Bioinformatics

The Morgan Kaufmann Series in Multimedia Information and Systems

Series Editor

Edward A. Fox,
Virginia Polytechnic University

Bioinformatics

Managing Scientific Data

Edited by

Zoé Lacroix
Arizona State University
Tempe, Arizona

And

Terence Critchlow
Lawrence Livermore National Laboratory
Livermore, California

With 34 Contributing Authors

MORGAN KAUFMANN PUBLISHERS

AN IMPRINT OF ELSEVIER SCIENCE

SAN FRANCISCO   SAN DIEGO   NEW YORK   BOSTON
LONDON   SYDNEY   TOKYO

Acquisitions Editor: Rick Adams
Developmental Editor: Karyn Johnson
Publishing Services Manager: Simon Crump
Project Manager: Jodie Allen
Designer: Eric Decicco
Production Services: Graphic World Publishing Services
Composition: International Typesetting and Composition
Illustration: Graphic World Illustration Studio
Printer: The Maple-Vail Book Manufacturing Group
Cover Printer: Phoenix

Designations used by companies to distinguish their products are often claimed as trademarks or registered trademarks. In all instances in which Morgan Kaufmann Publishers is aware of a claim, the product names appear in initial capital or all capital letters. Readers, however, should contact the appropriate companies for more complete information regarding trademarks and registration.

Morgan Kaufmann Publishers
An imprint of Elsevier Science
340 Pine Street, Sixth Floor
San Francisco, CA 94104-3205
www.mkp.com

07 06 05 04 03   5 4 3 2 1

Library of Congress Cataloging-in-Publication Data

Bioinformatics: managing scientific data / edited by Zoé Lacroix and Terence Critchlow.
    p. cm. -- (Morgan Kaufmann series in multimedia information and systems)
    Includes bibliographical references and index.
    ISBN 1-55860-829-X (pbk. : alk paper)
    1. Bioinformatics. I. Lacroix, Zoé. II. Critchlow, Terence. III. Series.

QH324.2.B55 2003
570'.285--dc21   2003044603

Library of Congress Control Number: 2003044603
ISBN: 1-55860-829-X

This book is printed on acid-free paper.

Contents

3   A Practitioner's Guide to Data Management and Data Integration in Bioinformatics   35

Barbara A. Eckman

4   Issues to Address While Designing a Biological Information System   75

Zoé Lacroix

5 SRS: An Integration Platform for Databanks and Analysis Tools in Bioinformatics 109

Thure Etzold, Howard Harris, and Simon Beaulah

8   The Information Integration System K2   225

Val Tannen, Susan B. Davidson, and Scott Harker

9   P/FDM Mediator for a Bioinformatics Database Federation   249

Graham J. L. Kemp and Peter M. D. Gray

Contributors

Patricia Baker
Department of Computer Science
University of Manchester
Manchester, United Kingdom

Simon Beaulah
LION Bioscience Ltd.
Cambridge, United Kingdom

Sean Bechhofer
Department of Computer Science
University of Manchester
Manchester, United Kingdom

Andy Brass
Department of Computer Science
University of Manchester
Manchester, United Kingdom

John Campbell
Gene Logic Inc.
Data Management Systems
Berkeley, California

I-Min A. Chen
Gene Logic Inc.
Data Management Systems
Berkeley, California

Jing Chen
geneticXchange Inc.
Menlo Park, California

Su Yun Chung
The Center for Research on Biological
Structure and Function
University of California, San Diego
La Jolla, California

Terence Critchlow
Lawrence Livermore National
Laboratory
Livermore, California

Susan B. Davidson
Department of Computer
and Information Science
University of Pennsylvania
Philadelphia, Pennsylvania

Barbara A. Eckman
IBM Life Sciences
West Chester, Pennsylvania

Thure Etzold
LION Bioscience Ltd.
Cambridge, United Kingdom

Carole Goble
Department of Computer Science
University of Manchester
Manchester, United Kingdom

Peter M. D. Gray
Department of Computing Science
University of Aberdeen
King's College
Aberdeen, Scotland, United Kingdom

Amarnath Gupta
San Diego Supercomputer Center
University of California, San Diego
San Diego, California

Laura M. Haas
IBM Silicon Valley Lab
San Jose, California

Howard Harris
LION Bioscience Ltd.
Cambridge, United Kingdom

Scott Harker
GlaxoSmithKline
King of Prussia, Pennsylvania

Graham J. L. Kemp
Department of Computing Science
Chalmers University of Technology
Göteborg, Sweden

Prasad Kodali
IBM Life Sciences
Somers, New York

Anthony Kosky
Gene Logic Inc.
Data Management Systems
Berkeley, California

Zoé Lacroix
Arizona State University
Tempe, Arizona

Eileen T. Lin
IBM Silicon Valley Lab
San Jose, California

Bertram Ludäscher
San Diego Supercomputer Center
University of California, San Diego
San Diego, California

Maryann E. Martone
Department of Neurosciences
University of California, San Diego
San Diego, California

Victor M. Markowitz
Gene Logic Inc.
Data Management Systems
Berkeley, California

Gary Ng
Network Inference Ltd.
London, United Kingdom

Krishna Palaniappan
Gene Logic Inc.
Data Management Systems
Berkeley, California

Norman W. Paton
Department of Computer Science
University of Manchester
Manchester, United Kingdom

Julia E. Rice
IBM Almaden Research Center
San Jose, California

Peter M. Schwarz
IBM Almaden Research Center
San Jose, California

Robert Stevens
Department of Computer Science
University of Manchester
Manchester, United Kingdom

Val Tannen
Department of Computer
and Information Science
University of Pennsylvania
Philadelphia, Pennsylvania

Thodoros Topaloglou
Gene Logic Inc.
Data Management Systems
Berkeley, California

Limsoon Wong
Institute for Infocomm Research
Singapore

John C. Wooley
Center for Research on Biological
Structure and Function
University of California, San Diego
La Jolla, California

About the Authors

Dr. Zoè Lacroix is currently a Research Assistant Professor at Arizona State University. She received a PhD in Computer Science in 1996 from the University of Paris XI (France). Her research interests cover various aspects of data management, and she has published more than 20 journal articles, conference papers, and book chapters. She also has served in numerous conference program committees, organized several panels and workshops, and was an active member in the working groups XML Query Language and XML Forms at the World Wide Web Consortium (W3C). Dr. Lacroix has been involved in bioinformatics for more than 7 years. She has interacted with the Center of Bioinformatics at the University of Pennsylvania and worked for two biotechnology companies, Gene Logic Inc. and SurroMed Inc. Her contributions in bioinformatics include publications, invited talks (Symposium on Bioinformatics organized at the National University of Singapore), and data integration middlewares, such as the Object-Web Wrapper, which is currently used at SmithKlineGlaxo.

Dr. Terence Critchlow is a computer scientist in the Center for Applied Scientific Computing at Lawrence Livermore National Laboratory (LLNL) and leads the DataFoundry project. His involvement in bioinformatics began more than 7 years ago as part of a collaboration between the University of Utah Computer Science department and the Utah Human Genome Center. Since completing his dissertation and joining LLNL in 1997, he has been an active member of the research community, publishing in both computer science and informatics forums, giving invited talks, participating in program committees, and organizing the XML Enabled Searches in Bioinformatics workshop.

Preface

Purpose and Goals

Bioinformatics can refer to almost any collaborative effort between biologists or geneticists and computer scientists and thus covers a wide variety of traditional computer science domains, including data modeling, data retrieval, data mining, data integration, data managing, data warehousing, data cleaning, ontologies, simulation, parallel computing, agent-based technology, grid computing, and visualization. However, applying each of these domains to biomolecular and biomedical applications raises specific and unexpectedly challenging research issues.

In this book, we focus on data management and in particular data integration, as it applies to genomics and microbiology. This is an important topic because data are spread across multiple sources, preventing scientists from efficiently obtaining the information required to perform their research (on average, a pharmaceutical company uses 40 data sources). In this environment, answering a single question may require accessing several data sources and calling on sophisticated analysis tools (e.g., sequence alignment, clustering, and modeling tools). While data integration is a dynamic research area in the database community, the specific needs of biologists have led to the development of numerous middleware systems that provide seamless data access in a results-driven environment (eight middleware systems are described in detail in this book).

The objective of the book is to provide life scientists and computer scientists with a complete view on biological data management by: (1) identifying specific issues in biological data management, (2) presenting existing solutions from both academia and industry, and (3) providing a framework in which to compare these systems.

Book Audience

This book is intended to be useful to a wide audience. Students, teachers, bioinformaticians, researchers, practitioners, and scientists from both academia and industry may all benefit from its material. It contains a comprehensive description of issues for biological data management and an overview of existing systems, making it appropriate for introductory and instructional purposes. Developers not yet familiar with bioinformatics will appreciate descriptions of the numerous challenges that need to be addressed and the various approaches that have been developed to solve them. Bioinformaticians may find the description of existing systems and the list of challenges that remain to be addressed useful. Decision makers will benefit from the evaluation framework, which will aide in their selection of the integration system that fits best the need of their research laboratory or company. Finally, life scientists, the ultimate users of these systems, may be interested in understanding how they are designed and evaluated.

Topics and Organization

The book is organized as follows: Four introductory chapters are followed by eight chapters presenting systems, an evaluation chapter, a summary, a glossary, and an appendix.

The introduction further refines the focus of this book and provides a working definition of bioinformatics. It also presents the steps that lead to the development of an information system, from its design to its deployment. Chapter 2 introduces the challenges faced by the integration of biological information. Chapter 3 refines these challenges into use cases and provides life scientists a translation of their needs into technical issues. Chapter 4 illustrates why traditional approaches often fail to meet life scientists' needs.

The following eight chapters each present an approach that was designed and developed to provide life scientists integrated access to data from a variety of distributed, heterogeneous data sources. The presented approaches provide a comprehensive overview of current technology. Each of these chapters is written by the main inventors of the presented system, specifies its requirements, and provides a description of both the chosen approach and its implementation. Because of the self-contained nature of these chapters, they may be read in any order. Chapter 13 provides users and developers with a methodology to evaluate presented systems. Such a methodology may be used to select the system most appropriate for an organization, to compare systems, or to evaluate a system developed in-house. The summary reiterates the state-of-the-art, existing solutions and new challenges that need to be addressed.

The appendix contains a list of useful biological resources (databases, organizations, and applications) organized in three tables. The acronyms commonly used to refer to them and used in the chapters of this book are spelled out, and current URLs are provided so that readers can access complete information.

Each of the chapters uses various technical terms. Because these terms involve expertise in life science and computer science, a glossary providing the spelling of acronyms or short definitions is provided at the end of the book.

Acknowledgments

Such a book requires hard work from a large number of individuals and organizations, and although we are not able to explicitly acknowledge everyone involved, we would like to thank as many as possible for their contributions.

We are obviously indebted to those individuals who contributed chapters, as this book would not have been as informative without them. Most of these contributions came in the form of detailed system descriptions. Whereas there are many bioinformatics data integration systems currently available, we selected several of the larger, better-known systems to include in this book. We are fortunate that key individuals working on these projects were willing and able to devote their time and energy to provide detailed descriptions of their systems. The fact that these contributors include the key architects of the systems makes them much more insightful than would otherwise be possible. We are also fortunate that Su Yun Chung, John Wooley, and Barbara Eckman were able to contribute their insights on a life scientist perspective of bioinformatics.

Beyond this obvious group, others contributed, directly and indirectly, to the final version of this book. We would like to thank our reviewers for their extremely helpful suggestions and our publishers for their support and tireless work bringing everything together. The manuscript reviewers included: Johann-Christoph Freytag, Humboldt-Universität zu Berlin; Mark Graves, Berlex; Michael Hucka, California Institute of Technology; Sean Mooney, Stanford University; and Shalom (Dick) Tsur, Ph.D., The Real-Time Enterprise Group. We would also like to thank Tom Slezak and Krishna Rajan for contributions that were not able to be included in the final version of this book.

Finally, Terence Critchlow would like to thank Carol Woodward for ongoing moral support, and Pete Eltgroth for providing the resources he used to perform this work. He would also like to extend his appreciation to Lawrence Livermore National Laboratory for their support of his effort and to acknowledge that this work was partially performed under the auspices of the U.S. DOE by LLNL under contract No. W-7405-ENG-48.

1 Introduction

Zoé Lacroix and Terence Critchlow

1.1 OVERVIEW

Bioinformatics and the management of scientific data are critical to support life science discovery. As computational models of proteins, cells, and organisms become increasingly realistic, much biology research will migrate from the wet-lab to the computer. Successfully accomplishing the transition to biology *in silico*, however, requires access to a huge amount of information from across the research community. Much of this information is currently available from publicly accessible data sources, and more is being added daily. Unfortunately, scientists are not currently able to identify easily and exploit this information because of the variety of semantics, interfaces, and data formats used by the underlying data sources. Providing biologists, geneticists, and medical researchers with integrated access to all of the information they need in a consistent format requires overcoming a large number of technical, social, and political challenges.

As a first step in helping to understand these issues, the book provides an overview of the state of the art of data integration and interoperability in genomics. This is accomplished through a detailed presentation of systems currently in use and under development as part of bioinformatics efforts at several organizations from both industry and academia. While each system is presented as a stand-alone chapter, the same questions are answered in each description. By highlighting a variety of systems, we hope not only to expose the different alternatives that are actively being explored, but more importantly, to give insight into the strengths and weaknesses of each approach. Given that an ideal bioinformatics environment remains an unattainable dream, compromises need to be made in the development of any real-world system. Understanding the tradeoffs inherent in different approaches, and combining that knowledge with specific organizational needs, is the best way to determine which alternative is most appropriate for a given situation.

Because we hope this book will be useful to both computer scientists and life scientists with varying degrees of familiarity with bioinformatics, three introductory chapters put the discussion in context and establish a shared vocabulary. The challenges faced by this developing technology for the integration of biological information are presented in Chapter 2. The complexity of use cases and the variety of techniques needed to support these needs are exposed in Chapter 3. This chapter also discusses the translation from specification to design, including the most common issues raised when performing this transformation in the life sciences domain. The difficulty of face-to-face communication between demanding users and developers is evoked in Chapter 4, in which examples are used to highlight the difficulty involved in directly transferring existing data management approaches to bioinformatics systems. These chapters describe the nuances that differentiate real-world bioinformatics from technology transferred from other domains. Whereas these nuances may be skeptically viewed as simple justifications for working on solved problems, they are important because bioinformatics occurs in the real world, complete with its ugly realities, not in an abstract environment where convenient assumptions can be used to simplify problems.

These introductory chapters are followed by the heart of this book, the descriptions of eight distinct bioinformatics systems. These systems are the results of collaborative efforts between the database community and the genomics community to develop technology to support scientists in the process of scientific discovery. Systems such as Kleisli (Chapter 6) were developed in the early stages of bioinformatics and matured through meetings on the Interconnection of Molecular Biology Databases (the first of the series was organized at Stanford University in the San Francisco Bay Area, August 9–12, 1994). Others, such as DiscoveryLink (Chapter 11), are recent efforts to adapt sophisticated data management technology to specific challenges facing bioinformatics. Each chapter has been written by the primary contributor(s) to the system being described. This perspective provides precious insight into the specific problem being addressed by the system, why the particular architecture was chosen, its strengths, and any weakness it may have. To provide an overall summary of these approaches, advantages and disadvantages of each are summarized and contrasted in Chapter 13.

1.2   PROBLEM AND SCOPE

In the last decade, biologists have experienced a fundamental revolution from traditional research and development (R&D) consisting in discovering and understanding genes, metabolic pathways, and cellular mechanisms to large-scale, computer-based R&D that simulates the disease, the physiology, the molecular mechanisms, and the pharmacology [1]. This represents a shift away from life science's empirical roots, in which it was an iterative and intuitive process. Today it is systematic and predictive with genomics, informatics, automation, and miniaturization all playing a role [2]. This fusion of biology and information science is expected to continue and expand for the foreseeable future. The first consequence of this revolution is the explosion of available data that biomolecular researchers have to harness and exploit. For example, an average pharmaceutical company currently uses information from at least 40 databases [1], each containing large amounts of data (e.g., as of June 2002, GenBank [3, 4] provides access to 20,649,000,000 bases in 17,471,000 sequences) that can be analyzed using a variety of complex tools such as FASTA [5], BLAST [6], and LASSAP [7].

Over the past several years, *bioinformatics* has become both an all-encompassing term for everything relating to computer science and biology, and a very trendy one.[1] There are a variety of reasons for this including: (1) As computational biology evolves and expands, the need for solutions to the data integration problems it faces increases; (2) the media are beginning to understand the implications of the genomics revolution that has been going on for the last 15 or more years; (3) the recent headlines and debates surrounding the cloning of animals and humans; and (4) to appear cutting edge, many companies have relabeled the work that they are doing as bioinformatics, and similarly many people have become bioinformaticians instead of geneticists, biologists, or computer scientists. As these events have occurred, the generally accepted meaning of the word *bioinformatics* has grown from its original definition of managing genomics data to include topics as diverse as patient record keeping, molecular simulations of protein sequences, cell and organism level simulations, experimental data analysis, and analysis of journal articles. A recent definition from the National Institutes of Health (NIH) phrases it this way:

> Bioinformatics is the field of science in which biology, computer science, and information technology merge to form a single discipline. The ultimate goal of the field is to enable the discovery of new biological insights as well as to create a global perspective from which unifying principles in biology can be discerned. [8]

This definition could be rephrased as: *Bioinformatics is the design and development of computer-based technology that supports life science.* Using this definition, bioinformatics tools and systems perform a diverse range of functions including: data collection, data mining, data analysis, data management, data integration, simulation, statistics, and visualization. Computer-aided technology directly supporting medical applications is excluded from this definition and is referred to as medical informatics. This book is not an attempt at authoritatively describing

---

1. The sentence claims that computer science is relating to biology. Whenever one refers to this "relationship," one uses the term *bioinformatics*.

the gamut of information contained in this field. Instead, it focuses on the area of genomics data integration, access, and interoperability as these areas form the cornerstone of the field. However, most of the presented approaches are generic integration systems that can be used in many similar scientific contexts.

This emphasis is in line with the original focus of bioinformatics, which was on the creation and maintenance of data repositories (flat files or databases) to store biological information, such as nucleotide and amino acid sequences. The development of these repositories mostly involved schema design issues (data organization) and the development of interfaces whereby scientists could access, submit, and revise data. Little or no effort was devoted to traditional data management issues such as storage, indexing, query languages, optimization, or maintenance. The number of publicly available scientific data repositories has grown at an exponential rate, to the point where, in 2000, there were thousands of public biomolecular data sources. In 2003, Baxevanis listed 372 key databases in molecular biology only [9]. Because these sources were developed independently, the data they contain are represented in a wide variety of formats, are annotated using a variety of methods, and may or may not be supported by a database management system.

1.3   BIOLOGICAL DATA INTEGRATION

Data integration issues have stymied computer scientists and geneticists alike for the last 20 years, and yet successfully overcoming them is critical to the success of genomics research as it transitions from a wet-lab activity to an electronic-based activity as data are used to drive the increasingly complicated research performed on computers. This research is motivated by scientists striving to understand not only the data they have generated, but more importantly, the information implicit in these data, such as relationships between individual components. Only through this understanding will scientists be able to successfully model and simulate entire genomes, cells, and ultimately entire organisms.

Whereas the need for a solution is obvious, the underlying data integration issues are not as clear. Chapter 4 goes into detail about the specific computer science problems, and how they are subtly different from those encountered in other areas of computer science. Many of the problems facing genomics data integration are related to data semantics—the meaning of the data represented in a data source—and the differences between the semantics within a set of sources. These differences can require addressing issues surrounding concept identification, data transformation, and concept overloading. *Concept identification and resolution* has two components: identifying when data contained in different data sources refer to the same object and reconciling conflicting information found in these sources. Addressing these issues should begin by identifying which abstract concepts are represented in each data source. Once shared concepts have been identified, conflicting information can be easily located. As a simple example, two sources may have different values for an attribute that is supposed to be the same. One of the wrinkles that genomics adds to the reconciliation process is that there may not be a "right" answer. Consider that a sequence representing the same gene should be identical in two different data sources. However, there may be legitimate differences between two sources, and these differences need to be preserved in the integrated view. This makes a seemingly simple query, "*return the sequence associated with this gene*," more complex than it first appears.

In the case where the differences are the result of alternative data formats, *data transformations* may be applied to map the data to a consistent format. Whereas mapping may be simple from a technical perspective, determining what it is and when to apply it relies on the detailed representation of the concepts and appropriate domain knowledge. For example, the translation of a protein sequence from a single-character representation to a three-character representation defines a corresponding mapping between the two representations. Not all transformations are easy to perform—and some may not be invertible. Furthermore, because of *concept overloading*, it is often difficult to determine whether or not two abstract concepts really have the same meaning—and to figure out what to do if they do not. For example, although two data sources may both represent genes as DNA sequences, one may include sequences that are postulated to be genes, whereas the other may only include sequences that are known to code for proteins. Whether or not this distinction is important depends on a specific application and the semantics that the unified view is supporting. The number of subtly distinct concepts used in genomics and the use of the same name to refer to multiple variants makes overcoming these conflicts difficult.

Unfortunately, the semantics of biological data are usually hard to define precisely because they are not explicitly stated but are implicitly included in the database design. The reason is simple: At a given time, within a single research community, common definitions of various terms are often well understood and have precise meaning. As a result, the semantics of a data source are usually understood by those within that community without needing to be explicitly defined. However, genomics (much less all of biology or life science) is not a single, consistent scientific domain; it is composed of dozens of smaller, focused research communities. This would not be a significant issue if researchers only accessed data from within a single domain, but that is not usually the case. Typically, researchers require integrated access to data from multiple domains, which requires resolving terms that have slightly different meanings across the communities. This is further complicated by the observations that the specific community whose terminology is being used by the data source is usually not explicitly identified and that the terminology evolves over time. For many of the larger, community data sources, the domain is obvious—the Protein Data Bank (PDB) handles protein structure information, the Swiss-Prot protein sequence database provides protein sequence information and useful annotations, etc.—but the terminology used may not be current and can reflect a combination of definitions from multiple domains. The terminology used in smaller data sources, such as the drosophila database, is typically selected based on a specific usage model. Because this model can involve using concepts from several different domains, the data source will use whatever definitions are most intuitive, mixing the domains as needed.

Biology also demonstrates three challenges for data integration that are common in evolving scientific domains but not typically found elsewhere. The first is the sheer number of available data sources and the inherent heterogeneity of their contents. The World Wide Web has become the preferred approach for disseminating scientific data among researchers, and as a result, literally hundreds of small data sources have appeared over the past 10 years. These sources are typically a "labor of love" for a small number of people. As a result, they often lack the support and resources to provide detailed documentation and to respond to community requests in a timely manner. Furthermore, if the principal supporter leaves, the site usually becomes completely unsupported. Some of these sources contain data from a single lab or project, whereas others are the definitive repositories for very specific types of information (e.g., for a specific genetic mutation). Not only do these sources complicate the concept identification issue previously mentioned (because they use highly specialized data semantics), but their number make it infeasible to incorporate all of them into a consistent repository.

Second, the data formats and data access methods (associated interfaces) change regularly. Many data providers extend or update their data formats approximately every 6 months, and they modify their interfaces with the same frequency. These changes are an attempt to keep up with the scientific evolution occurring in the community at large. However, a change in a data source representation can have a dramatic impact on systems that integrate that source, causing the integration to fail on the new format or worse, introducing subtle errors into the systems. As a result of this problem, bioinformatics infrastructures need to be more flexible than systems developed for more static domains.

Third, the data and related *analysis* are becoming increasingly complex. As the nature of genomics research evolves from a predominantly wet-lab activity into knowledge-based analysis, the scientists' need for access to the wide variety of available information increases dramatically. To address this need, information needs to be brought together from various heterogeneous data sources and presented to researchers in ways that allow them to answer their questions. This means providing access not only to the sequence data that is commonly stored in data sources today, but also to multimedia information such as expression data, expression pathway data, and simulation results. Furthermore, this information needs to be available for a large number of organisms under a variety of conditions.

1.4   DEVELOPING A BIOLOGICAL DATA INTEGRATION SYSTEM

The development of a biological data integration and management system has to overcome the difficulties outlined in Section 1.3. However, there is no obvious best approach to doing this, and thus each of the systems presented in this book addresses these issues differently. Furthermore, comparing and contrasting these systems is extremely difficult, particularly without a good understanding of how they were developed. This is because the goals of each system are subtly different, as reflected by the system requirements defined at the outset of the design process. Understanding the development environment and motivation behind the initial system constraints is critical to understanding the tradeoffs that were made later in the design process and the reasons why.

1.4.1   Specifications

The design of a system starts with collecting requirements that express, among other things:

- ✦ Who the users of the system will be
- ✦ What functionality the system is expected to have
- ✦ How this functionality is to be viewed by the users
- ✦ The performance goals for the system

System requirements (or specifications) describe the desired system and can be seen as a contract agreed upon by the target users (or their surrogates) and the developers. Furthermore, these requirements can be used to determine if a delivered system performs properly.

The user profile is a concise description of who the target users for a system are and what knowledge and experience they can be assumed to have. Specifying the user *profile* involves agreeing on the level of computer literacy expected of users (e.g., Are there programmers helping the scientists access the data? Are the users expected to know any programming language?), the type of interface the users will have (e.g., Will there be a visual interface? A user customizable interface?), the security issues that need to be addressed, and a multitude of other concerns.

Once the user profile is defined, the tasks the system is supposed to perform must be analyzed. This analysis consists in listing all the tasks the system is expected to perform, typically through *use cases*, and involves answering questions such as: What are the sources the system is expected to integrate? Will the system allow users to express queries? If so, in what form and how complex will they be? Will the system incorporate scientific applications? Will it allow users to navigate scientific objects?

Finally, technical issues must be agreed upon. These issues include the platforms the system is expected to work on (i.e., UNIX, Microsoft, Macintosh), its scalability (i.e., the amount of data it can handle, the number of queries it can simultaneously support, and the number of data sources that can be integrated), and its expected efficiency with respect to data storage size, communication overhead, and data integration overhead.

The collection of these requirements is traditional to every engineering task. However, in established engineering areas there are often intermediaries that initially evaluate the needs for new technology and significantly facilitate the definition of system specifications. Unfortunately, this is not the case in life sciences. Although technology is required to address complex user needs, the scientists generally directly communicate their needs to the system designers. While communication between specialists in different domains is inherently difficult, bioinformatics faces an additional challenge—the speed at which the underlying science is evolving. A common result of this is that both scientists and developers become frustrated. Scientists are frustrated because systems are not able to keep up with their ever-changing requirements, and developers are frustrated because the requirements keep changing on them. The only way to overcome this problem is to have an intermediary between the specialists. A common goal can be formulated and achieved by forging a bridge between the communities and accurately representing the requirements and constraints of both sides.

1.4.2   Translating Specifications into a Technical Approach

Once the specifications have been agreed upon, they can be translated into a set of approaches. This can be thought of as an optimization problem in which the hard constraints define a feasibility region, and the goal is to minimize the cost of the system while maximizing its usefulness and staying within that region. Each attribute in the system description can be mapped to a dimension. Existing data management approaches can then be mapped to overlapping regions in this space.

Once the optimal location has been identified, these approaches can be used as a starting point for the implementation.

Obviously, this problem is not always formally specified, but considering it in this way provides insight into the appropriate choices. For example, in the dimension of storage costs, two alternatives can be considered: materializing the data and not materializing it. The materialized approach collects data from various sources and loads them into a single system. This approach is often closely related to a data warehousing approach and is favored when the specifications include characteristics such as data curation, infrequent data updates, high reliability, and high levels of security. The non-materialized approach integrates all the resources by collecting the requested data from the distributed data sources at query execution time. Thus, if the specifications require up-to-date data or the ability to easily include new resources in the integration, a non-materialized approach would be more appropriate.

1.4.3   Development Process

The system development implements the approaches identified in Section 1.4.2, possibly extending them to meet specific constraints. System development is often an iterative process in which the following steps are repeatedly performed as capabilities are added to the system:

+ Code design: describing the various software components/objects and their respective capabilities
+ Implementation: actually writing the code and getting it to execute properly
+ Testing: evaluating the implementation, identifying and correcting bugs
+ Deployment: transferring the code to a set of users

The formal deployment of a system often includes an analysis of the tests and training the users. The final phases are the system migration and the operational process. More information on managing a programming project can be found in *Managing a Programming Project—Processes and People* [10].

1.4.4   Evaluation of the System

Two systems may have the same specifications and follow the same approach yet end up with radically different implementations. The eight systems presented in the book (Chapters 5 through 12) follow various approaches. Their design and implementation choices lead to vastly different systems. These chapters provide few details on the numerous design and implementation decisions and instead focus on the main characteristics of their systems. This will provide some insight into the vast array of tradeoffs that are possible while still developing feasible systems.

There are several metrics by which a system can be evaluated. One of the most obvious is whether or not it meets its requirements. However, once the specifications are satisfied, there are many characteristics that reflect a system's performance. Although similar criteria may be used to compare two systems that have the same specifications, these same criteria may be misleading when the specifications differ. As a result, evaluating systems typically requires insight into the system design and implementation and information on users' satisfaction. Although such a difficult task is beyond the scope of this book, in Chapter 13 we outline a set of criteria that can be considered a starting point for such an evaluation.

REFERENCES

[1]   M. Peitsch. "From Genome to Protein Space." Presentation at the Fifth Annual Symposium in Bioinformatics, Singapore, October 2000.

[2]   D. Valenta. "Trends in Bioinformatics: An Update." Presentation at the Fifth Annual Symposium in Bioinformatics, Singapore, October 2000.

[3]   D. Benson, I. Karsch-Mizrachi, D. Lipman, et al. "GenBank." *Nucleic Acids Research* 31, no. 1 (2003):23–27, http://www.ncbi.nlm.nih.gov/Genbank.

[4]   "Growth of GenBank." (2003): http://www.ncbi.nlm.nih.gov/Genbank/genbankstats.html.

[5]   W. Pearson and D. Lipman. "Improved Tools for Biological Sequence Comparison." *Proceedings of the National Academy of Sciences of the United States of America* 85, no. 8 (April 1988): 2444–2448.

[6]   S. Altschul, W. Gish, W. Miller, et al. "Basic Local Alignment Search Tool." *Journal of Molecular Biology* 215, no. 3 (October 1990): 403–410, http://www.ncbi.nlm.nih.gov/BLAST.

[7]   E. Glenet and J-J. Codani. "LASSAP: A Large Scale Sequence Comparison Package." *Bioinformatics* 13, no. 2 (1997): 137–143.

[8]   NCBI. "Just the Facts: A Basic Introduction to the Science Underlying NCBI Resources." *A Science Primer* (November 2002): http://www4.ncbi.nlm.nih.gov/About/primer/bioinformatics.html

[9]   A. Baxevanis. "The Molecular Biology Database Collection: 2003 Update." *Nucleic Acids Research* 31, no. 1 (2003): 1–12, http://nar.oupjournals.org/cgi/content/full/31/1/1.

[10] P. Metzger and J. Boddie. *Managing a Programming Project—Processes and People*. Upper Saddle River, NJ: Prentice Hall, 1996.

Challenges Faced in the Integration of Biological Information

Su Yun Chung and John C. Wooley

Biologists, in attempting to answer a specific biological question, now frequently choose their direction and select their experimental strategies by way of an initial computational analysis. Computers and computer tools are naturally used to collect and analyze the results from the largely automated instruments used in the biological sciences. However, far more pervasive than this type of requirement, the very nature of the intellectual discovery process requires access to the latest version of the worldwide collection of data, and the fundamental tools of bioinformatics now are increasingly part of the experimental methods themselves. A driving force for life science discovery is turning complex, heterogeneous data into useful, organized information and ultimately into systematized knowledge. This endeavor is simply the classic pathway for all science, Data $\Rightarrow$ Information $\Rightarrow$ Knowledge $\Rightarrow$ Discovery, which earlier in the history of biology required only brainpower and pencil and paper but now requires sophisticated computational technology.

In this chapter, we consider the challenges of information integration in biology from the perspective of researchers using information technology as an integral part of their discovery processes. We also discuss why information integration is so important for the future of biology and why and how the obstacles in biology differ substantially from those in the commercial sector—that is, from the expectations of traditional business integration. In this context, we address features specific to the biological systems and their research approaches. We then discuss the burning issues and unmet needs facing information integration in the life sciences. Specifically, data integration, meta-data specification, data provenance and data quality, ontology, and Web presentations are discussed in subsequent sections. These are the fundamental problems that need to be solved by the bioinformatics community so that modern information technology can have a deeper impact on the progress of biological discovery. This chapter raises the challenges rather than trying to establish specific, ideal solutions for the issues involved.

2.1   THE LIFE SCIENCE DISCOVERY PROCESS

In the last half of the 20th century, a highly focused, hypothesis-driven approach known as reductionist molecular biology gave scientists the tools to identify and characterize molecules and cells, the fundamental building blocks of living systems. To understand how molecules, and ultimately cells, function in tissues, organs, organisms, and populations, biologists now generally recognize that as a community they not only have to continue reductionist strategies for the further elucidation of the structure and function of individual components, but they also have to adopt a systems-level approach in biology. Systems analysis demands not just knowledge of the parts—genes, proteins, and other macromolecular entities—but also knowledge of the connection of these molecular parts and how they work together. In other words, the pendulum of bioscience is now swinging away from reductionist approaches and toward synthetic approaches characteristic of systems biology and of an integrated biology capable of quantitative and/or detailed qualitative predictions. A synthetic or integrated view of biology obviously will depend critically on information integration from a variety of data sources. For example, neuroinformatics includes the anatomical and physiological features of the nervous system, and it must interact with the molecular biological databases to facilitate connections between the nervous system and molecular details at the level of genes and proteins.[1] In phylogeny and evolution biology, comparative genomics is making new impacts on evolutionary studies. Over the past two decades, research in evolutionary biology has come to depend on sequence comparisons at the gene and protein level, and in the future, it will depend more and more on tracking not just DNA sequences but how entire genomes evolve over time [1]. In ecology there is an opportunity ultimately to study the sequences of all genomes involved in an entire ecological community. We believe integration bioinformatics will be the backbone of 21st-century life sciences research.

Research discovery and synthesis will be driven by the complex information arising intrinsically from biology itself and from the diversity and heterogeneity of experimental observations. The database and computing activities will need to be integrated to yield a cohesive information infrastructure underlying all of biology. A conceptual example of how biological research has increasingly come to depend on the integration of experimental procedures and computation activities is illustrated in Figure 2.1. A typical research project may start with a collection of known or unknown genomic sequences (see Genomics in Figure 2.1). For unknown sequences, one may conduct a database search for similar sequences

---

1. For information about neuroinformatics, refer to the Human Brain Project at the National Institute of Mental Health (*http://www.nimh.nih.gov/neuroinformatics/abs.cfm*).

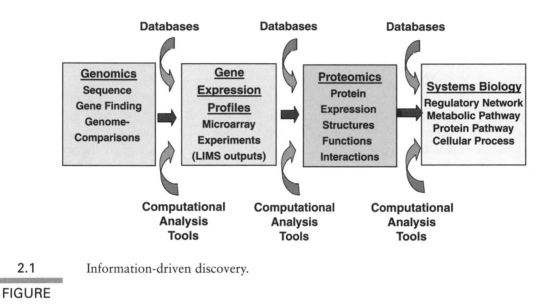

2.1   Information-driven discovery.

FIGURE

or use various gene-finding computer algorithms or genome comparisons to predict the putative genes. To probe expression profiles of these genes/sequences, high-density microarray gene expression experiments may be carried out. The analysis of expression profiles of up to 100,000 genes can be conducted experimentally, but this requires powerful computational correlation tools. Typically, the first level of experimental data stream output for a microarray experiment (laboratory information management system [LIMS] output) is a list of genes/sequences/identification numbers and their expression profile. Patterns or correlations within the massive data points are not obvious by manual inspection. Different computational clustering algorithms are used simultaneously to reduce the data complexity and to sort out relationships among genes/sequences according to their expression levels or changes in expression levels.

These clustering techniques, however, have to deal with a high-dimensional data element space; the possibility for correlation by chance is high because a set of genes clustered together does not necessarily imply participation in a common biological process. To back up the clustering results, one may proceed to proteomics (see Figure 2.1) to connect the gene expression results with available protein expression patterns, known protein structures and functions, and protein–protein interaction data. Ultimately, the entire collection of interrelated macromolecular information may be considered in the context of systems biology (see Figure 2.1), which includes analyses of protein or metabolic pathways, regulatory networks, and other, more complex cellular processes. The connections and interactions among areas of genomics, gene expression profiles, proteomics, and systems biology depend on the integration of experimental procedures with database searches and the applications of computational algorithms and analysis tools.

As one moves up in the degree of complexity of the biological processes under study, our understanding at each level depends in a significant way on the levels beneath it. In every step, database searches and computational analysis of the data are an integral part of the discovery process. As we choose complex systems for study, experimentally generated data must be combined with data derived from databases and computationally derived models or simulations for best interpretation. On the other hand, modeling and simulation of protein–protein interactions, protein pathways, genetic regulatory networks, biochemical and cellular processes, and normal and disease physiological states are in their infancy and need more experimental observations to fill in missing quantitative details for mature efforts. In this close interaction, the boundaries between experimentally generated data and computationally generated data are blurring. Thus, accelerating progress now requires multidisciplinary teams to conduct integrated approaches. Thus, *in silico* discovery, that is, experiments carried out with a computer, is fully complementary to traditional wet-laboratory experiments. One could say that an information infrastructure, coupled with continued advances in experimental methods, will facilitate computing an understanding of biology.

2.2   AN INFORMATION INTEGRATION ENVIRONMENT FOR LIFE SCIENCE DISCOVERY

Biological data sources represent the collective research efforts and products of the life science communities throughout the world. The growth of the Internet and the availability of biological data sources on the Web have opened up a tremendous opportunity for biologists to ask questions and solve problems in unprecedented ways. To harness these community resources and assemble all available information to investigate specific biological problems, biologists must be able to find, extract, merge, and synthesize information from multiple, disparate sources. Convergence of biology, computer science, and information technology (IT) will accelerate this multidisciplinary endeavor. The basic needs are:

1. On demand access and retrieval of the most up-to-date biological data and the ability to perform complex queries across multiple heterogeneous databases to find the most relevant information

2. Access to the best-of-breed analytical tools and algorithms for extraction of useful information from the massive volume and diversity of biological data

3. A robust information integration infrastructure that connects various computational steps involving database queries, computational algorithms, and application software

This multidisciplinary approach demands close collaboration and clear understanding between people with extremely different domain knowledge and skill sets. The IT professionals provide the knowledge of syntactic aspects of data, databases, and algorithms, such as how to search, access, and retrieve relevant information, manage and maintain robust databases, develop information integration systems, model biological objects, and support a user-friendly graphical interface that allows the end user to view and analyze the data. The biologists provide knowledge of biological data, semantic aspects of databases, and scientific algorithms. Interpreting biological relationships requires an understanding of the biological meaning of the data beyond the physical file or table layout. Particularly, the effective usage of scientific algorithms or analytical tools (e.g., sequence alignment, protein structure prediction, and other analysis software) depends on having a working knowledge of the computer programs and of biochemistry, molecular biology, and other scientific disciplines. Before we can discuss biological information integration, we need first to consider the specific nature of biological data and data sources.

2.3   THE NATURE OF BIOLOGICAL DATA

The advent of automated and high-throughput technologies in biological research and the progress in the genome projects has led to an ever-increasing rate of data acquisition and exponential growth of data volume. However, the most striking feature of data in life science is not its volume but its diversity and variability.

2.3.1   Diversity

The biological data sets are intrinsically complex and are organized in loose hierarchies that reflect our understanding of the complex living systems, ranging from genes and proteins, to protein–protein interactions, biochemical pathways and regulatory networks, to cells and tissues, organisms and populations, and finally the ecosystems on earth. This system spans many orders of magnitudes in time and space and poses challenges in informatics, modeling, and simulation equivalent to

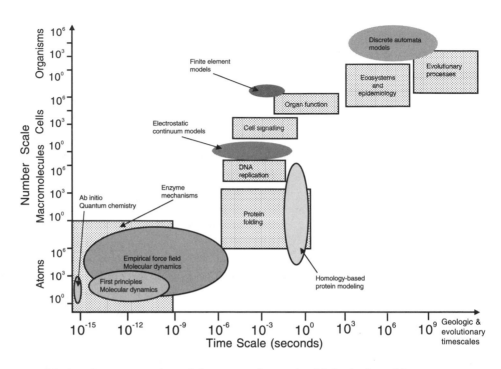

Notional representation of the vast and complex biological world.

or beyond any other scientific endeavor. A notional description of the vast scale of complexity, population, time, and space in the biological systems is given in Figure 2.2 [2]. Reflecting the complexity of biological systems, the types of biological data are highly diverse. They range from the plain text of laboratory records and literature publications, nucleic acid and protein sequences, three-dimensional atomic structures of molecules, and biomedical images with different levels of resolutions, to various experimental outputs from technology as diverse as microarray chips, gels, light and electronic microscopy, Nuclear Magnetic Resonance (NMR), and mass spectrometry. The horizontal abscissa in Figure 2.2 shows time scales ranging from femtoseconds to eons that represent the processes in living systems from chemical and biochemical reactions, to cellular events, to evolution. The vertical ordinate shows the numerical scale, the range of number of atoms involved in molecular biology, the number of macromolecules in cellular biology, the number of cells in physiological biology, and the number of organisms in population biology. The third dimension indicated by rectangles illustrates the hierarchical nature of biology from subcellular structures to ecosystems. The fourth dimension, indicated by ovals, represents the current state of computation biology in modeling and simulation of biological systems.

2.3.2 Variability

Different individuals and species vary tremendously, so naturally biological data does also. For example, structure and function of organs vary across age and gender, in normal and different disease states, and across species. Essentially, all features of biology exhibit some degree of variability. Biological research is in an expanding phase, and many fields of biology are still in the developing stages. Data for these systems are incomplete and very often inconsistent. This presents a great challenge in modeling biological objects.

2.4 DATA SOURCES IN LIFE SCIENCE

In response to current advances in technology and research scope, massive amounts of data are routinely deposited in public and private databases. In parallel, there is a proliferation of computational algorithms and analysis tools for data analysis and visualization. Because most databases are accompanied by specific computational algorithms or tools for analysis and presentation and vice versa, we use the term *data source* to refer to a database or computational analysis tool or both. There are more than 1000 life science data sources scattered over the Internet (see the Biocatalog and the Public Catalog of Databases), and these data sources vary widely in scope and content. Finding the right data sources alone can be a challenge. Searching for relevant information largely relies on a Web information retrieval system or on published catalog services. Each January, the *Journal of Nucleic Acid Research* provides a yearly update of molecular biology database collections. The current issue lists 335 entries in molecular biology databases alone [3]. Various Web sites provide a catalog and links to biological data sources (see "biocat" and "dbcat" cited previously). In addition to the public sources, there are numerous private, proprietary data sources created by biotechnology or pharmaceutical companies.

The scope of the public data sources ranges from the comprehensive, multidisciplinary, community informatics center, supported by government public funds and sustained by teams of specialists, to small boutique data sources by individual investigators. The content of databases varies greatly, reflecting the broad disciplines and sub-disciplines across life sciences from molecular biology and cell biology, to medicine and clinical trials, to ecology and biodiversity. A sampling of various public biological databases is given in the Appendix.

2.4.1   Biological Databases Are Autonomous

Biological data sources represent a loose collection of autonomous Web sites, each with its own governing body and infrastructure. These sites vary in almost every possible instance such as computer platform, access, and data management system. Much of the available biological data exist in legacy systems in which there are no structured information management systems. These data sources are inconsistent at the semantic level, and more often than not, there is no adequate attendant meta-data specification. Until recently, biological databases were not designed for interoperability [4].

2.4.2   Biological Databases Are Heterogeneous in Data Formats

Data elements in public or proprietary databases are stored in heterogeneous data formats ranging from simple files to fully structured database systems that are often *ad hoc,* application-specific, or vendor-specific. For example, scientific literature, images, and other free-text documents are commonly stored in unstructured or semi-structured formats (plain text files, HTML or XML files, binary files). Genomic, microarray gene expression, and proteomic data are routinely stored in conventional spreadsheet programs or in structured relational databases (Oracle, Sybase, DB2, Informix). Major data depository centers have implemented various data formats for operations; the National Center for Biotechnology Information (NCBI) has adopted the highly nested data system ASN.1 (Abstract Syntax Notation) for the general storage of gene, protein, and genomic information [5]; the United States Department of Agriculture (USDA) Plant Genome Data and Information Center has adopted the object-oriented, A C. elegans Data Base (Ace DB) data management systems and interface [6].

2.4.3   Biological Data Sources Are Dynamic

In response to the advance of biological research and technology, the overall features of biological data sources are subjected to continuous changes including data content and data schema. New databases spring up at a rapid rate and older databases disappear.

2.4.4   Computational Analysis Tools Require Specific Input/Output Formats and Broad Domain Knowledge

Computational software packages often require specific input and output data formats and graphic display of results, which pose serious compatibility and interoperability issues. The output of one program is not readily suitable as direct input for the next program or for a subsequent database search. Development of a standard data exchange format such as XML will alleviate some of the interoperability issues.

Understanding application semantics and the proper usages of computer software is a major challenge. Currently, there are more than 500 software packages or analysis tools for molecular biology alone (reviewed in the Biocatalog at the European Bioinformatics Institute [EBI] Web site given previously). These programs are extremely diverse, ranging from nucleic and protein sequence analysis, genome comparison, protein structure prediction, biochemical pathway and genetic network analysis, and construction of phylogenetic trees, to modeling and simulation of biological systems and processes. These programs, developed to solve specific biological problems, rely on input from other domain knowledge such as computer science, applied mathematics, statistics, chemistry, and physics. For example, protein folding can be approached using *ab initio* prediction based on first principles (physics) or on knowledge-based (computer science) threading methods [7]. Many of these software packages, particularly those available through academic institutions, lack adequate documentation describing the algorithm, functionality, and constraints of the program. Given the multidisciplinary nature and the scope of domain knowledge, proper usage of a scientific analysis program requires significant (human) expertise. It is a daunting task for the end users to choose and evaluate the proper software programs for analyses, so they will be able to understand and interpret the results.

2.5   CHALLENGES IN INFORMATION INTEGRATION

With the expansion of the biological data sources available across the World Wide Web, integration is a new, major challenge facing researchers and institutions that wish to explore these rich deposits of information. Data integration is an ongoing active area in the commercial world. However, information integration in biology must consider the characteristics of the biological data and data sources as discussed in the previous two sections (2.3 and 2.4): (1) diverse data are stored in autonomous data sources that are heterogeneous in data formats, data

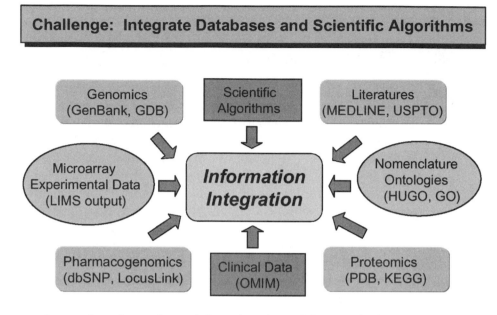

2.3

FIGURE

Integration of experimental data, data derived from multiple database queries, and applications of scientific algorithms and computational analysis tools (Refer to the Appendix for the definitions of acronyms).

management systems, data schema, and semantics; (2) analysis of biological data requires both database query activities and proper usage of computational analysis tools; (3) a broad spectrum of knowledge domains divide traditional biological disciplines.

For a typical research project, a user must be able to merge data derived from multiple, diverse, heterogeneous sources freely and readily. As illustrated in Figure 2.3, the LIMS output from microarray gene expression experiments must be interpreted and analyzed in the context of the information and tools available across the Internet, including genomic data, literature, clinical data, analysis algorithms, etc. In many cases, data retrieved from several databases may be selected, filtered, and transformed to prepare input data sets for particular analytic algorithms or applications. The output of one program may be submitted as input to another program and/or to another database search. The integration process involves an intricate network of multiple computational steps and data flow. Information integration in biology faces challenges at the technology level for data integration architectures and at the semantic level for meta-data specification, maintenance of data provenance and accuracy, ontology development for knowledge sharing and reuse, and Web presentations for communication and collaboration.

2.5.1   Data Integration

First-generation bioinformatics solutions for data integration employ a series of non-interoperable and non-scalable quick fixes to translate data from one format into another. This means writing programs, usually in programming language such as Perl, to access, parse, extract, and transform necessary data for particular applications. Writing a translation program requires intensive coding efforts and knowledge of the data and structures of the source databases. These *ad hoc* point-to-point solutions are very inefficient and are not scalable to the large number of data sources to be integrated. This is dubbed the N^2 *factor* because it would require N (N-1)/2 programs to connect N data sources. If one particular data source changes formats, all of the programs involved with this data source must be upgraded. Upgrades are inevitable because changes in Web page services and schema are very common for biological data sources.

The second generation of data integration solutions provides a more structured environment for code re-use and flexible, scalable, robust integration. Over the past decade, enormous efforts and progress have been made in many data integration systems. They can be roughly divided into three major categories according to access and architectures: the data warehousing approach, the distributed or federated approach, and the mediator approach. However, the following fundamental functions or features are desirable for a robust data integration system:

1. Accessing and retrieving relevant data from a broad range of disparate data sources

2. Transforming the retrieved data into designated data model for integration

3. Providing a rich common data model for abstracting retrieved data and presenting integrated data objects to the end user applications

4. Providing a high-level expressive language to compose complex queries across multiple data sources and to facilitate data manipulation, transformation, and integration tasks

5. Managing query optimization and other complex issues

The Data Warehouse Approach

The data warehouse approach assembles data sources into a centralized system with a global data schema and an indexing system for integration and navigation. The data warehouse world is dominated by relational database management systems (RDBMS), which offer the advantage of a mature and widely accepted database technology and a high level standard query language (SQL) [8]. These systems have proven very successful in commercial enterprises, health care, and government sectors for resource management such as payroll, inventory, and records. They require reliable operation and maintenance, and the underlying databases are under a controlled environment, are fairly stable, and are structured. The biological data sources are very different from those contained in the commercial databases. The biological data sources are much more dynamic and unpredictable, and few of the public biological data sources use structured database management systems. Given the sheer volume of data and the broad range of biological databases, it would require substantial effort to develop any monolithic data warehouses encompassing diverse biological information such as sequence and structure and the various functions of biochemical pathways and genetic polymorphisms. As the number of databases in a data warehouse grows, the cost of storage, maintenance, and updating data will be prohibitive. A data warehouse has an advantage in that the data are readily accessed without Internet delay or bandwidth limitation in network connections. Vigorous data cleansing to remove potential errors, duplications, and semantic inconsistency can be performed before entering data in the warehouse. Thus, limited data warehouses are popular solutions in the life sciences for data mining of large databases, in which carefully prepared data sets are critical for success [9].

The Federation Approach

The distributed or federated integration approaches do not require a centralized persistent database, and thus the underlying data sources remain autonomous. The federated systems maintain a common data model and rely on schema mapping to translate heterogeneous source database schema into the target schema for integration. A data dictionary is used to manage various schema components. In the life science arena, in which schema changes in data sources are frequent, the maintenance of a common schema for integration could be costly in large federated systems. As the database technology progresses from relational toward object-oriented technology [10], many distributed integration solutions employ object-oriented paradigms to encapsulate the heterogeneity of underlying data sources in life science. These systems typically rely on client–server architectures and software platforms or interfaces such as Common Object Request Broker Architecture (CORBA), an open standards by the Object Management Group (OMG) to facilitate interoperation of disparate components [11, 12].

The Mediator Approach

The most flexible data integration designs adopt a mediator approach that introduces an intermediate processing layer to decouple the underlying heterogeneous distributed data sources and the client layer of end users and applications.

The mediator layer is a collection of software components performing the task of data integration. The concept was first introduced by Wiederhold to provide flexible modular solutions for integration of large information systems with multiple knowledge domains [13, 14].

Most database mediator systems use a wrappers layer to handle the tasks of data access, data retrieval, and data translation. The wrappers access specified data sources, extract selected data, and translate source data formats into a common data model designated for the integration system.

The mediator layer performs the core function of data transformation and integration and communicates with the wrappers and the user application layer. The integration system provides an internal common data model for abstraction of incoming data derived from heterogeneous data sources. Thus, the internal data model must be sufficiently rich to accommodate various data formats of existing biological data sources, which may include unstructured text files, semi-structured XML and HTML files, and structured relational, object-oriented, and nested complex data models. In addition, the internal data model facilitates structuring integrated biological objects to present to the user application layer. The flat, tabular forms of the relational model encounter severe difficulty in model complex and hierarchical biological systems and concepts. XML and other object-oriented models are more natural in model biological systems and are gaining popularity in the community.

In addition to the core integration function, the mediator layer also provides services such as filtering, managing meta-data, and resolving semantic inconsistency in source databases. Ideally, instead of relying on low-level programming efforts, a full integration system supports a high-level query language for data transformation and manipulation. This would greatly facilitate the composition of complex queries across multiple data sources and the management of architecture layers and software components.

The advantage of the mediator approach is its flexibility, scalability, and modularity. The heterogeneity and dynamic nature of the data sources is isolated from the end user applications. Wrappers can readily handle data source schema changes. New data sources can be added to the system by simply adding new wrappers. Scientific analytical tools are simply treated as data sources via wrappers and can be seamlessly integrated with database queries. This approach is most suitable for scientific investigations that need to access the most up-to-date data and issue queries against multiple heterogeneous data sources on demand.

There are many flavors of mediator approaches in life science domains, which differ in database technologies, implementations, internal data models, and query languages. The Kleisli system provides an internal, nested, complex data model and a high-power query and transformation language for data integration [15-17].

The K2 system shares many design principles with Kleisli in supporting a complex data model, but it adopts more object-oriented features [18, 19] (see Chapter 8). The Object-Protocol Model (OPM) supports a rich object model and a global schema for data integration [20, 21]. The IBM DiscoveryLink middleware system is rooted in the relational database technology and supports a full SQL3 [22, 23] (see Chapter 11). The Transparent Access to Multiple Bioinformatics Information Sources (TAMBIS) provides a global ontology to facilitate queries across multiple data sources [24, 25] (see Chapter 7). The Stanford-IBM Manager of Multiple Information Sources (TSIMMIS) is a mediation system for information integration with its own data model, the Object-Exchange Model (OEM), and query language [26].

2.5.2   Meta-Data Specification

Meta-data is data describing data, that is, data that provides documentation on other data managed within an application or environment.

In a structured database environment, the meta-data are formally included in the data schema and type definition. However, few of the biomedical databases use commercial, structured database management systems. The majority of biological data are stored and managed in collections of flat files in which the structure and meaning of the data are not well documented. Furthermore, most biological data are presented to the end users as loosely structured Web pages, even with those databases that have underlying structured database management systems (DBMS).

Many biological data sources provide keyword-search querying interfaces with which a user can input specified Boolean combinations of search terms to access the underlying data. Formulating effective Boolean queries requires domain expertise and knowledge of the contents and structure of the databases. Without meta-data specification, users are likely to formulate queries that return no answers or return an excessively large number of irrelevant answers. In such unstructured or semi-structured data access environments, the introduction of meta-data in the databases across the Web would be important for information gathering and to enhance the user's ability to capture the relevant information independent of data formats.

The need for adequate meta-data specification for scientific analytical algorithms and software tools is particularly acute. Very little attention has been given to meta-data specification in existing programs, especially those available in the public domain from academic institutions. In general, they lack adequate documentation on algorithms, data formats, functionality, and constraints. This could lead to potential misunderstanding of computational tools by the end users. For example, sequence comparison programs are the most commonly used tools to search similar sequences in databases. There are many such programs in the public and private domains. The Basic Local Alignment Tool (BLAST) uses heuristic approximation algorithms to search for related sequences against the databases [27]. BLAST has the advantage of speed in searching very large databases and is a widely used tool. Very often it is an overly used tool in the molecular biology community. The BLAST program trades speed for sensitivity and may not be the best choice for all purposes. The Smith–Waterman dynamic programming algorithm, which strives for optimal global sequence alignment, is more sensitive in finding distantly related sequences [28]. However, it requires substantial computation power and a much slower search speed (50-fold or more). Recently, a number of other programs have been developed using hidden Markov models, Bayesian statistics, and neural networks for pattern matching [29]. In addition to algorithmic differences, these programs vary in accuracy, statistical scoring system, sensitivity, and performance. Without an adequate meta-data specification, it would be a challenge for users to choose the most appropriate program for their application, let alone to use the optimal parameters to interpret the results properly and evaluate the statistical significance of the search results.

In summary, with the current proliferation of biological data sources over the Internet and new data sources constantly springing up around the world, there is an urgent need for better meta-data specification to enhance our ability to find relevant information across the Web, to understand the semantics of scientific application tools, and to integrate information. Ultimately, the communication and sharing of biological data will follow the concept and development of the Semantic Web [30].[2] The Resource Description Format (RDF) schema developed by the Semantic Web offers a general model for meta-data applications such that data sources on the Web can be linked and be understood by both humans and computers.[3]

2.5.3 Data Provenance and Data Accuracy

As databases move to the next stage of development, more and more secondary databases with value-added annotations will be developed. Many of the data providers will also become data consumers. Data provenance and data accuracy become major issues as the boundaries among primary data generated experimentally, data generated through application of scientific analysis programs, and data

---

2. See also *http://www.w3.org/2001/sw.*
3. The RDF Schema is given and discussed at *http://www.w3.org/RDF/overview.html* and *http://www.w3.org/DesignIssues/Semantic.html.*

derived from database searches will be blurred. When users find and examine a set of data from a given database, they will have to be concerned about where the data came from and how the data were generated.

One example of this type of difficulty can be seen with the genome annotation pipeline. The raw experimental output of DNA sequences needs to be characterized and analyzed to turn into useful information. This may involve the application of sequence comparison programs or a sequence similarity search against existing sequence databases to find similar sequences that have been studied in other species to infer functions. For genes/sequences with unknown function, gene prediction programs can be used to identify open reading frames, to translate DNA sequences into protein sequences, and to characterize promoter and regulatory sequence motifs. For genes/sequences that are known, database searches may be performed to retrieve relevant information from other databases for protein structure and protein family classification, genetic polymorphism and disease, literature references, and so on. The annotation process involves computational filtering, transforming, and manipulating of data, and it frequently requires human efforts in correction and curation.

Thus, most curated databases contain data that have been processed with specific scientific analysis programs or extracted from other databases. Describing the provenance of some piece of data is a complex issue. These annotated databases offer rich information and have enormous value, yet they often fail to keep an adequate description of the provenance of the data they contain [31].

With increasingly annotated content, databases become interdependent. Errors caused by data acquisition and handling in one database can be propagated quickly into other databases, or data updated in one database may not be immediately propagated to the other related databases. At the same time, differences in annotations of the same object may arise in different databases because of the application of different scientific algorithms or to different interpretations of results.

Scientific analysis programs are well known to be extremely sensitive to input datasets and the parameters used in computation. For example, a common practice in annotation of an unknown sequence is to infer that similar sequences share common biochemical function or a common ancestor in evolution. The use of different algorithms and different cut-off values for similarity could potentially yield different results for remotely related sequences. Other forms of evidence are required to resolve the inconsistency. This type of biological reasoning also points to another problem. Biological conclusions derived by inference in one database will be propagated and may no longer be reliable after numerous transitive assertions.

Data provenance touches the issue of data accuracy and reliability. It is critical that databases provide meta-data specification on how the data are generated and derived. This has to be as rigorous as the traditional standards for experimental data for which the experimental methods, conditions, and material are provided. Similarly, computationally generated data should be documented with the computational conditions involved, including algorithms, input datasets, parameters, constraints, and so on.

2.5.4   Ontology

On top of the syntactic heterogeneity of data sources, one of the major stumbling blocks in information integration is at the semantic level. In naming and terminology alone, there are inconsistencies across different databases and within the same database. In the major literature database MEDLINE, multiple aliases for genes are the norm, rather than the exception. There are cases in which the same name refers to different genes that share no relationship with each other. Even the term *gene* itself has different meanings in different databases, largely because it has different meanings in various scientific disciplines; the geneticists, the molecular biologists, and the ecologists have different concepts at some levels about genes.

The naming confusion partly stems from the isolated, widely disseminated nature of life science research work. At the height of molecular cloning of genes in the 1980s and 1990s, research groups that cloned a new gene had the privilege of naming the gene. Very often, laboratories working on very different organisms or biological systems independently cloned genes that turned out to encode the same protein. Consequently, various names for the same gene are populated in the published scientific literature and in databases. Biological scientists have grown accustomed to the naming differences. This becomes an ontology issue when information and knowledge are represented in electronic form because of the necessity of communication between human and computers and between computer and computer. For the biological sciences community, the idea and the use of the term *ontology* is relatively new, and it generates controversy and confusion in discussions.

What Is an Ontology?

The term *ontology* was originally a philosophical term that referred to "the subject of existence." The computer science community borrowed the term *ontology* to refer to a "specification of a conceptualization" for knowledge sharing in artificial intelligence [32]. An *ontology* is defined as a description of concepts and relationships that exist among the concepts for a particular domain of knowledge. In the world of structured information and databases, ontologies in life science provide controlled vocabularies for terminology as well as specifying object classes, relations, and functions. Ontologies are essential for knowledge sharing and communications across diverse scientific disciplines.

Throughout the history of the field, the biology community has made a continuous effort to strive for consensus in classifications and nomenclatures. The Linnaean system for naming of species and organisms in taxonomy is one of the oldest ontologies. The nomenclature committee for the International Union of Pure and Applied Chemistry (IUPAC) and the International Union of Biochemistry and Molecular Biology (IUBMB) make recommendations on organic, biochemical, and molecular biology nomenclature, symbols, and terminology. The National Library of Medicine Medical Subject Headings (MeSH) provides the most comprehensive controlled vocabularies for biomedical literature and clinical records. The Systematized Nomenclature of Medicine International, a division of the College of American Pathologists, oversees the development and maintenance of a comprehensive and multi-axial controlled terminology for medicine and clinical information known as SNOMED.

Development of standards is and always has been complex and contentious because getting agreement has been a long and slow process. The computer and IT communities dealt with software standards long before the life science community. Recently, the Object Management Group (OMG), an established organization in the IT community, established a life sciences research group (LSR) to improve communication and interoperability among computational resources in life sciences.[4] LSR uses the OMG technology adoption process to standardize models and interfaces for software tools, services, frameworks, and components in life sciences research.

Because of its longer history and diverse scientific disciplines and constituents, developing standards in the life science community is harder than doing so in the information technology community. Besides the great breadth of academic and research communities in the life sciences, some fields of biology are a century or more older than molecular biology. Thus, the problems are sociological and technological. Standardization further requires a certain amount of stability and certainty in the knowledge content of the field. In contrast, the level, extent, and nature of biological knowledge is still extensively, even profoundly, dynamic in content. The meaning attached to a term may change over time as new facts are discovered that are related to that term. So far, the attempts to standardize the gene names alone have met a tremendous amount of resistance across different biological communities. The Gene Nomenclature Committee (HGNC) led by the Human Genome Organization (HUGO) made tremendous progress to standardize

---

4. This is discussed on the OMG Web site: *http://lsr.omg.org*. OMG is an open-membership, not-for-profit consortium that produces and maintains computer industry specifications for interoperable enterprise application.

gene names for humans with the support of the mammalian genetics community [33]. However, the attempt to expand the naming standard across other species turned out to be more difficult [34]. Researchers working in different organisms or fields have their own established naming usages, and it takes effort to convert to a new set of standards.

An ontology is domain-knowledge specific and context dependent. For example, the term *vector* differs (not surprisingly or problematically) in meaning between its usage in biology and in the physical sciences, as in a mathematical vector. However, within biology, the specific meaning of a term also can be quite different: Molecular biologists use *vector* to mean a vehicle, as in cloning vector, whereas parasitologists use *vector* to refer to an organism as an agent in transmission of disease. Thus, the development of ontologies is a community effort and the adoption of a successful ontology must have wide endorsement and participation of the users. The ecological and biodiversity communities have made major efforts in developing meta-data standards, common taxonomy, and structural vocabulary for their Web site with the help of the National Science Foundation and other government agencies [35].[5] The molecular biology community encompasses a much more diverse collection of sub-disciplines, and for researchers in the molecular biology domain, reaching a community-wide consensus is much harder. To circumvent these issues, there is a flurry of grassroots movements to develop ontologies in specific areas or research such as sequence analysis, gene expression, protein pathways, and so on [36].[6] These group or consortium efforts usually adopt a use case and open source approach for community input. The ontologies are not meant to be mandatory, but instead they serve as a reference framework to go forward for further development. For example, one of the major efforts in molecular biology is the Gene Ontology (GO) consortium, which stems from the annotation projects for the fly genome and the human genome. Its goal is to design a set of structured, controlled vocabularies to describe genes and gene products in organisms [37]. Currently, the GO consortium is focused on building three ontologies for molecular function, biological process, and cellular components, respectively. These ontologies will greatly facilitate queries across genetic and genome databases. The GO consortium started with the core group from the genome databases for the fruit fly, FlyBase; budding yeast, Saccharomyces Genome Database (SGD); and mouse genome database (MGD). It is gaining momentum with growing participants from other genome databases. With such a grassroots

---

5. See also *http://www.nbii.gov/disciplines/systematics.html*, a general systematics site, and *http://www.fgdc.gov*, for geographic data.
6. See the work by the gene expression ontology working group at *http://www.mged.org*.

approach, interactions between different domain ontologies are critical in future development. For example, brain ontology will inevitably relate to ontologies of other anatomical structures or at the molecular level will share ontologies for genes and proteins [38]. A sample collection of ontology resources in life science is listed in the Appendix.

A consistent vocabulary is critical in querying across multiple data sources. However, given the diverse domains of knowledge and specialization of scientific disciplines, it is not foreseeable that in the near future a global, common ontology covering broad biological disciplines will be developed. Instead, in biomedical research alone, there will be multiple ontologies for genomes, gene expression, proteomes, and so on. Semantic interoperability is an active area of research in computer science [39]. Information integration across multiple biological disciplines and sub-disciplines would depend on the close collaborations of domain experts and IT professionals to develop algorithms and flexible approaches to bridge the gaps among multiple biological ontologies.

2.5.5   Web Presentations

Much of the biological data is delivered to end users via the Web. Currently, the biological Web sites resemble a collection of rival medieval city-states, each with its own design, accession methods, query interface, services, and data presentation format [40]. Much of the data retrieval efforts in information integration rely on brittle, screen scraping methods to parse and extract data from HTML files. In an attempt to reduce redundancy and share efforts, an open source movement in the bioinformatics community has began to share various scripts for parsing HTML files from popular data sources such as GenBank report [3], Swiss-Prot report [41], and so forth.

Recently, the biological IT community has been picking up momentum to adopt the merging XML technology for biological Web services and for exchange of data. Many online databases already make their data available in XML format.[7] Semi-structured XML supports user-defined tags to hold data, and thus an XML document contains both data and meta-data. The ability for data sources to exchange information in an XML document strictly depends on their sharing a special document known as Data Type Declaration (DTD), which defines the terms (names for tags) and their data types in the XML document [42]. Therefore, DTD serve as data schema and can be viewed as a very primitive ontology in which DTD defines a set of terms, but not the relationship between terms. XML will

---

7. See the Distributed System Annotation, *http://www.biodas.org*, and the Protein Information Resource, *http://nbrfa.georgetown.edu/pir/databases/pir_xml*.

ease some of the incompatibility problems of data sources, such as data formats. However, semantic interoperability and consistency remain a serious challenge. With the autonomous nature of life science Web sites, one can envision that the naming space of DTD alone could easily create an alphabet soup of confusing terminology as encountered in the naming of genes. Recently, there has been a proliferation of XML-based markup languages to represent models of biological objects and to facilitate information exchange within specific research areas such as microarray and gene expression markup language,[8] systems biology markup language,[9] and bio-polymer markup language.[10] Many of these are available through the XML open standard organization.[11] However, we caution that development of such documents must be compatible with existing biological ontologies or viewed as a concerted community effort.

CONCLUSION

IT professionals and biologists have to work together to address the level of challenges presented by the inherent complexity and vast scales of time and space covered by the life sciences. The opportunities for biological science research in the 21st century require a robust, comprehensive information integration infrastructure underlying all aspects of research. As discussed in the previous sections, substantial progress has been made for data integration at the technical and architectural level. However, data integration at the semantic level remains a major challenge. Before we will be able to seize any of these opportunities, the biology and bioinformatics communities have to overcome the current limitations in metadata specification, maintenance of data provenance and data quality, consistent semantics and ontology, and Web presentations. Ultimately, the life science community must embrace the concept of the Semantic Web [30] as a web of data that is understandable by both computers and people. The bio-ontology efforts for the life sciences represent one important step toward this goal. The brave, early efforts to build computational solutions for biological information integration are discussed in subsequent chapters of this book.

---

8. The MicroArray and Gene Expression (MAGE) markup language is being developed by the Microarray Gene Expression Data Society (see *http://www.mged.org/Workgroups/mage.html*).

9. The Systems Biology Workbench (SBW) is a modular framework designed to facilitate data exchange by enabling different tools to interact with each other (see *http://www.cds.caltech.edu/erato*).

10. The Biopolymer Markup Language (BioML) is an XML encoding schema for the annotation of protein and nucleic acid sequence (see *http://www.bioml.com*).

11. OASIS is an international, not-for-profit consortium that designs and develops industry standard specifications for interoperability based on XML.

REFERENCES

[1] E. Pennisi. "Genome Data Shake Tree of Life." *Science* 280, no. 5364 (1998): 672–674.

[2] J. C. Wooley. "Trends in Computational Biology." *Journal of Computational Biology* 6, no. 314 (1999): 459–474.

[3] A. D. Baxevanis. "The Molecular Biology Database Collection: 2002 Update." *Nucleic Acid Research* 30, no. 1 (2002): 1–12.

[4] P. D. Karp. "Database Links are a Foundation for Interoperability." *Trends in Biotechnology* 14, no. 7 (1996): 273–279.

[5] D. L. Wheeler, D. M. Church, A. E. Lash, et al. "Database Resources of the National Center of Biotechnology Information: 2002 Update." *Nucleic Acids Research* 30, no. 1 (2002): 13–16.

[6] J. Thierry-Meig and R. Durbin. "Syntactic Definitions for the ACeDB Data Base Manager." *AceDB-A* C. elegans *Database*, 1992, http://genome.cornell.edu/acedocs/syntax.html.

[7] T. Head-Gordon and J. C. Wooley. "Computational Challenges in Structural Genomics." *IBM Systems Journal* 40, no. 2 (2001): 265–296.

[8] J. D. Ullmann and J. Widom. *A First Course in Database Systems*. Upper Saddle River, NJ: Prentice Hall, 1997.

[9] R. Resnick. "Simplified Data Mining." *Drug Discovery and Development* October (2000): 51–52.

[10] R. G. G. Cattell. *Object Data Management: Object-Oriented and Extended Relational Database Systems*, revised ed. Reading, MA: Addison-Wiley, 1994.

[11] K. Jungfer, G. Cameron, and T. Flores. "EBI: CORBA and the EBI Databases." In *Bioinformatics: Databases and Systems*, edited by S. Letovsky, 245–254. Norwell, MA: Kluwer Academic Publishers, 1999.

[12] A. C. Siepel, A. N. Tolopko, A. D. Farmer, et al. "An Integration Platform for Heterogeneous Bioinformatics Software Components." *IBM Systems Journal* 40, no. 2 (2001): 570–591.

[13] G. Wiederhold. "Mediators in the Architecture of Future Information Systems." *IEEE Computer* 25, no. 3 (1992): 38–49.

[14] G. Wiederhold and M. Genesereth. "The Conceptual Basis for Mediation Services." *IEEE Expert, Intelligent Systems and Their Applications* 12, no. 5 (1997): 38–47.

[15] S. Davidson, C. Overton, V. Tannen, et al. "BioKleisli: A Digital Library for Biomedical Researchers." *International Journal of Digital Libraries* 1, no. 1 (1997): 36–53.

[16] L. Wong. "Kleisli, A Functional Query System." *Journal of Functional Programming* 10, no. 1 (2000): 19–56.

[17] S. Y. Chung and L. Wong. "Kleisli: A New Tool for Data Integration in Biology." *Trends in Biotechnology* 17 (1999): 351–355.

[18] J. Crabtree, S. Harker, and V. Tannen. "The Information Integration System K2," 1998, http://db.cis.upenn.edu/K2/K2.doc.

[19] S. B. Davidson, J. Crabtree, B. P. Brunk, et al. "K2/Kleisli and GUS: Experiments in Integrated Access to Genomic Data Sources." *IBM Systems Journal* 40, no. 2 (2001): 512–531.

[20] I-M. A. Chen and V. M. Markowitz. "An Overview of the Object-Protocol Model (OPM) and OPM Data Management Tools." *Information Systems* 20, no. 5 (1995): 393–418.

[21] I-M. A. Chen, A. S. Kosky, V. M. Markowitz, et al. "Constructing and Maintaining Scientific Database Views in the Framework of the Object-Protocol Model." In *Proceedings of the Ninth International Conference on Scientific and Statistical Database Management*, 237–248. New York: IEEE, 1997.

[22] L. M. Haas, P. M. Schwartz, P. Kodali, et al. "DiscoveryLink: A System for Integrated Access to Life Science Data Sources." *IBM Systems Journal* 40, no. 2 (2001): 489–511.

[23] L. M. Haas, R. J. Miller, B. Niswonger, et al. "Transforming Heterogeneous Data With Database Middleware: Beyond Integration." *IEEE Data Engineering Bulletin* 22, no. 1 (1999): 31–36.

[24] N. W. Patton, R. Stevens, P. Baker, et al. "Query Processing in the TAMBIS Bioinformatics Source Integration System." In *Proceedings of the 11th International Conference on Scientific and Statistical Database Management*, 138–147. New York: IEEE, 1999.

[25] R. Stevens, P. Baker, S. Bechhofer, et al. "TAMBIS: Transparent Access to Multiple Bioinformatics Information Sources." *Bioinformatics* 16, no. 2 (2000): 184–186.

[26] Y. Papakonstantinou, H. Garcia-Molina, and J. Widom. "Object Exchange Across Heterogeneous Information Sources." In *Proceedings of the IEEE Conference on Data Engineering*, 251–260. New York: IEEE, 1995.

[27] S. F. Altschul, W. Gish, W. Miller, et al. "Basic Local Alignment Search Tool." *Journal of Molecular Biology* 215, no. 3 (1990): 403–410.

[28] T. F. Smith and M. S. Waterman. "Identification of the Common Molecular Subsequences." *Journal of Molecular Biology* 147, no. 1 (1981): 195–197.

[29] D. W. Mount. *Bioinformatics: Sequence and Genome Analysis*. Cold Spring Harbor, NY: Cold Spring Harbor Laboratory Press, 2001.

[30] T. Berners-Lee, J. Hendler, and O. Lassila. "The Semantic Web." *Scientific American* 278, no. 5 (May 2001): 35–43.

[31] P. Buneman, S. Khanna, and W-C. Tan. "Why and Where: A Characterization of Data Provenance." In J. Vander Bussche and V. Vianu. *Proceedings of the Eighth International Conference on Database Theory (ICDT)*, 316–330. Heidelberg, Germany: Springer-Verlag, 2001.

[32] T. R. Gruber. "A Translation Approach to Portable Ontology Specification." *Knowledge Acquisition* 5, no. 2 (1993): 199–220.

[33] H. M. Wain, M. Lush, F. Ducluzeau, et al. "Genew: The Human Gene Nomenclature Database." *Nucleic Acids Research* 30, no. 1 (2002): 169–171.

[34] H. Pearson. "Biology's Name Game." *Nature* 411, no. 6838 (2001): 631–632.

[35] J. L. Edwards, M. A. Lane, and E. S. Nielsen. "Interoperability of Biodiversity Databases: Biodiversity Information on Every Desk." *Science* 289, no. 5488 (2000): 2312–2314.

[36] D. E. Oliver, D. L. Rubin, J. M. Stuart, et al. "Ontology Development for a Pharmacogenetics Knowledge Base." In *Pacific Symposium on Biocomputing*, 65–76. Singapore: World Scientific, 2002.

[37] M. Ashburner, C. A. Ball, J. A. Blacke, et al. "Gene Ontology: Tool for the Unification of Biology." *Nature Genetics* 25, no. 1 (2000): 25–29.

[38] A. Gupta, B. Ludäscher, M. E. Martone. "Knowledge-Based Integration of Neuroscience Data Source." In *Proceedings of the 12th International Conference on Scientific and Statistical Database Management (SSDBM)*, 39–52. New York: IEEE, 2000.

[39] P. Mitra, G. Wiederhold, and M. Kersten. "A Graph-Oriented Model for Articulation of Ontology Interdependencies." In *Proceedings of the Conference on Extending Database Technology (EDBT)*, 86–100. Heidelberg, Germany: Springer-Verlag, 2000.

[40] L. Stein. "Creating a Bioinformatics Nation." *Nature* 417, no. 6885 (2002): 119–120.

[41] A. Bairoch and R. Apweiler. "The SWISS-PROT Protein Sequence Database and Its Supplement TrEMBL in 2000." *Nucleic Acids Research* 28, no. 1 (2000): 45–48.

[42] E. T. Oay. *Learning XML: Guide to Creating Self-Describing Data*. San Jose, CA: O'Reilly, 2001.

3

A Practitioner's Guide to Data Management and Data Integration in Bioinformatics

Barbara A. Eckman

3.1   INTRODUCTION

Integration of a large and widely diverse set of data sources and analytical methods is needed to carry out bioinformatics investigations such as identifying and characterizing regions of functional interest in genomic sequence, inferring biological networks, and identifying patient sub-populations with specific beneficial or toxic reactions to therapeutic agents. A variety of integration tools are available, both in the academic and the commercial sectors, each with its own particular strengths and weaknesses. Choosing the right tools for the task is critical to the success of any data integration endeavor. But the wide variety of available data sources, integration approaches, and vendors makes it difficult for users to think clearly about their needs and to identify the best means of satisfying them. This chapter introduces use cases for biological data integration and translates them into technical challenges. It introduces terminology and provides an overview of the landscape of integration solutions, including many that are detailed in other chapters of this book, along with a means of categorizing and understanding individual approaches and their strengths and weaknesses.

This chapter is written from the point of view of a bioinformatician practicing database integration, with the hope that it will be useful for a wide variety of readers, from biologists who are unfamiliar with database concepts to more computationally experienced bioinformaticians. A basic familiarity with common biological data sources and analysis algorithms is assumed throughout.

The chapter is organized as follows. Section 3.2 introduces traditional database terms and concepts. Those already familiar with these concepts may want to skim that section and begin reading at Section 3.3, which introduces multiple dimensions to integration, thus intermediate terminology. Section 3.4 presents various use cases for integration solutions. Strengths and weaknesses of integration approaches are given in Section 3.5. Section 3.6 is devoted to tough integration problems. Therefore, computer scientists and information technologists may benefit from the advanced problems evoked in Section 3.6.

The goal of this chapter is to convey a basic understanding of the variety of data management problems and needs in bioinformatics; an understanding of the variety of integration strategies currently available, and their strengths and weaknesses; an appreciation of some difficult challenges in the integration field; and the ability to evaluate existing or new integration approaches according to six general categories or dimensions. Armed with this knowledge, practitioners will be well prepared to identify the tools that are best suited to meet their individual needs.

3.2   DATA MANAGEMENT IN BIOINFORMATICS

Data is arguably the most important commodity in science, and its management is of critical importance in bioinformatics. One introductory textbook defines bioinformatics as "the science of creating and managing biological databases to keep track of, and eventually simulate, the complexity of living organisms" [1]. If the central task of bioinformatics is the computational analysis of biological sequences, structures, and relationships, it is crucial that biological sequence and all associated data be accurately captured, annotated, and maintained, even in the face of rapid growth and frequent updates. It is also critical to be able to retrieve data of interest in a timely manner and to define and retrieve data of interest precisely enough to separate effectively its signal from the distracting noise of irrelevant or insignificant data.

3.2.1   Data Management Basics

To begin the discussion of data management in bioinformatics, basic terms and concepts will be introduced by means of *use cases*, examples or scenarios of familiar data management activities. The term *database* will be used both as "a collection of data managed by a database management system" (DBMS) and, more generally, when concepts of data representation are presented, regardless of how the data is managed or stored. Otherwise, the term *data collection* or *data source* will be used for collections of data not managed by a DBMS. For a more detailed explanation of basic data management than is possible in this chapter, see Ullman and Widom's *A First Course in Database Systems* [2].

Use Case: A Simple Curated Gene Data Source

Consider a simple collection of data about known and predicted human genes in a chromosomal region that has been identified as likely to be related to a genetic predisposition for a disease under investigation. The properties stored for each gene are as follows:

✦ GenBank accession number (accnum) [3]

✦ Aliases in other data sources (e.g., Swiss-Prot accession number) [4]

✦ Description of the gene

✦ Chromosomal location

✦ Protein families database (Pfam) classification [5]

✦ Coding sequence (CDS)

✦ Peptide sequence

✦ Gene Ontology (GO) annotation [6]

✦ Has expression results? (Are there expression results for this gene?)

✦ Has Single Nucleotide Polymorphisms (SNPs)? (Are there known SNPs for this gene?)

✦ Date gene was entered

✦ Date gene entry was last modified

The complement of properties stored in a database, along with the relationships among them, is called the database's *schema*. Individual properties, GenBank accession number, are *attributes*. Attributes can be *single-valued*, like peptide sequence, or *multi-valued*, like aliases. Attributes can be *atomic*, like peptide sequence, which is a simple character string, or *nested*, like aliases, which themselves have structure (data source + identifier).

Data accuracy is critically important in scientific data management. Single attributes or groups of attributes must satisfy certain rules or *constraints* for the data to be valid and useful. When entering data into the database, or *populating* it, care must be taken to ensure that these constraints are met. Examples of constraints in the simple gene data source are:

✦ The chromosomal location of each gene must lie within the original region of interest.

✦ The CDS and peptide sequences must contain only valid nucleotide and amino acid symbols, respectively.

✦ The CDS sequence must have no internal in-frame stop codons (which would terminate translation prematurely).

✦ The peptide sequence must be a valid translation of the CDS sequence.

✦ The Pfam classification must be a valid identifier in the Pfam data source.

This simple gene data collection is subject to continual curation, in which new data is inserted, old data is updated, and erroneous data is deleted. A user might make changes to existing entries as more information becomes known, such as more accurate sequence or exon boundaries of a predicted gene, refined GO classification, SNPs discovered, or expression results obtained. A user might also make changes to the source's schema, such as adding new attributes like mouse orthologues, or links to LocusLink, RefSeq [7], or KEGG pathways [8]. New linkage studies may result in a widening or narrowing of the chromosomal region of interest, requiring a re-evaluation of which genes are valid members of the collection and the addition or deletion of genes. Finally, multiple curators may be working on the data collection simultaneously. Care must be taken that an individual curator's changes are completed before a second curator's changes are applied, lest inconsistencies result (e.g., if one curator changes the CDS sequence and the other changes the peptide sequence so they are no longer in the correct translation relationship to one another). The requirement for correctly handling multiple, simultaneous curators' activities is called *multi-user concurrency*.

Databases are only useful, of course, if data of interest can be retrieved from them when needed. In a small database, a user might simply need to retrieve all the attributes at once in a report. More often, however, users wish to retrieve subsets of a database by specifying conditions, or *search predicates*, that the data retrieved should meet. Examples of queries from the curated gene data collection described previously are: "*Retrieve the gene whose GenBank accession number is AA123456*"; "*retrieve only genes that have expression results*"; "*retrieve only genes that contain in their description the words 'serotonin receptor.'*" Search predicates may be combined using logical AND and OR operators to produce more complex conditions; for example, in the query "*Retrieve genes which were entered since 09/01/2002 and lie in a specified sub-region of the chromosomal region of interest*," the conjunction of the two search predicates will be expressed by an operator AND.

Use Case: Retrieving Genes and Associated Expression Results

Along with the simple curated gene data collection, a user may wish to view expression data on the genes that have been gathered through microarray experiments. For example, an expression data source might permit the retrieval of genes that show equal to or greater than two-fold difference in expression intensities between ribonucleic acid (RNA) isolated from normal and diseased tissues. To retrieve all genes with known SNPs with at least two-fold differential expression between normal and diseased tissue, search predicates would need to be applied to each of the two data sources. The result would be the genes that satisfy both of the conditions.

There are many different but equivalent methods of retrieving the genes that satisfy both of these predicates, and an important task of a database system is to identify and execute the most efficient of these alternate methods. For example, the system could first find all genes with SNPs from among the curated genes, and then check the expression values for each of them one by one in the expression data source. Alternatively, the system could find all genes in the expression data source with two-fold expression in normal versus diseased tissue, then find all genes in the curated data source that have SNPs, and finally merge the two lists, retaining only the genes that appear in both lists. Typically, methods differ significantly in their speed due to such factors as the varying speeds of the two databases, the volume of data retrieved, the specificity of some predicates, the lack of specificity of others, and the order in which predicates are satisfied. They may also differ in their usage of computer system resources such as central processing unit (CPU) or disk. Depending on individual needs, the execution *cost* may be defined either as execution time or resource usage (see Chapter 13). The process of estimating costs of various alternative data retrieval strategies and identifying the lowest one among them is known as *cost-based query optimization.*

3.2.2 Two Popular Data Management Strategies and Their Limitations

Two approaches that have commonly been used to manage and distribute data in bioinformatics are spreadsheets and semi-structured text files.

Spreadsheets

Spreadsheets are easy to use and handy for individual researchers to browse their data quickly, perform simple arithmetic operations, and distribute them to collaborators. The cell-based organization of a spreadsheet enables the structuring of data into separate items, by which the spreadsheet may then be sorted. The Microsoft Excel spreadsheet software [9] provides handy data entry features for replicating values in multiple cells, populating a sequence of rows with a sequence of integer identifiers, and entering values into a cell that have appeared in the same column previously.

A disadvantage of spreadsheets, at least as they are typically used, is that very little data validation is performed when data is entered. It is certainly possible, by programming in Microsoft Visual Basic or using advanced Excel features, to perform constraint checking such as verifying that data values have been taken from an approved list of values or *controlled vocabulary*, that numeric data fall in the correct range, or that a specific cell has not been left blank; but in practice this is not often done. Furthermore, while advanced features exist to address this problem, in practice spreadsheets typically include a great deal of repeated or redundant data. For example, a spreadsheet of gene expression data might include the following information, repeated for each tissue sample against which the gene was tested: `GenBank accession number`, `gene name`, `gene description`, `LocusLink Locus ID`, and `UniGene Cluster ID`. If an error should be found in any of these redundant fields, the change would have to be made in each row corresponding to the gene in question. If the change is not made in all relevant rows, an inconsistency arises in the data. In database circles, this inconsistency caused by unnecessary data duplication is called an *update anomaly*.

Another problem with spreadsheets is they are fundamentally single-user data sources. Only one user may enter data into a spreadsheet at a time. If multiple users must contribute data to a data source housed in a spreadsheet, a single curator must be designated. If multiple copies of a spreadsheet have been distributed, and each has been edited and added to by a different curator, it will be a substantial task to harmonize disagreements among the versions when a single canonical version is desired. The spreadsheet itself offers no help in this matter.

Finally, search methods over data stored in spreadsheets are limited to simple text searches over the entire spreadsheet; complex combinations of search conditions, such as *"return serotonin receptors that have SNPs but do not have gene expression results"* are not permitted. Additional limitations of text searches are presented in the next section.

Semi-Structured Text Files

Semi-structured text files, that is, text files containing a more or less regular series of labels and associated values, have data management limitations similar to spreadsheets. A prominent example is the GenBank sequence annotation flat files [10]. It should be noted that the National Center for Biotechnology Information (NCBI) does not store its data in flat file format; rather, the GenBank flat file format is simply a report format based on the structured ASN.1 data representation [11].

An advantage of the semi-structured text format is that it permits more complex, hierarchical (tree-like) structures to be represented. A sequence has multiple references, each of which has multiple authors. Text files are also perhaps the most portable of formats—anyone with a text editor program can view and edit them (unless the file size exceeds the limit of the editor's capability). However, most text editors provide no data validation features. Like spreadsheets, they are not oriented toward use by multiple concurrent users and provide little help in merging or harmonizing multiple copies that have diverged from an original canonical version. Without writing an indexing program, searching a text file is very inefficient because the entire file must be read sequentially, looking for a match to the user's input. Further, it is impossible to specify which part of the flat file entry is to be matched. If a user wants to find mammalian sequences, there is no way to limit the search to the section `organism` of the file to speed the search. As with spreadsheets, full-text searches over text files do not support complex combinations of search conditions. Full-text searches may also result in incorrect data retrieval. For example, consider a flat-file textual data source of human genes and their mouse orthologues, both of which have chromosomal locations. Suppose the user wants to "*find all human genes related to mouse orthologues on mouse chromosome 10*"; simple text-searching permits no way of specifying that the match to `chromosome 10` should refer to the human gene and not the mouse gene. Finally, text editors provide no easy means of retrieving associated data from two related text data sources at once, for example, a GenBank entry and its associated Swiss-Prot entry. More sophisticated search capability over semi-structured, text-formatted data sources is provided by systems like LION Biosciences' Sequence Retrieval System (SRS) [12] (presented in Chapter 5); however, such read-only indexing systems do not provide tools for data validation during curation or solve the multi-user concurrency problem, and they have limited power to compensate for data irregularities in the underlying text files.

3.2.3 Traditional Database Management

This discussion of the limitations of spreadsheets and flat files points toward the advantages of traditional data management approaches. The most mature of these, relational technology, was conceived in 1970 in a seminal paper by E. F. Codd [13]. In the succeeding 30 years, the technology has become very mature and robust, and a great deal of innovative thought has been put into making data retrieval faster and faster. For example, a great step forward was cost-based optimization, or planning a query based on minimizing the expense to execute it, invented in 1979 by Patricia Selinger [14]. Similarly, because relational technology was originally developed for business systems with a high volume of simultaneous inserts, updates and deletes, its ability to accommodate multiple concurrent users is highly advanced.

The Relational Model

A *data model* is the fundamental abstraction through which data is viewed. Although the terms are often confused, a *data model* is not the same as a *schema,* which represents the structure of a particular set of data. The basic element of the *relational data model* is a table (or *relation*) of rows (or *tuples*) and columns (or *attributes*). A representation of gene expression data in tabular fashion means the relational data model is being used. A particular relational schema might contain a gene table whose columns are GenBank accession number, Swiss-Prot accession number, description, chromosomal location, Pfam classification, CDS sequence, peptide sequence, GO annotation, gene expression results, SNPs, date_entered, and date_modified; and a gene expression table whose columns are GenBank_accession number, tissue_ID, and intensity_value.

A number of basic operations are defined on relations, expressed by the *relational algebra operators* [2]:

+ **Projection** (π) produces from a relation R a new relation (noted πR) that has only some of R's columns. In the example, the projection operator might return only the GenBank and Swiss-Prot accession numbers of the genes in the table.

+ **Selection** (σ) produces from a relation R a new relation (noted σR) with a subset of R's rows. For example, this could be the genes that have a Pfam protein kinase domain.

+ The **union** ($\cup$) of two relations R and S (noted $R \cup S$) is the set of rows that are in R or S or both. (R and S must have identical sets of attributes.) For example, if there were 24 separate tables of genes, one for each human chromosome, the union operator could be used to yield a single table containing all the genes in the genome.

+ The **difference** operation noted R–S of two relations R and S is the set of elements that are in R but not in S; for example, this could be the set of GenBank accession numbers that appear in the genes table but are not present in the gene expression table.

+ The **join** ($\bowtie$) of two relations R and S (noted $R \bowtie S$) is a relation consisting of all the columns of R and S, with rows from R and S paired if they agree on particular attribute(s) common to R and S, called the *join attribute(s)*. For example, a user might join the genes table and the expression table on GenBank accession number, pairing genes with their expression results.

The relational algebra operations are the building blocks that may be combined to form more complex expressions, or *queries*, that enable users to ask complex questions of scientific interest. For example, the following query involves projection, selection, union, and join: "*Retrieve the GenBank accession numbers, peptide sequence, and tissue IDs*" [projection] "*for all genes on any chromosome*" [union] "*that have associated SNPs*" [selection] "*and show expression*" [join] "*in central nervous system tissue*" [selection].

A key element of the relational approach is enabling users to describe the behavior they want to ensure or the results they want to retrieve, rather than requiring them to write a program that specifies, step by step, how to obtain the results or ensure the behavior. The Structured Query Language (SQL) [2], the language through which users pose questions to a relational database and specify constraints on relational data, is thus *declarative* rather than *procedural*. Through declarative statements, users can specify that a column value may not be null, that it must be unique in its table, that it must come from a predefined set or range of values, or that it must already be present in a corresponding column of another table. For example, when adding an expression result, the gene used must already be registered in the gene table. Through declarative queries, users can ask complex questions of the data involving many different columns in the database at once, and because relational tables may be indexed on multiple columns, such searches are fast. Advanced search capabilities permit defining subsets of the database and then counting or averaging numeric values over the subset. An example would be listing all tissues sampled and the average expression value in each over a set of housekeeping genes. Performing such computations over subsets of tables is called *aggregation*, and functions like count, average, minimum, and maximum are *aggregate functions*.

Finally, because it is easy to define multiple related tables in a relational database, a user may define separate tables for genes and their aliases, permitting fast searches over multiple aliases and eliminating the need for users to know what type of alias they are searching with, that is, where it comes from (Swiss-Prot, GenBank, etc.). There are two main disadvantages of relational databases when compared to flat file data sources and spreadsheets: Specialized software is required to query the data, and free text searches of the entire entry are not supported in a traditional relational database.

A criticism sometimes made of the relational data model is that it is not natural to model complex, hierarchically structured biological objects as flat, relational tables. For example, an annotated sequence, as represented in GenBank, is a rich structure. The systems in the BioKleisli family (see Chapter 6) address this issue by defining their basic operations on nested relations, that is, relations whose attributes can themselves be relations. Another approach to management of hierarchically structured data is to represent it in eXtensible Markup Language (XML) [15], a structured text data exchange format based on data values combined with *tags* that indicate the data's structure. Special-purpose XML query languages are in development that will enable users to pose complex queries against XML databases and specify the desired structure of the resulting data [16].

Use Case: Transforming Database Structure

Often, transformation of database structures is necessary to enable effective querying and management of biological data. Many venerable data sources no longer represent biological objects optimally for the kinds of queries investigators typically want to pose. For example, it has often been noted that GenBank is sequence-centric, not gene-centric, so queries concerning the structure of individual genes are not easy to express. In contrast, Swiss-Prot is sequence-centric, not domain-centric, so it is rather awkward to ask for proteins with carbohydrate features in a certain domain because all these features are represented in terms of the sequence as a whole.

To illustrate one method of handling data transformations, consider a very simple gene table with attributes `GenBank accession number`, `SwissProt accession number`, and `sequence`. It might be advantageous to enable users to retrieve sequences by accession number without knowing where the accession number originated (GenBank or Swiss-Prot). Creating a separate table for aliases is one solution, particularly if each gene has many different accession numbers, including multiple accession numbers from the same original data source. Another way to permit this search is to transform the database into the following schema: `accession number` and `sequence`. This transformation can be accomplished by retrieving all the `GenBank accession numbers` and their associated sequences, then retrieving all the `SwissProt accession numbers` and their associated `sequences`, and finally doing a union of those two sets. The formula or expression that defines this transformed relation is called a *view*. This expression may be used to create a new table, called a *materialized* view, which exists separately from the original table, so that changes to the original are not applied to the new table. If the expression is not used to create a new table, but only to retrieve data from the original table and transform it on the fly, it is a *non-materialized* view, or simply a view.

Recall the critique that it is not natural to model complex, hierarchically structured biological objects as flat, relational tables. A user might choose relational database technology for storing and managing data due to its efficiency, maturity, and robustness but still wish to present a hierarchical view of the data to the user, one that more closely matches biological concepts. This (non-materialized) view may be accomplished by means of a *conceptual schema* layered on top of the relational database. The biological object layers of Transparent Access to Multiple Bioinformatics Information Sources (TAMBIS) (see Chapter 7) and the Acero Genome Knowledge Platform [17] are efforts in this direction.

3.3 DIMENSIONS DESCRIBING THE SPACE OF INTEGRATION SOLUTIONS

There is nearly universal agreement in the bioinformatics and genomics communities that scientific investigation requires an integrated view of all relevant data. A general discussion of the scope of biological data integration, as well as the obstacles that currently exist for integration efforts, is presented in Chapter 1 of this book. The typical bioinformatics practitioner encounters data in a wide variety of formats, as Chapter 2 presents, including relational databases, semi-structured flat files, and XML documents. In addition, the practicing bioinformatician must integrate the results of analytical applications performing such tasks as sequence comparison, domain identification, motif search, and phylogenetic classification. Finally, Internet sites are also critical due to the traditional importance of publicly funded, public domain data at academic and government Web sites, whether they are central resources or boutique data collections targeting specific research interests. These Internet resources often provide specialized search functionality as well as data, such as the Basic Local Alignment Search Tool (BLAST) at NCBI [18] and the Simple Modular Architecture Research Tool (SMART) at the European Molecular Biology Laboratory (EMBL) [19]. A bioinformatics integration strategy must make sure this specialized search capability is retained.

3.3.1 A Motivating Use Case for Integration

To motivate the need for an integration solution, consider the following use case: "*Retrieve sequences for all human expressed sequence tags (ESTs) that by BLAST are >60% identical over >50 amino acids to mouse channel genes expressed in central nervous system (CNS) tissue.*" For those less familiar with biological terms, a *channel gene* is a gene coding for a protein that is resident in the membrane of a cell and that controls the passage of ions (potassium, sodium, calcium) into and out of the cell. The channels open and close in response to appropriate signals and establish ion levels within the cell. This is particularly important for neural network cells. The data sources used in this query are: the Mouse Genome Database (MGD) at the Jackson Laboratory in Bar Harbor, Maine [20]; the Swiss-Prot protein sequence data source at the Swiss Institute for Bioinformatics, and the BLAST search tool and the GenBank nucleotide sequence data collection at NCBI. The data necessary to satisfy this query are split, or *distributed*, across multiple data sources at multiple sites. One way to integrate these data sources is to enable the user to access them as if they were all components of a single, large database with a single schema. This large *global schema* is an integrated view of all the *local schemas* of the individual data sources. Producing such a global schema is the task of *schema integration*.

This example illustrates six dimensions for categorizing integration solutions:

+ Is data accessed via *browsing* or *querying*?

+ Is access provided via *declarative* or *procedural* code?

+ Is the access code *generic* (used for all similar data sources) or *hard-coded* for the particular source?

+ Is the focus on overcoming *semantic* heterogeneity (heterogeneity of meaning) or *syntactic* heterogeneity (heterogeneity of format)?

+ Is integration accomplished via a *data warehouse* or a *federated* approach?

+ Is data represented in a *relational* or a *non-relational* data model?

As will become evident, some approaches will be better suited to addressing this particular use case than others; this is not intended to prejudice but to clarify the differences among the approaches. The rest of Section 3.3 discusses various alternative approaches to addressing this motivating use case.

3.3.2   Browsing vs. Querying

The relationship between browsing and querying is similar to the relationship in library research between browsing the stacks and conducting an online search. Both are valid approaches with distinct advantages. Browsing, like freely wandering in the stacks, permits relatively undirected exploration. It involves a great deal of leg work, but it is the method of choice when investigators want to explore the domain of interest to help sharpen their focus. It is also well suited to retrieval of a single Web page by its identifier or a book by its call number. On the other hand, querying, like online searching, permits the formulation of a complex search request as a single statement, and its results are returned as a single collated set. Both browsing and querying allow the user to select a set of documents from a large collection and retrieve them. However, browsing stops at retrieval, requiring manual navigation through the resulting documents and related material via static hyperlinks. Querying goes further than retrieval: It accesses the content [21] of the resulting documents, extracts information and manipulates it, for example, dropping some items and performing computations on others. Querying thus makes very efficient use of human time and is the method of choice when an investigator's interests are already focused, especially if aggregations over subsets of data are involved.

While the motivating use case may be successfully addressed using the browsing approach, it is tedious, error-prone, and very cumbersome, involving an average of 70 BLAST result sets consisting of up to 500 EST hits each. In the browsing approach, the user searches for channel sequences expressed in CNS tissues using the MGD query form. The result is 14 genes from 17 assays. The user then visits each gene's MGD page. Assume that the user is only interested in Swiss-Prot sequences and that each gene has an average of five associated Swiss-Prot sequence entries. The user has to visit each sequence's Swiss-Prot page, from which a BLAST search against gbest (the EST portion of GenBank) is launched. Each BLAST result must be inspected to eliminate non-human sequence hits and alignments that do not meet the inclusion criteria (>60% identity over >50 amino acids) and to eliminate duplicate ESTs hit by multiple Swiss-Prot sequences. Finally, the full EST sequences for all the hits that survive must be retrieved from GenBank. If the browsing approach was used to satisfy this query, these steps would then be repeated for each of the 14 genes returned by the initial query (Figure 3.1).

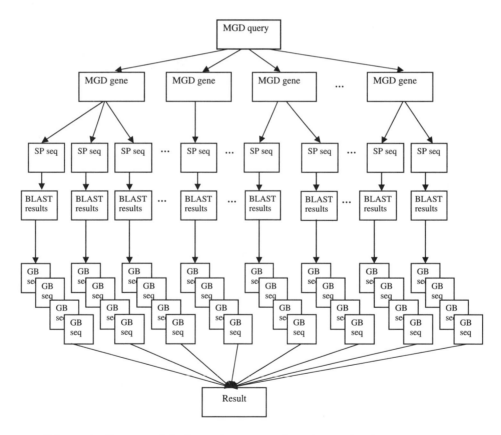

3.1

FIGURE

Schematic diagram of the browsing approach to the motivating use case.

Querying Approach

"Show me all human EST sequences that are >60% identical over 50 AA to mouse channel genes expressed in CNS"

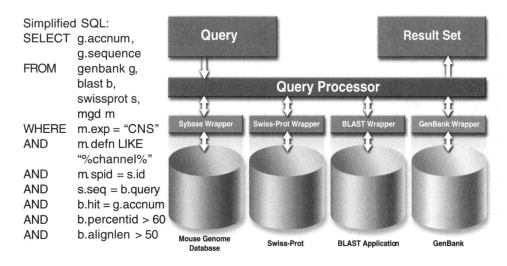

Simplified SQL:
```
SELECT  g.accnum,
        g.sequence
FROM    genbank g,
        blast b,
        swissprot s,
        mgd m
WHERE   m.exp = "CNS"
AND     m.defn LIKE
        "%channel%"
AND     m.spid = s.id
AND     s.seq = b.query
AND     b.hit = g.accnum
AND     b.percentid > 60
AND     b.alignlen > 50
```

3.2

FIGURE

The querying approach to the motivating use case.

In a querying approach to this problem, a short SQL query is submitted to the query processor. The query processor visits MGD to identify channel genes expressed in CNS, and the Swiss-Prot Web site to retrieve their sequences. For each of these sequences, it launches a BLAST search against gbest, gathers the results, applies the stringency inclusion criteria, and finally retrieves the full-length EST sequences from GenBank (Figure 3.2).

3.3.3   Syntactic vs. Semantic Integration

As stated previously, syntactic integration addresses heterogeneity of form. Gen-Bank is a structured file, MGD is a Sybase (relational) database, and BLAST is an analytical application. These differences in form are overcome in the browsing strategy by providing a Web-based front end to the sources and in the querying strategy by providing SQL access to all the sources. Contrariwise, semantic integration addresses heterogeneity of meaning. In GenBank, a gene is an annotation on a sequence, while in MGD a gene is a locus conferring phenotype (e.g., black hair, blindness). Neither of the integration approaches in this example specifically focuses on resolving this heterogeneity of meaning. They rely instead on the user's knowledge of the underlying data sources to combine data from the sources in scientifically meaningful ways.

3.3.4 Warehouse vs. Federation

In a warehousing approach to integration, data is migrated from multiple sources into a single DBMS, typically a relational DBMS. As it is copied, the data may be cleansed or filtered, or its structure may be transformed to match the desired queries more closely. Because it is a copy of other data sources, a warehouse must be refreshed at specified times—hourly, daily, weekly, monthly, or quarterly. A data warehouse may contain multiple *data marts*, subset warehouses designed to support a specific activity or inquiry.

While a warehouse replicates data, a federated approach leaves data in its native format and accesses it by means of the native access methods. In the previous example, the querying approach is a federated approach—it accesses MGD as a Sybase database, Swiss-Prot and GenBank as Web sites, and BLAST via run-time searches, and it integrates their results using complex software known as *middleware*. An alternative demonstration of the querying approach could have imported GenBank, Swiss-Prot, MGD, and the results of BLAST searches into Sybase, Oracle, or IBM DB2 database systems and executed the retrievals and filtering there. This would have been an example of the warehousing approach.

3.3.5 Declarative vs. Procedural Access

As discussed previously, declarative access means stating what the user wants, while procedural access specifies how to get it. The typical distinction opposes the use of a query language (e.g., SQL) and writing access methods or sub-routines in Perl, Java, or other programming languages to access data. In the motivating use case in Section 3.3.1, the querying approach uses the SQL query language. Alternatively, Perl [22] sub-routines or object methods that extract data from MGD, Swiss-Prot, and GenBank and run the necessary BLAST searches could have performed the task.

3.3.6 Generic vs. Hard-Coded

The federated approach in the previous example was generic; it assumed the use of a general-purpose query execution engine and general purpose *wrappers* (software modules tailored to a particular family of data sources) for data access. An example of a hard-coded approach to the problem would be writing a special purpose Perl script to retrieve just the information needed to answer this particular question. A generic system enables users to ask numerous queries supporting a variety of scientific tasks, while a hard-coded approach typically answers a single query and supports users in a single task. Generic approaches generally involve higher up-front development costs, but they can pay for themselves many times over in flexibility and ease of maintenance because they obviate the need for extensive programming every time a new research question arises.

3.3.7   Relational vs. Non-Relational Data Model

Recall that a data model is not a specific database schema, but rather something more abstract: the way in which data are conceptualized. For example, in the relational data model, the data are conceptualized as a set of tables with rows and columns. Oracle, Sybase, DB2, and MySQL are all DBMSs built on the relational model. In data management systems adhering to a non-relational data model, data may be conceptualized in many different ways, including hierarchical (tree-like) structures, ASCII text files, or Java or Common Object Request Broker Architecture (CORBA) [23] objects. In the motivating example, MGD is relational, and the other sources are non-relational.

3.4   USE CASES OF INTEGRATION SOLUTIONS

The motivating use case in Section 3.3.1 permitted a brief outline of the six dimensions for categorizing integration solutions. To further elucidate these dimensions and demonstrate their use, this section describes each dimension in greater detail, presents a prototypical featured solution, and categorizes the featured integration solution on all six dimensions.

3.4.1   Browsing-Driven Solutions

As in the previous example, in a browsing approach users are provided with interactive access to data, allowing them to step sequentially through the exploratory process. A typical browsing session begins with a query form that supports a set of pre-defined, commonly posed queries. After the user has specified the parameters of interest and the query is executed, a summary screen is typically returned. From here the user may drill down, one by one, into the individual objects meeting the search criteria and from there view related objects by following embedded links, such as hypertext markup language (HTML) or XML hyperlinks. The data source(s) underlying a browsing application may be warehoused or federated, and relational or non-relational. Browsing applications are ubiquitous on the Internet; examples are Swiss-Prot and the other data collections on the Expert Protein Analysis System (ExPASy) server at the Swiss Institute of Bioinformatics [24], the FlyBase Web site for Drosophila genetics [25, 26], and the featured example, the Entrez Web site at NCBI [10].

Browsing Featured Example: NCBI Entrez

As an example of the browsing approach, consider the following query: "*Find in PubMed articles published in 2002 that are about human metalloprotease genes and retrieve their associated GenBank accession numbers and sequences.*" The sequence of steps in answering this query is shown in Figures 3.3 through 3.7. First the user enters the Entrez Boolean search term "metalloprotease AND human AND 2002 [pdat]" in the PubMed online query form [7]. The result is a summary of qualifying hits; there were 1054 in December 2002 (Figure 3.3). From here, the user can visit individual PubMed entries (Figure 3.4), read their abstracts, check for a GenBank sequence identifier in the secondary source ID attribute (Figure 3.5), and visit the associated GenBank entry to retrieve the sequence.

3.3 PubMed articles published in 2002 on human metalloprotease genes.

FIGURE

3.4

FIGURE

One of the qualifying PubMed abstracts.

Alternatively, the user can take advantage of the LinkOut option on the PubMed entry page (Figures 3.4 and 3.6), which enables access to the sequence information provided by LocusLink (Figure 3.7). Notice that there are many more navigation paths to follow via hyperlinks than are described here; a strength of the browsing approach is that it supports many different navigation paths through the data.

The categorization of Entrez based on the six dimensions is given in Table 3.1 on page 56.

3.4.2   Data Warehousing Solutions

In the data warehousing approach, data is integrated by means of replication and storage in a central repository. Often data is cleaned and/or transformed during the loading process. While a variety of data models are used for data warehouses, including XML and ASN.1, the relational data model is the most popular choice (e.g., Oracle, Sybase, DB2, MySQL). Examples of the integration solutions following the data warehousing approach include Gene Logic's GeneExpress Database

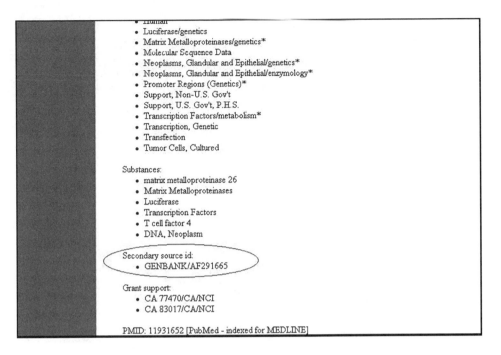

3.5   Checking for GenBank references in the PubMed entry.

FIGURE

(presented in Chapter 10) [27], the Genome Information Management System of the University of Manchester [28], the data source underlying the GeneCards Web site at the Weizmann Institute in Israel [29, 30], and AllGenes [31], which will serve as the featured example.

Warehousing Featured Example: AllGenes

A research project of the Computational Biology and Informatics Laboratory at the University of Pennsylvania, AllGenes is designed to provide access to a database integrating every known and predicted human and mouse gene, using only publicly available data. Predicted human and mouse genes are drawn from transcripts predicted by clustering and assembling EST and messenger RNA (mRNA) sequences. The focus is on integrating the various types of data (e.g., EST sequences, genomic sequence, expression data, functional annotation). Integration is performed in a structured manner using a relational database and controlled vocabularies and ontologies [32]. In addition to clustering and assembly, significant cleansing and transformation are done before data is loaded onto AllGenes, making data warehousing an excellent choice.

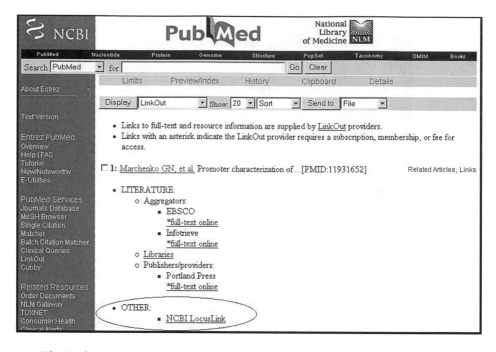

3.6

FIGURE

The LinkOuts page enables access to LocusLink.

A sample query for AllGenes is the following: "*Show me the DNA repair genes that are known to be expressed in central nervous system tissue.*" The query is specified and run using a flexible query-builder interface (Figure 3.8), yielding a summary of qualifying assemblies (Figure 3.9). From the query result page the user can visit a summary page for each qualifying assembly, which includes such valuable information as predicted GO functions; hyperlinks to GeneCards, the Mouse Genome Database (MGD), GenBank, ProDom, and so on; Radiation Hybrid (RH) Map locations; the 10 best hits against the GenBank non-redundant protein database (nr); and the 10 best protein domain/motif hits.

The categorization of AllGenes based on the six dimensions is given in Table 3.2 on page 58.

3.4.3   Federated Database Systems Approach

Recall that in a federated approach, data sources are not migrated from their native source formats, nor are they replicated to a central data warehouse. The data sources remain autonomous, data is integrated on the fly to support specific queries

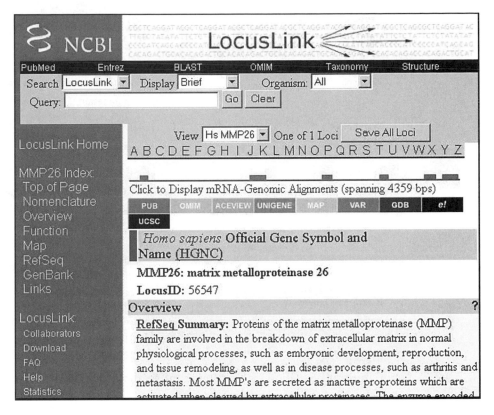

3.7

FIGURE

Sequences may be obtained from LocusLink entries corresponding to PubMed articles.

or applications, and access is typically through a declarative query language. Examples of federated systems and their data models include complex-relational systems, such as BioKleisli/K2 (Chapter 8) and its cousin GeneticXchange's K1 (see Chapter 6), object-relational systems (OPM/TINet) [33], and IBM's relational system DiscoveryLink, which is detailed in Chapter 10 and will serve as the featured example [34].

Federated Featured Example: DiscoveryLink

The motivating use case of Section 3.3.1 is a good fit for a federated approach like DiscoveryLink's. DiscoveryLink provides transparency: The federation of diverse types of data from heterogeneous sources appears to the user or the application as a single large database, in this case a relational database. The SQL query language

Browsing	**Querying**
Interactive Web browser access to data	No querying capability
Semantic	**Syntactic**
No semantic integration	Provides access to nucleotide and protein sequence, annotation, MEDLINE abstracts, etc.
Warehouse	**Federation**
Provides access to data sources at NCBI	No federation
Declarative Access	**Procedural Access**
No declarative access	Access via Entrez Programming Utilities (E-utilities)
Generic	**Hard-Coded**
Not generic	Hard-coded for NCBI sources only
	Links are hard-coded indices
Relational Data Model	**Non-Relational Data Model**
Relational data model not used	Data stored in the ASN.1 complex-relational data model

3.1

TABLE

Entrez categorization with respect to the six dimensions of integration.

is supported over all the federated sources, even if the underlying sources' native search capabilities are less full-featured than SQL; a single federated query, as in the earlier motivating example, typically combines data from multiple sources. Similarly, specialized non-SQL search capabilities of the underlying sources are also available as DiscoveryLink functions.

The architecture of DiscoveryLink appears in Chapter 10 (Figure 10.1). At the far right are the data sources. To these sources, DiscoveryLink looks like an application—they are not changed or modified in any way. DiscoveryLink talks to the sources using *wrappers*, which use the data source's own client-server mechanism to interact with the sources in their native dialect. DiscoveryLink has a local catalog in which it stores information (*meta-data*) about the data accessible (both local data, if any, and data at the back end data sources). Applications of DiscoveryLink manipulate data using any supported SQL Application Programming Interface (API); for example, Open Database Connectivity (ODBC) or Java DataBase Connectivity (JDBC) are supported, as well as embedded SQL. Thus a DiscoveryLink application looks like any normal database application.

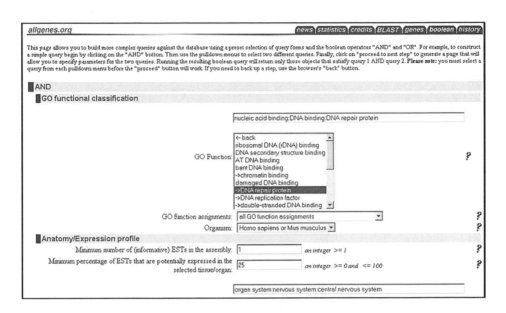

3.8 The AllGenes query builder.

FIGURE

3.9 Results of the AllGenes query.

FIGURE

Browsing	**Querying**
Interactive Web browser access to data	Limited querying capability via parameterized query builder
Semantic	**Syntactic**
Ontologies for semantic integration	Data warehousing for syntactic integration
Warehouse	**Federation**
Data stored in relational warehouse	Not a federation
Declarative Access	**Procedural Access**
Under the covers; users use parameterized query builder	No procedural access
Generic	**Hard-Coded**
Information not available	Information not available
Relational Data Model	**Non-Relational Data Model**
Data stored in Oracle DBMS	Not used

3.2   AllGenes categorization with respect to the six dimensions of integration.

TABLE

The categorization of DiscoveryLink based on the six dimensions is given in Table 3.3.

3.4.4   Semantic Data Integration

Recall that semantic data integration focuses on resolving heterogeneity of meaning, while syntactic data integration focuses on heterogeneity of form. In a volume on management of heterogeneous database systems, Kashyap and Sheth write:

> In any approach to interoperability of database systems [database integration], the fundamental question is that of identifying objects in different databases that are semantically related and then resolving the schematic [schema-related] differences among semantically related objects. [35]

This is the fundamental problem of semantic data integration. The same protein sequence is known by different names or accession numbers (synonyms) in GenBank and Swiss-Prot. The same mouse gene may be represented as a genetic map locus in MGD, the aggregation of multiple individual exon entries in GenBank, and a set of EST sequences in UniGene; in addition, its protein product may be an entry in

Browsing	Querying
No browsing capability	Full *ad hoc* SQL query language
Semantic	**Syntactic**
No semantic integration	• Maps heterogeneous sources into relational model
	• Maps SQL into native query languages of sources
Warehouse	**Federation**
Not available, though warehouses may be members of a DiscoveryLink federation	Integrates heterogeneous sources through wrappers and middleware
Declarative Access	**Procedural Access**
SQL query language	No procedural access
Generic	**Hard-Coded**
Query processor, most wrappers	Some access wrappers (e.g., BLAST)
Relational Data Model	**Non-Relational Data Model**
Built on top of DB2	Not used

3.3

TABLE

DiscoveryLink categorization with respect to the six dimensions of integration.

Swiss-Prot and its human orthologues may be represented as a disease-associated locus in Online Mendelian Inheritance in Man (OMIM) [36]. Semantic integration also deals with how different data sources are to be linked together. For example, according to documentation at the Jackson Lab Web site [37], MGD links to Swiss-Prot through its marker concept, to RatMap [38] through orthologues, to PubMed through references, and to GenBank through either markers (for genes) or molecular probes and segments (for anonymous DNA segments). Finally, a schema element with the same names in two different data sources can have different semantics and therefore different data values. For example, retrieving orthologues to the human BRCA1 gene in model organisms from several commonly used Web sites yields varying results: GeneCards returns the BRCA1 gene in mouse and *C. elegans*; MGD returns the mouse, rat, and dog genes; the Genome DataBase (GDB) [39, 40] returns the mouse and drosophila genes; and LocusLink returns only the mouse gene.

Approaches to semantic integration in the database community generally center on schema integration: understanding, classifying, and representing schema differences between two disparate databases. For example, in capturing the semantics of the relationships between objects in multiple databases, Kashyap and Sheth describe work on understanding the context of the comparison, the abstraction relating the domains of the two objects, and the uncertainty in the relationship [35].

Bioinformatics efforts at semantic integration have largely followed the approach of the artificial intelligence community. Examples of such semantic integration efforts are the Encyclopedia of *Escherichia coli* genes and metabolism (EcoCyc) [41], GO, and TAMBIS, the featured example and the subject of Chapter 7 of this book.

Semantic Integration Featured Example: TAMBIS

The TAMBIS system is the result of a research collaboration between the departments of computer science and biological sciences at the University of Manchester in England. Its chief components are an ontology of biological and bioinformatics terms managed by a terminology server and a wrapper service that, as in DiscoveryLink, handles access to external data sources. An *ontology* is a rigorous formal specification of the conceptualization of a domain. The TAMBIS ontology (TaO) [42] describes the biologist's knowledge in a manner independent of individual data sources, links concepts to their real equivalents in the data sources, mediates between (near) equivalent concepts in the sources, and guides the user to form appropriate biological queries. The TaO contains approximately 1800 asserted biological concepts and their relationships and is capable of inferring many more. Coverage currently includes proteins and nucleic acids, protein structure and structural classification, biological processes and functions, and taxonomic classification.

The categorization of TAMBIS based on the six dimensions is given in Table 3.4.

3.5   STRENGTHS AND WEAKNESSES OF THE VARIOUS APPROACHES TO INTEGRATION

This chapter has described multiple approaches to database integration in the bioinformatics domain and provided examples of each. Each of these approaches has strengths and weaknesses and is best suited to a particular set of integration needs.

Browsing	**Querying**
Interactive browser	Limited querying capability via parameterized query builder
Semantic	**Syntactic**
According to TAMBIS' authors, its "big win" lies in the ontology	Integrates via its wrapper service
Warehouse	**Federation**
Not used	Uses BioKleisli for federated integration
Declarative Access	**Procedural Access**
Uses the CPL query language, but users see only the parameterized query builder	No procedural access
Generic	**Hard-Coded**
Information not available	Information not available
Relational Data Model	**Non-Relational Data Model**
Relational data model not used	Object/complex-relational data model

3.4

TABLE

TAMBIS categorization with respect to the six dimensions of integration.

3.5.1   Browsing and Querying: Strengths and Weaknesses

The strengths of a browsing approach are many. As noted previously, its interactive nature makes it especially well suited to exploring the data landscape when an investigator has not yet formulated a specific question. It is also well suited to retrieval of information about single objects and for optionally drilling down to greater levels of detail or for following hyperlinks to related objects. The ubiquity of the Internet makes Web browsers familiar to even the most inexperienced user.

The weaknesses of a browsing approach are the flip side of its strengths. Because it is fundamentally based on visiting single pages containing data on a single object, it is not well suited to handling large data sets or to performing a large, multi-step workflow including significant processing of interim results. Its flexibility is also limited, as the user is confined to the query forms and navigation paths the application provides.

The strengths of a querying approach are the natural opposite of those of the browsing approach. Because it is based on specifying attributes of result sets via a query language, often with quite complex search conditions, the querying approach is well suited to multi-step workflows resulting in large result sets. This approach is also flexible, allowing the user to specify precisely inclusion and exclusion criteria and noting which attributes to include in the final result set. Contrariwise, the querying approach is not as well suited to the exploration or manual inspection of interim results, and the need to specify desired results using query language syntax requires more computational sophistication than many potential users possess.

3.5.2　Warehousing and Federation: Strengths and Weaknesses

A major strength of a data warehousing approach is that it permits cleansing and filtering of data because an independent copy of the data is being maintained. If the original data source is not structured optimally to support the most commonly desired queries, a warehousing approach may transform the data to a more amenable structure. Copying remote data to a local warehouse can yield excellent query performance on the warehouse, all other things being equal. Warehousing exerts a load on the remote sources only at data refresh times, and changes in the remote sources do not directly affect the warehouse's availability.

The primary weakness of the data warehousing approach is the heavy maintenance burden incurred by maintaining a cleansed, filtered, transformed copy of remote data sources. The warehouse must be refreshed frequently to ensure users' access to up-to-date data; the warehousing approach is probably not the method of choice for integrating large data sources that change on a daily basis. Adding a data source to a warehouse requires significant development, loading, and maintenance overhead; therefore this approach is unlikely to scale well beyond a handful of data sources. Warehousing data may lose the specialized search capability of the native data sources; an example would be specialized text searching over documents or sub-structure searching over chemical compound data collections.

A major strength of the federated approach is that the user always enjoys access to the most up-to-date data possible. While connectivity to remote sources requires some maintenance, the burden of adding and maintaining a new data source is considerably less than in the warehousing case. The federated approach scales well, even to very large numbers of data sources, and it readily permits new sources to be added to the system on a prototype or trial basis to evaluate their potential utility to users. In a fast-paced, ever-changing field like bioinformatics, this nimbleness is invaluable. The federated approach meshes well with a landscape of many individual, autonomous data sources, which the bioinformatics community currently boasts. Finally, a federated system can provide access to data that cannot be easily copied into a warehouse, such as data only available via a Web site.

Any data cleansing must be done on the fly, for a federation accesses remote data sources in their native form. The members of the federation must be able to handle the increased load put on them by federated queries, and if network bandwidth is insufficient, performance will suffer.

3.5.3 Procedural Code and Declarative Query Language: Strengths and Weaknesses

Procedural code may be tuned very precisely for a specific task. There are virtually no limitations on its expressive power; however, this very strength can make it difficult to optimize. *Ad hoc* inquiries can be difficult to support, and extending the system to handle additional sources or additional queries can be difficult.

Declarative languages are flexible and permit virtually unlimited *ad hoc* querying. Queries expressed in a declarative language are relatively easy to program and maintain due to their small size and economy of expression. Sometimes, however, their simplicity is misleading; for example, it is easy to write a syntactically correct SQL query, but the results returned may not be what was intended because the query was written using the wrong constructs for the desired meaning. Finally, some programming tasks are much more easily written in a procedural language than a declarative one; the classic example is recursive processing over tree-like structures.

3.5.4 Generic and Hard-Coded Approaches: Strengths and Weaknesses

Generic coding is generally acknowledged to be desirable, where practicable, due to its extensibility and maintainability and because it facilitates code re-use. It does, however, yield a greater up-front cost than programming hard-coded for a specific task, and sometimes schedules do not permit this up-front expenditure. If the instances being generalized are not sufficiently similar, the complexity of generic code can be prohibitive.

Hard-coding permits an application to be finely tuned to optimize for a specific critical case, potentially yielding very fast response times; this approach may be the preferred strategy when only a limited set of queries involving large datasets is required. In the absence of an already existing generic system, it is generally quicker to prototype rapidly by hard coding. On the other hand, code with many system-specific assumptions or references can be difficult to maintain and extend. Adding a new data source or even a new query often means starting from scratch.

3.5.5   Relational and Non-Relational Data Models: Strengths and Weaknesses

The relational data model is based on a well-understood, theoretically rock-solid foundation. Relational technology has been maturing for the past 30 years and can provide truly industrial-strength robustness and constant availability. Relational databases prevent anomalies while multiple users are reading and writing concurrently, thus safeguarding data integrity. Optimization of queries over relational databases has been developed and honed for decades. The SQL query language is powerful and widely used, so SQL programmers are relatively easy to find. However, the relational model is based on tables of rows and columns, and several individual tables are typically required to represent a single complex biological object.

Hierarchical non-relational data models seem to be a more natural fit for complex scientific objects. However, this technology is still quite immature, and standard database *desiderata* such as cost-based query optimization, data integrity, and multi-user concurrency have been hard to attain because of the increased complexity of the non-relational systems.

3.5.6   Conclusion: A Hybrid Approach to Integration Is Ideal

Considering the variety of integration needs in a typical organization, a hybrid approach to database integration is generally the best strategy. For data that it is critical to clean, transform, or hand curate, and for which only the best query performance is adequate, data warehousing is probably the best approach. If the warehouse is derived from data outside the organization, it is best if the original data source changes infrequently, so the maintenance burden in merging updates is not too onerous. Otherwise, the federated model is an excellent choice because of its relatively low maintenance cost and its extensibility and scalability. Federations allow easy prototyping and swapping of new data sources for old in evaluation mode, and they permit integration of external data that is not accessible for duplicating internally, such as data only available via Web sites. They also permit the integration of special purpose search algorithms such as sequence comparison, secondary structure prediction, text mining, clustering, chemical structure searching, and so forth. Wherever possible, strategies should be generic, except for one-time, one-use programs or where hard-coding is needed to fine tune a limited set of operations over a limited set of data.

Both browsing and querying interfaces are important for different levels of users and different needs. For access to data in batch mode, the most common queries can be pre-written and parameterized and offered to users via a Web form-based interface. Both semantic and syntactic data integration are needed, although semantic integration is just beginning to be explored and understood.

Due to the maturity of the technology and its industrial strength, the relational data model is currently the method of choice for large integration efforts, both warehousing and federation. A middle software layer may be provided to expose biological objects to users, as mentioned previously. But based on the current state of the industry, the underlying data curation, storage, querying planning, and optimization are arguably best done in relational databases.

3.6   TOUGH PROBLEMS IN BIOINFORMATICS INTEGRATION

In spite of the variety of techniques and approaches to data integration in bioinformatics, many tough integration problems remain. These include query processing in a federated system when some members of the federation are inaccessible; universally accepted standards of representation for central biological concepts such as gene, protein, transcript, sequence, polymorphism, and pathway; and representing and querying protein and DNA interaction networks. This section discusses two additional examples of tough problems in bioinformatics integration: semantic query planning and schema management.

3.6.1   Semantic Query Planning Over Web Data Sources

While the TAMBIS and GO projects have made an excellent start in tackling the semantic integration problem, more remains to be done. TAMBIS and GO have focused on building ontologies and controlled vocabularies for biological concepts. Another fruitful area of investigation in semantic integration is using knowledge of the semantics of data sources to generate a variety of alternative methods of answering a question of scientific interest, thus freeing the user from the need to understand every data source in detail [43].

Recall that in accessing multiple data sources there are usually multiple ways of executing a single query, or multiple *query execution plans*. Each may have a different execution cost, as discussed in Section 3.2.1. Similarly, there may be multiple data sources that can be used to arrive at an answer to the same general question, though the semantics of the result may differ slightly. A semantic query planner considers not only the cost of different execution plans but also their semantics and generates alternate paths through the network of interconnected data sources. The goal is to help the user obtain the best possible answers to questions of scientific interest.

Web sources are ubiquitous in bioinformatics, and they are connected to each other in a complex tangle of relationships. Links between sources can be either explicit hypertext links or constructed calls in which an identifier for a remote data source may be extracted from a Web document and used to construct a Uniform Resource Identifier (URI) to access the remote source.

Not all inter-data source links are semantically equivalent. For example, there are two ways of navigating from PubMed to GenBank: through explicit occurrences of GenBank accession numbers within the secondary source identifier (SI) attribute of the MEDLINE formatted entry and through the Entrez Nucleotide Link display option. Following these two navigation paths does not always produce the same set of GenBank entries: For example, for the PubMed entry with ID 8552191, there are four embedded GenBank accession numbers, while the Nucleotide Links option yields 10 sequence entries (the four embedded entries plus related RefSeq entries) [43].

To generate alternate plans, a semantic query planner requires knowledge of certain characteristics of Web sources, including their query and search *capabilities*, the links between sources, and overlaps between the contents of data sources. An example of modeling a subset of the search capabilities of PubMed is as follows:

1. Search by key (`PubMedID` or `MedlineID`), returning a single entry. Single or multiple bindings for `PubMedID` or `MedlineID` are accepted.

2. Search by phrase, returning multiple entries. For example, the search term `gene expression` performs an untyped text search; `Science [journal]` returns all articles from *Science*; and `2001/06:2002[pdat]` returns articles published since June 2002.

To further illustrate semantic query planning, consider the following query: "*Given a Human Genome Organization (HUGO) name, retrieve all associated PubMed citations.*" There are at least three plans for this query:

1. Search PubMed directly for the HUGO name using the second search capability above.

2. Find the GeneCards entry for the HUGO name and follow its link to PubMed publications.

3. From the GeneCards entry for the HUGO name, follow the links to the Entrez RefSeq entry and extract the relevant PubMed identifiers.

These three plans all return different answers. For example, given the HUGO name BIRC1 (neuronal apoptosis inhibitory protein), plan 1 returns no answer, plan 2 returns two answers (PubMed identifiers 7813013 and 9503025), and plan 3 returns five answers (including the two entries returned by plan 2) [43].

In summary, a query planner who took advantage of semantic knowledge of Web data sources and their search capabilities would first identify that there are multiple alternate sources and capabilities to answer a query. Then, semantic knowledge would be used to determine if the results of each alternate plan would be identical. Finally, such a planner might suggest these alternate plans to a user, whose expert judgment would determine which plan was the most suitable to the scientific task. The user would be freed to focus on science instead of on navigating the often treacherous waters of data source space. Semantic query planning will be addressed in more detail in Section 4.4.2 in Chapter 4.

3.6.2   Schema Management

A *schema management* system supports databases and information systems as they deal with a multitude of schemas in different versions, structure, semantics, and format. Schema management is required whenever data is transformed from one structure to another, such as publishing relational data as XML on a Web site, restructuring relational data in hierarchical form for a biological object conceptual view layer, and integrating overlapping data sets with different structures, as needed in a merger of two large pharmaceutical companies. The system developed by the Clio research project [44] is an example of a basic schema management system with plans for development in the direction to be described.

The six building blocks of a schema management system are listed below. They will be defined and illustrated through three use cases.

+ Schema association/schema extraction
+ Schema versioning/schema evolution
+ Schema mapping/query decomposition
+ View building/view composition
+ Data transformation
+ Schema integration

Use Case: Data Warehousing

As described earlier, data warehousing is often used as an integration approach when the data must be extensively cleaned, transformed, or hand-curated. A warehouse may be built from a variety of data sources in different native formats. *Schema association* determines if these heterogeneous documents match a schema already stored in the schema manager. If no existing schema is found, *schema extraction* determines a new schema based on the data, for example, an XML document, and adds it to the schema manager. *Schema integration* helps develop a warehouse global schema accommodating all relevant data sources. As the warehouse evolves, *schema mapping* determines how to map between the schemas of newly discovered data sources and the warehouse's global schema. Finally, *data transformation* discovers and executes the complex operations needed to clean and transform the source data into the global warehouse schema. The transformations generated would be specified in the XML query language (XQuery) or Extensible Stylesheet Language Transformations (XSLT) for XML data, and SQL for relational data.

Use Case: Query and Combine Old and New Data

Because bioinformatics is a young, research-oriented field, database schemas to hold lab notebook data change frequently as new experimental techniques are developed. Industry standards are still emerging, and they evolve and change rapidly. Suppose two Web sites at a large pharmaceutical company were using two different versions of the same database schema, but scientists wanted to query the old-version and new-version data sources in uniform fashion, without worrying about the schema versions. *Schema evolution* keeps track of schema versions and their differences. *Query decomposition* allows the user to query old and new documents in a single query, as if they all conformed to the latest schema version, using knowledge of the differences between versions.

Use Case: Data Federation

Assume a federated database system integrates relational (e.g., MGD, GO, GeneLynx [45]) and XML data sources (e.g., PubMed and the output of bioinformatics algorithms such as BLAST) and provides integrated SQL access to them. *View building* allows users to build customized views on top of relational and XML schemas using a graphical interface. *Schema mapping* provides knowledge about correspondences among the different sources. To respond efficiently to queries against these sources, *view composition* and *query decomposition* must use the correspondences gained through schema mapping and view building to issue the right queries to the right sub-systems. Finally, knowledge about the data sources' capabilities and global query optimization allow the processor to push expensive operations to local sources as appropriate.

3.7 SUMMARY

Effective data management and integration are critical to the success of bioinformatics, and this chapter has introduced key concepts in these technical areas. While the wide and varied landscape of integration approaches can seem overwhelming to the beginner, this chapter has offered six dimensions by which to characterize current and new integration efforts: browsing/querying, declarative/procedural code, generic/hard-coded code, semantic/syntactic integration, data warehousing/federation, and relational/non-relational data model. Basic definitions and the relative strengths and weaknesses of a variety of approaches were explored through a series of use cases, which are summarized in Table 3.5. The optimal strategy for a given organization or research project will vary with its individual needs and constraints, but it will likely be a hybrid strategy, based on a careful consideration of the relative strengths and weaknesses of the various approaches. While many areas of data integration are solved or nearly so, tough, largely unsolved problems still remain. The chapter concluded by highlighting two of them: semantic query planning and schema management.

Use Case	Preferred Approach
3.2.1.1 Simple curated gene data source	Relational technology
3.2.1.2 Retrieving genes and associated expression results	Relational technology
3.2.3.2 Transforming database structure	Data warehousing, views
3.3.1 Multi-source heterogeneous integration	Federation, querying (DiscoveryLink, 3.4.3)
3.4.1.1 Exploring sequences associated with recent articles about metalloproteases	Browsing (Entrez)
3.4.2.1 Database of known and predicted human and mouse genes and transcripts	Warehouse (AllGenes)
3.4.4 Querying through unified biological concepts	Semantic integration (TAMBIS)

3.5
Summary of use cases and approaches.

TABLE

ACKNOWLEDGMENTS

Warm thanks are offered to my colleagues at IBM for many stimulating discussions and collegial support: Laura Haas, Peter Schwarz, Julia Rice, and Felix Naumann. Thanks also to Carole Goble and Robert Stevens of the University of Manchester for kindly providing materials on the TAMBIS project; Howard Ho and the IBM Clio team for their generous contributions to the schema management section; and Bill Swope for help on data warehousing. Finally, I thank the editors and reviewers for their patience and helpful suggestions.

REFERENCES

[1]　H. H. Rashidi, L. K. Buehler. *Bioinformatics Basics: Applications in Biological Science and Medicine.* Boca Raton, FL: CRC Press, 2000.

[2]　J. D. Ullman and J. Widom. *A First Course in Database Systems.* Upper Saddle River, NJ: Prentice Hall, 1997.

[3]　D. Benson, I. Karsch-Mizrachi, D. Lipman, et al. "GenBank." *Nucleic Acids Research* 31, no. 1 (2003):23–27, http://www.ncbi.nlm.nih.gov/Genbank.

[4]　B. Boeckmann, A. Bairoch, R. Apweiler, et al. "The SWISS-PROT Protein Knowledgebase and Its Supplement TrEMBL in 2003." *Nucleic Acids Research* 31, no. 1 (2003): 365–370.

[5]　A. Bateman, E. Birney, L. Cerruti, et al. "The Pfam Protein Families Database." *Nucleic Acids Research* 30, no. 1 (2002): 276–280.

[6]　M. Ashburner, C. A. Ball, J. A. Blake, et al. "Gene Ontology: Tool for the Unification of Biology. The Gene Ontology Consortium." *Nature Genetics* 25, no. 1 (2000): 25–29.

[7]　D. L. Wheeler, D. M. Church, A. E. Lash, et al. "Database Resources of the National Center for Biotechnology Information: 2002 Update." *Nucleic Acids Research* 30, no. 1 (2002): 13–16.

[8]　M. Kanehisa, S. Goto, S. Kawashima, et al. "The KEGG Databases at GenomeNet." *Nucleic Acids Research* 30, no. 1 (2002): 42–46.

[9]　Microsoft Corporation. Microsoft Excel 2000, 1999.

[10]　National Center for Biotechnology Information. The Entrez Search and Retrieval System. http://www.ncbi.nlm.nih.gov/Entrez, 2002.

[11]　J. M. Ostell, S. J. Wheelan, and J. A. Kans. "The NCBI Data Model." *Methods of Biochemical Analysis* 43 (2001): 19–43.

[12] T. Ezold and P. Argos. "SRS: An Indexing and Retrieval Tool for Flat File Data Libraries." *Computer Applications in the Biosciences* 9, no. 1 (1993): 49–57.

[13] E. F. Codd. "A Relational Model of Data for Large Shared Data Banks." *Communications of the ACM* 13, no. 6 (1970): 377–387.

[14] P. G. Selinger, M. M. Astrahan, D. D. Chamberlin, et al. "Access Path Selection in a Relational Database Management System." Proceedings of the 1979 ACM SIGMOD International Conference on Management of Data, Boston, MA, May 30–June 1. *ACM* (1979): 23–34.

[15] T. Bray, J. Paoli, C. M. Sperberg-McQueen, et al. *Extensible Markup Language (XML): World Wide Web Consortium (W3C) Recommendation*, 2nd edition, October 6, 2000, http://www.w3.org/TR/REC-xml/html.

[16] http://www.w3c.org/xme/query.

[17] The Acero Genome Knowledge Platform, http://www.acero.com.

[18] S. F. Altschul, W. Gish, W. Miller, et al. "Basic Local Alignment Search Tool." *Journal of Molecular Biology* 215, no. 3 (1990): 403–410.

[19] I. Letunic, L. Goodstadt, N. J. Dickens, et al. "Recent Improvements to the SMART Domain-Based Sequence Annotation Resource." *Nucleic Acids Research* 30, no.1 (2002): 242–244.

[20] J. A. Blake, J. E. Richardson, C. J. Bult, et al. "The Mouse Genome Database (MGD): The Model Organism Database for the Laboratory Mouse." *Nucleic Acids Research* 30, no. 1 (2002): 113–115.

[21] Z. Lacroix, A. Sahuget, and R. Chandrasekar. "Information Extraction and Database Techniques: A User-Oriented Approach to Querying the Web." Proceedings of the 1998 10th International Conference on Advanced Information Systems Engineering (CAiSE '98), Pisa, Italy, June 8–12, 1998: 289–304.

[22] L. Wall, T. Christiansen, and R. Schwartz. *Programming Perl*, 2nd edition. Sebastopol, CA: O'Reilly and Associates, 1996.

[23] T. J. Mowbray and R. Zahavi. *The Essential CORBA: Systems Integration Using Distributed Objects*. New York: Wiley, 1995.

[24] The Expert Protein Analysis System (ExPASy) server at the Swiss Institute of Bioinformatics, http://www.expasy.ch.

[25] The FlyBase Web site for Drosophila genetics, http://flybase.bio.indiana.edu.

[26] W. M. Gelbart, M. Crosby, B. Matthews, et al. "FlyBase: A Drosophilia Database. The FlyBase Consortium." *Nucleic Acids Research* 25, no. 1 (1997): 63–66.

[27] Gene Logic's GeneExpress Database, http://www.genelogic.com/genexpress.cfm.

[28] M. Cornell, N. W. Paton, S. Wu, et al. "GIMS—A Data Warehouse for Storage and Analysis of Genome Sequence and Functional Data." In *Proceedings of the 2nd*

        IEEE International Symposium on Bioinformatics and Bioengineering (BIBE).
        Rockville, MD: IEEE Press, 2001, 15–22.

[29] GeneCards Web site at the Weizmann Institute in Israel,
      http://bioinfo.weizmann.ac.il/cards.

[30] M. Rebhan, V. Chalifa-Caspi, J. Prilusky, et al. "GeneCards: A Novel Functional
      Genomics Compendium with Automated Data Mining and Query Reformulation
      Support." *Bioinformatics* 14, no. 8 (1998): 656–664.

[31] The Computational Biology and Informatics Library. "AllGenes: A Web Site
      Providing Access to an Integrated Database of Known and Predicted Human and
      Mouse Genes. (version 5.0, 2002). Center for Bioinformatics, Unisversity of
      Pennsylvania. http://www.allgenes.org.

[32] S. B. Davidson, J. Crabtree, B. P. Brunk, et al. "K2/Kleisli and GUS: Experiments in
      Integrated Access to Genomic Data Sources." *IBM Systems Journal* 40, no. 2
      (2001): 512–531.

[33] B. A. Eckman, A. S. Kosky, and L. A. Laroco Jr. "Extending Traditional
      Query-Based Integration Approaches for Functional Characterization of
      Post-Genomic Data." *Bioinformatics* 17, no. 7 (2001): 587–601.

[34] L. M. Haas, P. M. Schwartz, P. Kodali, et al. "DiscoveryLink: A System for
      Integrating Life Sciences Data." *IBM Systems Journal* 40, no. 2 (2001): 489–511.

[35] V. Kashyap and A. Sheth. "Semantic Similarities Between Objects in Multiple
      Databases." In *Management of Heterogeneous and Autonomous Database Systems,*
      3rd edition, by A. Elmagarmid, M. Rusinkiewicz, and A. Sheth, 57–89. San
      Francisco: Morgan Kaufmann, 1999.

[36] A. Hamoush, A. F. Scott, J. Amberger, et al. "Online Mendelian Inheritance in Man
      (OMIM), A Knowledge Base of Human Genes and Genetic Disorders." *Nucleic
      Acids Research* 30, no. 1 (2002): 52–55.

[37] The Jackson Lab Web site,
      http://www.informatics.jax.org/mgihome/overview.shtml.

[38] RatMap, http://ratmap.gen.gu.se/.

[39] Genome DataBase (GDB), http://www.gdb.org.

[40] C. C. Talbot Jr. and A. J. Cuticchia. "Human Mapping Databases." In *Current
      Protocols in Human Genetics,* 1.13.1–1.13.12. New York: Wiley, 1999.

[41] A. Pellegrini-Toole, C. Bonavides, and S. Gama-Castro. "The EcoCyc Database."
      Nucleic Acids Research 30, no. 1 (2002): 56–58.

[42] P. G. Baker, C. A. Gobel, S. Bechhofer, et al. "An Ontology for Bioinformatics
      Applications." *Bioinformatics* 15, no. 6 (1999): 510–520.

[43] B. A. Eckman, Z. Lacroix, and L. Raschid. "Optimized, Seamless Integration
      of Biomolecular Data." In *In proceedings of the 2nd IEEE International*

Symposium on Bioinformatics and Bioengineering (BIBE). Rockville, MD: IEEE, 2001, 23–32.

[44] R. J. Miller, L. M. Haas, L. Yan, et al. "The Clio Project: Managing Heterogeneity." *ACM SIGMOD Record* 30, no. 1 (2001): 78–83.

[45] B. Lenhard, W. S. Hayes, and W. W. Wasserman. "GeneLynx: A Gene-Centric Portal to the Human Genome." *Genome Research* 11, no. 12 (2001): 2151–2157.

Issues to Address While Designing a Biological Information System

Zoé Lacroix

Life science has experienced a fundamental revolution from traditional in vivo discovery methods (understanding genes, metabolic pathways, and cellular mechanisms) to electronic scientific discovery consisting in collecting measurement data through a variety of technologies and annotating and exploring the resulting electronic data sets. To cope with this dramatic revolution, life scientists need tools that enable them to access, integrate, mine, analyze, interpret, simulate, and visualize the wealth of complex and diverse electronic biological data. The development of adequate technology faces a variety of challenges. First, there exist thousands of biomedical data sources: There are 323 relevant public resources in molecular biology alone [1]. The number of biological resources increases at great pace. Previous lists of key public resources in molecular biology contained 203 data sources in 1999 [2], 226 in 2000 [3], and 277 in 2001 [4]. Access to these data repositories is fundamental to scientific discovery. The second challenge comes from the multiple software tools and interfaces that support electronic-based scientific discovery. An early report from the 1999 U.S. Department of Energy Genome Program meeting [5] held in Oakland, California identified these challenges with the following statement:

> Genome-sequencing projects are producing data at a rate exceeding current analytical and data-management capabilities. Additionally, some current computing problems are expected to scale up exponentially as the data increase. [5]

The situation has worsened since, whereas the need for technology to support scientific discovery and bioengineering has significantly increased. Chapter 1 covers the reasons why, ultimately, all of these resources must be combined to form a comprehensive picture. Chapter 2 claims that this challenge may well constitute the backbone of 21st century life science research. In the past, the specific research and development of geographical and spatial data management systems led to the emergence of an important and very active geographic information systems (GIS) community. Likewise, the field of biological information systems (BIS) aiming to support life scientists is now emerging.

To develop biological information systems, computer scientists must address the specific needs of life scientists. The identification of the specifications of computer-aided biology is often impeded by difficulty of communication between life scientists and computer scientists. Two main reasons can be identified. First, life and computer scientists have radically different perspectives in their development activities. These discrepancies can be explained by comparing the design process in engineering and in experimental sciences. A second reason for misunderstanding results from their orthogonal objectives. A computer scientist aims to build a system, whereas a life scientists aims to corroborate an hypothesis. These viewpoints are illustrated in the following.

Engineering vs. Experimental Science

Software development has an approach similar to engineering. First, the specifications (or requirements) of the system to be developed are identified. Then, the development relies on a long initial design phase when most of the cases, if not all, are identified and offered a solution. Only then is a prototype implemented. Later, iterations of the loop *design ↔ implementation* aim to correct the implementation's failure to perform effectively the requirements and to extend, significantly, the implementation to new requirements. These iterations are typically expressed through codified versioning of the prototype. In practice, initial design phases are typically shortened because of drastic budget cuts and a hurry to market the product. However, short design phases often cause costly revisions that could have been avoided with appropriate design effort.

Bioinformatics aims to support life scientists in the discovery of new biological insights as well as to create a global perspective from which unifying principles in biology can be discerned. Scientific discovery is experimental and follows a progress track blazed by experiments designed to corroborate or fail hypotheses. Each experiment provides the theory with additional material and knowledge that builds the entire picture. An hypothesis can be seen as a *design* step, whereas an experiment is an *implementation* of the hypothesis. Learning thus results from multiple iterations of *design ↔ implementation*, where each refinement of an hypothesis is motivated by the failure of the previous implementation.

These two approaches seem very similar, but they vary by the number of iterations of *design ↔ implementation*. When computer scientists are in the design phase of developing a new system for life scientists, they often have difficulties in collecting use cases and identifying the specifications of the system prior to implementation. Indeed, life scientists are likely able to provide just enough information to build a prototype, which they expect to evaluate to express more requirements for a better prototype, and so on. This attitude led a company proposing to build a biological data management system to offer to "build a little, test a little" to ensure meeting the system requirements. Somehow the prototype corresponds to an experiment for a life scientist. Understanding these two dramatically different approaches to design is mandatory to develop useful technology to support life scientists.

Generic System vs. Query-Driven Approach

Computer scientists aim to build systems. A system is the implementation of an approach that is generic to many applications having similar characteristics. When provided with use cases or requirements for a new system, computer scientists typically abstract them as much as possible to identify the intrinsic characteristics and therefore design the most generic approach that will perform the requirements in various similar applications.

Life scientists, in their discovery process, are motivated by an hypothesis they wish to validate. The validation process typically involves some data sets extracted from identified data sources and follows a pre-defined manipulation of the collected data. In a nutshell, a validation approach corresponds to a complex *query* asked against multiple and often heterogeneous data sources. Life scientists have a query-driven approach.

These two approaches are orthogonal but not contradictory. Computer scientists present the value of their approach by illustrating the various queries the system will answer, whereas life scientists value their approach by the quality of the data set obtained and the final validation of the hypothesis. This orthogonality also explains the legacy in bioinformatics implementations, which mostly consist of hard-coded queries that do not offer the flexibility of a system as explained and illustrated in Chapters 1 and 3. This legacy problem is addressed in Section 4.1.2.

This chapter does not aim to present or compare the systems that will be described in the later chapters of this book. Instead it is devoted to issues specific to data management that need to be addressed when designing systems to support life science. As with any technology, data management assumes an ideal world upon which most systems are designed. They appear to suit the needs of large corporate usage such as banking; however, traditional technology fails to adjust properly to many new technological challenges such as Web data management and scientific data management. The following sections introduce these issues.

Section 4.1 presents some of the characteristics of available scientific data and technology. Section 4.2 is devoted to the first issue that traditional data management technology needs to address: changes. Section 4.3 addresses issues related to biological queries, whereas Section 4.4 focuses on query processing. Finally, data visualization is addressed in Section 4.5. Bioinformaticians should find a variety of illustrations of the reasons why BIS need innovative solutions.

4.1   LEGACY

Scientific data has been collected in electronic form for many years. While new data management approaches are designed to provide the basis for future homogeneous collection, integration, and analysis of scientific data, they also need to integrate existing large data repositories and a variety of applications developed to analyze them. Legacy data and tools may raise various difficulties for their integration that may affect the design of BIS.

4.1.1   Biological Data

Scientific data are disseminated in myriad different data sources across disparate laboratories, available in a wide variety of formats, annotated, and stored in flat files and relational or object-oriented databases. Access to heterogeneous biological data sources is mandatory to scientists. A single query may involve flat files (stored locally or remotely) such as GenBank [6] or Swiss-Prot [7], Web resources such as the Saccharomyces Genome Database [8], GeneCards [9], or the references data source PubMed [10]. A list of useful biological data sources is given in the Appendix. These sources are mostly textual and of restricted access facilities. Their structure varies from ASN.1 data exchange format to poorly structured hypertext markup language (HTML) and extensible markup language (XML) formats. This variety of repositories justifies the need to evoke *data sources* rather than databases. This chapter only refers to a *database* when the underlying system is a database management system (a relational database system, for example). A system based on flat files is not a database.

Unlike data hosted in a database system, scientific data is maintained by life scientists through user-friendly interfaces that offer great flexibility to add and revise data in these data sources. However, this flexibility often affects data quality dramatically. First, data sources maintained by a large community such as GenBank contain large quantities of data that need to be curated. This explains the numerous overlapping data sources sometimes found aiming to complete or correct data from existing sources. As importantly, data organization also suffers from this wide and flexible access: Data fields are often completed with different goals and objectives, fields are missing, and so on. This flexibility is typically provided when the underlying data representation is a formatted file with no types or constraints checking. Unfortunately this variety of data representation makes it difficult to use traditional approaches as explained in Section 4.2.3.

The quantity of data sources to be exploited by life scientists is overwhelming. Each year the number of publicly available data sources increases significantly: It rose 43% between 1999 and 2002 for the key molecular data sources [1, 2]. In addition to this proliferation of sources, the quantity of data contained in each data source is significantly large and also increasing. For example, as of January 1, 2001, GenBank [6] contained 11,101,066,288 bases in 10,106,023 sequences, and its growth continues to be exponential, doubling every 14 months [11]. While the number of distinct human genes appears to be smaller than expected, in the range of 30,000–40,000 [12, 13], the distinct human proteins in the proteome are expected to number in the millions due to the apparent frequency of alternative splicing, ribonucleic acid (RNA) editing, and post-translational modification [14, 15]. As of May 4, 2001, Swiss-Prot contained 95,674 entries, whereas PubMed contained more than 11 million citations. Managing these large data sets efficiently will be critical in the future. Issues of efficient query processing are addressed in Section 4.4.

Future collaborations between computer and life scientists may improve the collection and storage of data to facilitate the exploitation of new scientific data. However, it is likely that scientific data management technology will need to address issues related to the characteristics of the large existing data sets that constitute part of the legacy of bioinformatics.

4.1.2 Biological Tools and Workflows

Scientific resources include a variety of tools that assist life scientists in searching, mining, and analyzing the proliferation of data. Biological tools include basic biosequence analyses such as FASTA, BLAST, Clustal, Mfold, Phylip, PAUP, CAP, and MEGA. A data management system not integrating these useful tools would offer little support to life scientists. Most of these tools can be used freely by loading their code onto a computer from a Web site. A list of 160 free applications supporting biomolecular biology is provided in Misener and Krawetz's *Bioinformatics: Methods and Protocols* [16]. The first problem is the various platforms used by life scientists: In 1998, 30-50% of biologists used Macintosh computers, 40-70% used a PC running any version of Microsoft Windows, and less than 10% used UNIX or LINUX [16]. Out of the 160 applications listed in Misener and Krawetz's book, 107 run on a PC, 88 on a Macintosh, and 42 on other systems such as UNIX. Although some applications (27 listed in the previously mentioned book) are made available for all computer systems, most of them only run on a single system. The need for integration of applications, despite the system for which they may be designed, motivated the idea of the *grid* as explained in Section 4.3.4.

Legacy tools also include a variety of hard-coded scripts in languages such as Perl or Python that implement specific queries, link data repositories, and perform a pre-defined sequence of data manipulation. Scripting languages were used extensively to build early bioinformatics tools. However, they do not offer expected flexibility for re-use and integration with other functions. Most legacy integration approaches were developed using *workflows*. Workflows are used in business applications to assess, analyze, model, define, and implement the core business processes of an organization (or other entity). A workflow approach automates the business procedures where documents, information, or tasks are passed between participants according to a defined set of rules to achieve, or contribute to, an overall business goal. In the context of scientific applications, a workflow approach may address overall collaborative issues among scientists, as well as the physical integration of scientific data and tools. The procedural support a workflow approach provides follows the query-driven design of scientific problems presented in the Introduction. In such an approach, the data integration problem follows step-by-step the single user's query execution, including all necessary "business rules" such as security and semantics. A presentation of workflows and their model is provided online by the Workflow Management Coalition [17].

The integration of these tools and query pipelines into a BIS poses problems that are beyond traditional database management as explained in Section 4.3.4.

4.2   A DOMAIN IN CONSTANT EVOLUTION

A BIS must be designed to handle a constantly changing domain while managing legacy data and technology. Traditional data management approaches are not suitable to address constant changes (see Section 4.2.1). Two problems are critical to address for scientific data management: changes in data representation (see Section 4.2.2) and data identification (see Section 4.2.4). The approach presented in Chapter 10 addresses specifically these problems with gene expression data.

4.2.1   Traditional Database Management and Changes

The main assumption of traditional data management approaches relies on a predefined, unchangeable system of data organization. Traditional database management systems are of three kinds: relational, object-relational, or object-oriented.

Relational database systems represent data in relations (tables). Object-relational systems provide a more user-friendly data representation through classes, but they rely on an underlying relational representation. Object-oriented databases organize data through classes. For the sake of simplicity, only relational databases are considered in this section because most of the databases currently used by life scientists are relational, and similar remarks could be made for all traditional systems.

Data organization includes the relations and attributes that constitute a relational database schema. When the database schema is defined, it can be populated by data to create an instance of the schema, in other words, a database. Each row of a relation is called a *tuple*. When a database has been defined, transactions may be performed to update the data contained in the database. They consist of insertions (adding new tuples in relations), deletions (removing tuples from relations), and updates (transforming one or more components of tuples in relations). All traditional database systems are designed to support transactions on their data; however, they support few changes in the data organization. Changes in the data organization include renaming relations or attributes, removing or adding relations or attributes, merging or splitting relations or attributes, and so on. Some transformations such as renaming are rather simple, and others are complex. Traditional database systems are not designed to support complex schema transactions. Typically, a change in the data organization of a database is performed by defining a new schema and loading the data from the database to an instance of the new schema, thus creating a new database. Clearly this process is tedious and not acceptable when such changes have to be addressed often. Another approach to the problem of restructuring is to use a view mechanism that offers a new schema to users as a virtual schema when the underlying database and schema have not changed. All user interfaces provide access to the data as they are defined in the view and no longer as they are defined in the database. This approach offers several advantages, including the possibility of providing customized views of databases. A view may be limited to part of the data for security reasons, for example. However, this approach is rather limited as the transactions available through the view may be restricted.

Another aspect of traditional database systems relies on a pre-defined identity. Objects stored in a relational database can be identified by a set of attributes that, together, characterize the object. For example, the three attributes—first name, middle name, and last name—can characterize a person. The set of characterizing attributes is called a primary key. The concept of primary key relies on a characterization of identity that will not change over time. No traditional database system is designed to address changes in identification, such as tracking objects that may have changed identity over time.

4.2.2　Data Fusion

Data fusion corresponds to the need to integrate information acquired from multiple sources (sensors, databases, information gathered by humans, etc.). The term was first used by the military to qualify events, activities, and movements to be correlated and analyzed as they occurred in time and space to determine the location, identity, and status of individual objects (equipment and units), assess the situation, qualitatively or quantitatively determine threats, and detect patterns in activity that would reveal intent or capability.

Scientific data may be collected through a variety of instruments and robots performing microarrays, mass spectrometry, flow cytometry, and other procedures. Each instrument needs to be calibrated, and the calibration parameters may affect the data significantly. Different instruments may be used to perform similar tasks and collect data to be integrated in a single data set for analysis. The analysis is performed over time upon data sets disparate by the context of their collection. The analysis must be tempered by parameters that directly affect the quality of the data. A similar problem is presented in Section 10.4.3 in the context of probe arrays and gene expression. A traditional database approach requires the complete collection of measurement data and all parameters to allow the expression of the complex queries that enable the analysis of the disparate data set. Should any information be missing (NULL in a table), the system ignores the corresponding data, an unacceptable situation for a life scientist. In addition, the use of any new instrument that requires the definition of new parameters may affect the data organization, as well as make the fusion process more complex. The situation is made even more complex by the constant evolution of the protocols. Their new specifications often change the overall data organization: Attributes may be added, split, merged, removed, or renamed. Traditional database systems' difficulty with these issues of data fusion explains the current use of Microsoft Excel spreadsheets and manual computation to perform the integration prior to analysis. The database system is typically used as a storage device.

Can a traditional database system be adjusted to handle these constant and complex changes in the data organization? It is unlikely. Indeed, all traditional approaches rely strongly on a pre-defined and stale data organization. A BIS shall offer great flexibility in the data organization to meet the needs of life scientists. New approaches must be designed to enable scientific data fusion. A solution is to relax the constraint on the data representation, as presented in the following section.

4.2.3　Fully Structured vs. Semi-Structured

Traditional database approaches are too structured: When the schema is defined, it is difficult to change it, and they do not support the integration of similar, but disparate, data sets. A solution to this need for adherence to a structure is offered by the *semi-structured* approach. In the semi-structured approach, the data organization allows changes such as new attributes and missing attributes. Semi-structured data is usually represented as an edge-labeled, rooted, directed graph [18-20]. Therefore, a system handling semi-structured data does not assume a given, pre-defined data representation: A new attribute name is a new labeled edge, a new attribute value is a new edge in the graph, and so on. Such a system should offer greater flexibility than traditional database systems. An example of representation of semi-structured data is XML, the upcoming standard for data exchange on the Web designed by the World Wide Web Consortium (W3C). XML extends the basic tree-based data representation of the semi-structured model by ordering elements and providing various levels of representation such as XML Schema [21-23]. These additional characteristics make XML data representation significantly less flexible than the original semi-structured data model. Fully structured data representation, semi-structured data representation, and XML are presented in *Data on the Web* [24].

There are currently two categories of XML management systems: XML-enabled and native XML. The first group includes traditional database systems extended to an XML interface for collection and publication. However, the underlying representation is typically with tables. Examples of XML-enabled systems are Oracle9i[1] and SQL Server 2000.[2] These systems were mostly designed to handle Business-to-Business (B2B) and Business-to-Customer (B2C) business tasks on the Web. They have not yet proven useful in scientific contexts. Native XML systems such as Tamino,[3] ToX,[4] and Galax[5] rely on a real semi-structured approach and should provide a flexibility interesting in the context of scientific data management.

Because of XML's promising characteristics, and because it is going to be the *lingua franca* for the Web, new development for BIS should take advantage of this new technology. A system such as KIND, presented in Chapter 12, already exploits XML format. However, the need for semantic data integration in addition to syntactic data integration (as illustrated in Section 4.2.5) limits the use of XML and its query language in favor of approaches such as description logics.

---

1. Oracle9i was developed by the Oracle corporation (see *http://www.oracle.com*).
2. SQL Server 2000 is a product of the Microsoft Corporation (see *http://www.microsoft.com*).
3. Tamino XML server is a commercial XML management system from SoftwareAG.
4. ToX is an academic XML management system being developed at the University of Toronto.
5. Galax was developed at the Bell Laboratory of Lucent Technology (see *http://db.bell-labs.com/galax*).

4.2.4   Scientific Object Identity

Scientific objects change identification over time. Data stored in data sources can typically be accessed with knowledge of their identification or other unique characterization initially entered into the data bank. Usually, each object of interest has a name or an identifier that characterizes it. However, a major problem arises when a given scientific object, such as a gene, may possess as many identifiers as there are data sources that contain information about it. The challenge is to manage these semantic heterogeneities at data access, as the following example illustrates.

Gene names change over time. For example, the Human Gene Nomenclature Database (HUGO) [25] contains 13,594 active gene symbols, 9635 literature aliases, and 2739 withdrawn symbols. In HUGO, SIR2L1 (withdrawn) is a synonym to SIRT1 (the current approved HUGO symbol) and sir2-like 1. P53 is a withdrawn HUGO symbol and an alias for TP53 (current approved HUGO symbol). When a HUGO name is removed, not all data sources containing the name are updated. Some information, such as the content of PubMed, will actually never be updated.

Table 4.1 illustrates the discrepancies found when querying biological data sources with equivalent (but withdrawn or approved) HUGO names in November 2001. The Genome DataBase (GDB) is the official central repository for genomic mapping data resulting from the Human Genome Initiative [26]. GenAtlas [27, 28] provides information relevant to the mapping of genes, diseases, and markers. Online Mendelian Inheritance in Man (OMIM) [29, 30] is a catalog of human genes and genetic disorders. GeneCards [9] is a data source of human genes, their products, and their involvement in diseases. LocusLink [31, 32] provides curated sequence and descriptive information about genetic loci.

Querying GDB with TP53 or its alias P53 does not affect the result of the query. However, SIRT1 returns a single entry, whereas its alias does not return

HUGO name	GDB	GenAtlas	OMIM	GeneCards	LocusLink
TP53	1	33	52	22	13
P53	1 (same)	17	188	69	63
SIRT1	1	0	5	1	2
SIR2L1	0	0	1	1 (same)	2 (same)

4.1

TABLE

Number of entries retrieved with HUGO names from GDB, GenAtlas, OMIM, GeneCards, and LocusLink.

any entry. GenAtlas returns more entries for the approved symbol TP53 than the withdrawn symbol P53. The question, then, is to determine if the entries corresponding to a withdrawn symbol are always contained in the set of entries returned for an approved symbol. This property does not hold in OMIM. OMIM returns many more entries for the withdrawn symbol P53 than the approved symbol TP53. However, it shows opposite behavior with SIRT1 and its alias. This demonstrates that even with the best understood and most commonly accepted characteristic of a gene—its identification—alternate identifier values need to be used when querying multiple data sources to get complete and consistent results. This problem requires significant domain expertise to resolve but is critical to the task of obtaining a successful, integrated biological information system.

The problem of gene identity is made more complex when the full name, alternative titles, and description of a gene are considered. Depending on the number of data sources that describe the gene, it may have that many equivalent source identifiers. For example, SIRT1 is equivalent to the full name (from HUGO) of sirtuin (silent mating type information regulation 2, *S. cerevisiae*, homolog) 1. This is also its description in LocusLink, but it has the following alternative title in OMIM:SIR2, S. CEREVISIAE, HOMOLOG-LIKE 1. These varying qualifications can often be easily discerned by humans, but they prove to be very difficult when automated. Here, too, extensive domain expertise is needed to determine that these descriptions each represent the very same gene, SIRT1. Although this is difficult, it does not describe the entire problem. There are as many identifiers to SIRT1 as there are data sources describing the gene. For example, SIRT1 also corresponds to 604479 (OMIM number), AF083106 (GenBank accession number), and 9956524 (GDB ID). Even UniGene clusters may have corresponding aliases. For example, Hs.1846 (the UniGene cluster for P53) [33] is an alias for Hs.103997 (primary cluster for TP53).

Existing traditional approaches do not address the complex issues of scientific object identity. However, recent work on ontologies may provide solutions to these issues.

4.2.5   Concepts and Ontologies

An *ontology* is a collection of vocabulary words that define a community's understanding of a domain. *Terms* are labels for concepts, which reside in a lattice of relationships between concepts. There have been significant contributions to the specification of standards and ontologies for the life science community as detailed in Chapter 2. In addition, some BIS were designed to provide users' access to data as close as possible to their understanding. The system [34] based on the Object Protocol Model (OPM), developed at Lawrence Berkeley Laboratory and later extended and maintained at Gene Logic, provides data organization through classes and relationships to the user (see Chapter 10). The most successful such approach is TAMBIS (see Chapter 7), which was developed to allow users to access and query their data sets through an ontology. Such approaches are friendly because they allow life scientists to visualize the data sets through their understanding of the overall organization (concepts and relationships) as opposed to an arbitrary and often complex database representation with tables or a long list of tags of a flat file.

A solution to the problem of capturing equivalent representations of objects consists in using concepts. For example, a concept gene can have a primary identity (its approved HUGO symbol) and equivalent representations (withdrawn HUGO names, aliases, etc.). There are many ways to construct these equivalent classes. One way consists in collecting these multiple identities and materializing them within a new data source. This approach was partially completed in LENS, which was developed at the University of Pennsylvania, and GeneCards. This first approach captures the expertise of life scientists, and these data sources are usually well curated. To make this task scalable to all the scientific objects of interest, specific tools need to be developed to enhance and assist scientists in the task of managing the identity of scientific objects. While this expertise could be materialized within a new data source, it is critical that it is used by a BIS to alert the biologist when it recognizes alternate identifiers that could lead to incomplete or inconsistent results. This approach could use entity matching tools [35, 36] that capture similarities in retrieved objects and are appropriate for matching many of the functional attributes such as description or alternate titles.

Recent work in the context of the Semantic Web activity of the W3C may develop more advanced technology to provide users a sound ontology layer to integrate their underlying biological resources. However, these approaches do not yet provide a solution to the problem of capturing equivalent representations of scientific objects, as presented in Section 4.2.4.

4.3   BIOLOGICAL QUERIES

The design of a BIS strongly depends on how it is going to be used. Section 1.4 of Chapter 1 presents the successive design steps, and Chapter 3 illustrates various design requirements with use cases. Traditional database approaches assume that the relational algebra, or query languages such as the Structured Query Language (SQL) or the Object Query Language (OQL), enable users to express all their queries. Life science shows otherwise. Similar to geographical information systems that aim to let users express complex geometric, topological, or algebraic queries, BIS should enable scientists to express a variety of queries that go beyond the relational algebra. The functionalities required by scientists include sophisticated search mechanisms (see Section 4.3.1) and navigation (see Section 4.3.2), in addition to standard data manipulation. In traditional databases, the semantics of queries are usually bi-valued: true or false. Practice shows that scientists wish to access their data sets through different semantic layers and would benefit from the use of probabilistic or other logical methods to evaluate their queries, as explained in Section 4.3.3. Finally, the complexity of scientific use cases and the applications that support them may drive the design of a BIS to *middleware* as opposed to a traditional data-driven database approach.

4.3.1   Searching and Mining

Searching typically consists in retrieving entries similar to a given string of characters (phrase, keyword, wildcard, DNA sequence, etc.). In most cases, the data source contains textual documents, and when searching the data source, users expect to retrieve documents that are similar to a given phrase or keyword. Search engines, such as Glimpse[6] used to provide search capabilities to GeneCards, use an index to retrieve documents containing the keywords and a ranking system to display ordered retrieved entries. Searching against a sequence data source is performed by honed sequence similarity search engines such as FASTA [37], BLAST [38], and LASSAP [39]. To search sequences, the input string is a sequence, and the ranking of retrieved sequences is customized with a variety of parameters.

Data mining aims to capture patterns out of large data sets with various statistical algorithms. Data mining can be used to discover new knowledge about a data set or to validate an hypothesis. Mining algorithms are often combined with association rules, neural networks, or genetic algorithms. Unlike searching, the data mining approach is not driven by a user's input expressed through a phrase nor does it apply to a particular data format. Mining a database distinguishes itself from querying a database by the fact that a database query is expressed in a language such as SQL and therefore only captures information organized in the schema. In contrast, a mining tool may exploit information contained in the database that was not organized in the schema and therefore not accessible by a traditional database query.

Most of the query capabilities expected by scientists fall under searching and mining. This was confirmed by the results of a survey of biologists in academia

---

6. The Glimpse search engine was developed and is maintained at the University of Arizona and is available at *http://glimpse.cs.arizona.edu/*.

and industry in 2000 [40], where 315 tasks and queries were collected from the answers to the following questions:

1. What tasks do you most perform?
2. What tasks do you commonly perform, that should be easy, but you feel are too difficult?
3. What questions do you commonly ask of information sources and analysis tools?
4. What questions would you like to be able to ask, given that appropriate sources and tools existed, that may not currently exist?

Interestingly, 54% of the collected tasks could be organized into three categories: (1) similarity search, (2) multiple pattern and functional motif search, and (3) sequence retrieval. Therefore, more than half of the identified queries of interest to biologists involve searching or mining capabilities.

Traditional database systems provide SQL as a query language, based on the relational algebra composed of selection (σ), projection (π), Cartesian product ($\times$), join ($\bowtie$), union ($\cup$), and intersection ($\cap$). These operators perform data manipulation and provide semantics equivalent to that of first order logic [41]. In addition, SQL includes all arithmetic operations, predicates for comparison and string matching, universal and existential quantifiers, summary operations for max/min or count/sum, and GROUP BY and HAVING clauses to partition tables by groups [42]. Commercial database systems extended the query capabilities to a variety of functionalities, such as manipulation of complex datatypes (such as numeric, string, date, time, and interval), OLAP, and limited navigation. However, none of these capabilities can perform the complex tasks specified by biologists in the 2000 survey [40].

Other approaches provide search capabilities only, and while failing to support standard data manipulation, they are useful for handling large data sets. They are made available to life scientists as Web interfaces that provide textual search facilities such as GeneCards [9], which uses the powerful Glimpse textual search engine [43], the Sequence Retrieval Service (SRS) [44], and the Entrez interface [45]. GeneCards provides textual search facilities for curated data warehoused in files. SRS, described in detail in Chapter 5, integrates data sources by indexing attributes. It enables queries composed of combinations of textual keywords on most attributes available at integrated databases. Entrez proposes an interesting approach to integrating resources through their similarities. It uses a variety of similarity search tools to index the data sources and facilitate their access through search queries. For instance, the neighbors of a sequence are its *homologs*, as identified by a similarity score using the BLAST algorithm [38]. On the other hand, the neighbors of a PubMed citation are the articles that use similar terms in their title and abstract [46]. These approaches are limited because they do not allow customized access to the sources. A user looking for PubMed references that have direct protein links will not be able to express the query through Entrez because the interface is designed to retrieve the protein linked from a given citation, not to retrieve all citations linked to a protein. This example, as well as others collected in a 1998 *Access* article by L. Wong [47], illustrates the weakness of these approaches. A real query language allows customization, whereas a selection of capabilities limits significantly the range of queries biologists are able to ask.

Biologists involve in their queries a variety of search and mining tools and employ traditional data manipulation operators to support their queries. In fact, they often wish to combine them all within a single query. For example, a typical query could start with searching PubMed and only retrieve the references that have direct protein links [47]. Most systems do not support this variety of functionalities yet. Systems such as Kleisli and DiscoveryLink, presented in Chapters 6 and 11, specifically address this issue.

4.3.2 Browsing

Biologists aim to perform complex queries as described in Section 4.3.1, but they also need to browse and navigate the data sets. Systems such as OPM and TAMBIS (see Chapter 7) are designed to provide a user-friendly interface that allows queries through ontologies or object classes, as presented in Section 4.2.5. But they do not provide navigational capabilities that enable access to other scientific objects through a variety of hyperlinks. Web interfaces such as GeneCards offer a large variety of hyperlinks that enable users to navigate directly to other resources such as GenBank, PubMed, and European Molecular Biology Laboratory (EMBL) [34, 48, 76]. Entrez, the Web interface to PubMed, GenBank, and an increasing number of resources hosted at the National Center for Biotechnology Information (NCBI) offer the most sophisticated navigational capabilities. All 15 available resources (as of July 2002) are linked together. For example, a citation in PubMed is linked to related citations in PubMed via the `Related Articles`, linked to relevant sequences or proteins, respectively, via the `Nucleotide Link` or the `Protein Link`. The links are completed by a variety of hyperlinks available in the display of retrieved entries. These navigational capabilities complete the query capabilities and assist the biologists in fulfilling their needs.

The recent development of XML and its navigational capabilities make XPath [49], the language designed to handle navigational queries, and XQuery [50], its extension to traditional data management queries as well as to document queries, good candidates for query languages to manipulate scientific data. There are additional motivations for choosing XML technology to handle scientific data. XML is designed as the standard for data exchange on the Web, and life scientists publish and collect large amounts of data on the Web. In addition, the need for a flexible data representation already evoked the choice of XML in Section 4.2.3. Scientific data providers such as NCBI already offer data in XML format.

Although clearly needed, the development of a navigational foreground for biological data raises complex issues of semantics, as will be presented in Section 4.5.2.

4.3.3   Semantics of Queries

Traditional database approaches use bi-valued semantics: true or false. When a query is evaluated, should any data be missing, the output is NULL. Such semantics are not appropriate for many biological tasks. Indeed, biologists often attempt to collect data with exploring queries, despite missing information. An attribute NULL does not always mean that the value is null but rather that the information is not available yet or is available elsewhere. The rigid semantics of traditional databases may be frustrating for biologists who aim to express queries with different layers of semantics.

Knowledge-based approaches [51] may be used to provide more flexibility. In knowledge bases, *reasoning* about the possible courses of action replaces the typical database evaluation of a query. Knowledge bases rely on large amounts of expertise expressed through statements, rules, and their associated semantics. Extending BIS with knowledge-based reasoning provides users with customized semantics of queries. BIS can be enhanced by the use of temporal logic that assumes the world to be ordered by time intervals and allows users to reason about time (e.g., "*Retrieve all symptoms that occurred before event A*") or fuzzy logic that allows degrees of truth to be attached to statements. Therefore, a solution consists in providing users with a hybrid query language that allows them to express various dependency information, or lack thereof, between events and build a logical reasoning framework on top of such statements of probability. Such an approach has been evaluated for temporal databases [52] and object databases [53]. BIS also could benefit from approaches that would cover the need for addressing object identity as presented in Section 4.2.4, as well as semantic issues such as those to be addressed in Sections 4.4.2 and 4.5.2.

4.3.4   Tool-Driven vs. Data-Driven Integration

Most existing BIS are *data-driven*: They focus on the access and manipulation of data. But should a BIS really be data-driven? It is not that clear. A traditional database system does not provide any flexibility in the use of additional functionalities. The query language is fixed and does not change. Public or commercial platforms aim to offer integrated software; however, their approach does not provide the ability to integrate easily and freely new softwares as they become available or improve. Commercial integrated platforms also are expensive to use. For scientists with limited budgets, free software is often the only solution.

Some systems provide APIs to use external programs, but the system is no longer the central query processing system; it only processes SQL queries, and an external program executes the whole request. The problem with this approach is that the system, which uses a database system as a component, no longer benefits from the database technology, including efficient query processing (as will be presented in Section 4.4).

Distributed object technology has been developed to cope with the heterogeneous and distributed computing environment that often forces information to be moved from one machine to another, disks to be cross-mounted so different programs can be run on multiple systems, and programs to be re-written in a different programming language to be compiled and executed on another architecture. The variety of scientific technology presented in Section 4.1.2 often generates significant waste of time and resources. Distributed object technology includes Common Object Request Broker Architecture (CORBA), Microsoft Distributed Component Object Model (DCOM), and Java Remote Method Invocation (RMI). This technology is tools-driven and favors a computational architecture that interoperates efficiently and robustly. Unlike traditional databases, it allows flexible access to computational resources with easy registration and removal of tools. For these reasons, many developers of BIS are currently using this technology. For sake of efficiency, a new version of TAMBIS no longer uses CPL but provides a user-friendly ontology of biological data sources using CORBA clients to retrieve information from these sources [54] (see Chapter 5). CORBA appears to be suitable for creating wrappers via client code generation from interface definition language (IDL) definitions. The European Bioinformatics Institute (EBI) is leading the effort to make its data sources CORBA compliant [48, 55]. Unfortunately, most data providers do not agree with this effort. Concurrent to the CORBA effort, NCBI, the American institute, and EBI provide their data sources in XML format.

A lot of interest has been given to grid architectures. A grid architecture aims to enable computing as well as data resources to be delivered and accessed seamlessly, transparently, and dynamically, when needed, on the Internet. The name *grid* was inspired by the electricity power grid. A biologist should be able to plug into the grid like an appliance is plugged into an outlet and use resources available on the grid transparently. The grid is an approach to a new generation of BIS. Examples of grids include the Open Grid Services Architecture (OGSA) [56] and the open source problem-solving environment Cactus [57]. TeraGrid [58] and DataGrid [59] are international efforts to build grids.

These tool-driven proposals do not yet solve the many problems of resource selection, query planning, optimization, and other semantics issues as will be presented in Sections 4.4 and 4.5.

4.4   QUERY PROCESSING

In the past, query processing often received less attention from designers of BIS. Indeed, BIS developers devoted most of their effort to meeting the needs for integration of data sets and applications and providing a user-friendly interface for scientists. However, as the data sets get larger, the applications more time-consuming, and the queries more complex, the specification for fast query processing becomes critical.

4.4.1   Biological Resources

A BIS must adequately capture and exploit the diverse, and often complex, query processing, or other computational capabilities of biological resources by specifying them in a catalog and using them at both query formulation and query evaluation. The W3C Semantic Web Activity [60, 61] aims to provide a meta-data layer to permit people and applications to share data on the Web. Recent efforts within the bioinformatics community address the use of OIL [62] to capture alternative representations of data to extend biomolecular ontologies [63]. Such efforts focusing on data representation of the contents of the sources must be extended to capture meta-data along several dimensions, including (1) the coverage of the information sources, (2) the capabilities, links, and statistical patterns, (3) the data delivery patterns of the resources, and (4) data representation and organization at the source.

The *coverage* of information sources is useful in solving the so-called *source relevance problem*, which involves deciding which of the myriad sources are relevant for the user and to evaluate the submitted query. Directions to characterize and exploit coverage of information sources include local, closed world assumptions, which state that the source is complete for a specific part of the database

[64, 65]; quantifications of coverage (e.g., the database contains at least 90% of the sequences), or intersource overlaps (e.g., the EMBL Nucleotide Sequence Database has a 75% likely overlap with DDBJ for sequences annotated with "calcium channel") [66–68]. Characterizing coverage enables the exploitation of coverage positioning of data sources from complement to partial or complete overlap (*mirror sites*).

Source *capabilities* capture the types of queries supported by the sources, the access pattern limitations, the ability to handle limited disjunction, and so on. Although previous research has addressed capabilities [69–74], it has not addressed the diverse and complex capabilities of biological sources. Recent work aims to identify the properties of sophisticated source capabilities including text search engines, similarity sequence search engines such as BLAST, and multiple sequence alignment tools such as Cluster [75, 76]. Their characteristics are significantly more complex than the capabilities addressed up to now, and their use is dramatically costly in terms of processing time. In addition, many of the tools are closely coupled to the underlying source, which requires the simultaneous identification of capabilities and coverage.

Statistical patterns include the description of information clusters and the selectivities of all or some of the data access mechanisms and capabilities. The use of simple statistical patterns has been studied [65, 66, 77]. However, BIS should exploit statistical patterns of real data sources that are large, complex, and constantly evolving (as opposed to their simplified simulations).

Delivery patterns include the response time, that is, units of time needed to receive the first block of answers, the size of these blocks and so on. Delivery patterns may affect the query evaluation process significantly. Depending on the availability of the proper indices, a source may either return answers in decreasing order of matching (from best to worst) or in an arbitrary (unordered) manner. Other delivery profiles include whether information can be provided in a sorted manner for certain attributes or not. These types of profiles will be essential in identifying sources to get the first, best answers. This is useful when a user expects to get the answers to a query sorted in a pre-defined relevant order (the more relevant answers are the first returned). Delivery patterns can also be exploited to provide users with a faster, relevant, but maybe incomplete answer to a query. BIS should exploit delivery patterns in conjunction with the actual capabilities supported by the sources.

The access and exploitation of the previously mentioned meta-knowledge of biological resources offers several advantages. First, it enables the comparison of diverse ways to evaluate a query as explained in the next section. Further, it can characterize the most efficient way to evaluate a query, as will be presented in Section 4.4.3.

4.4.2   Query Planning

Query planning consists in considering the many potential combinations of accesses to evaluate a query. Each combination is a *query evaluation plan*. Consider the query (Q) defined as follows:

> *(Q)"Return accession numbers and definitions of GenBank EST sequences that are similar (60% identical over 50AA) to 'Calcium channel' sequences in Swiss-Prot that have references published since 1995 and mention 'brain.'"* [78]

There exist many plans to evaluate the query. One possible plan for this query is illustrated in Figure 4.1 and described as follows: (1) access PubMed and retrieve references published since 1995 that mention `brain`; (2) extract from all these references the Swiss-Prot identifiers; (3) obtain the corresponding sequences from Swiss-Prot whose function is `calcium channel`; and (4) execute a BLAST search using a wrapped BLAST application to retrieve similar sequences from GenBank (`gbest` sequences).

Figure 4.2 presents an alternative approach that first accesses Swiss-Prot and retrieves sequences whose function is `calcium channel`. In parallel, it retrieves the citations from PubMed that mention `brain` and are published since 1995 and extracts sequences from them. Then it determines which sequences are in common. Finally, it executes a BLAST search to retrieve similar sequences from GenBank (`gbest` sequences).

Scientific resources overlap significantly. The variety of capabilities, as well as the coverages and statistical patterns presented in Section 4.4.1, offer many alternative evaluation plans for a query. The number of evaluation plans is exponential to the size of similar resources. Therefore not all plans should be evaluated to answer a query. To select the plan to evaluate a given query, first the semantics of the plan should be captured accurately. Indeed, two plans may be similar and yet not semantically equivalent. For example, suppose a user is interested in retrieving the sequences relevant to the article entitled "Suppression of Apoptosis in Mammalian Cells by `NAIP` and a Related Family of IAP Genes" published in *Nature* and referenced by `8552191` in PubMed. A first plan is to extract the GenBank identifiers explicitly provided in the MEDLINE format of the reference. A second plan consists in using the capability `Nucleotide Link`, provided at NCBI. The two plans are not semantically equivalent because the first plan returns four GenBank identifiers when the Nucleotide Link returns eight GenBank identifiers (as of August 2001).

Verifying whether two plans are semantically equivalent, that is, if the answers that are returned from the two plans are identical, is non-trivial and depends on the meta-data of the particular resources used in each plan. This issue is closely

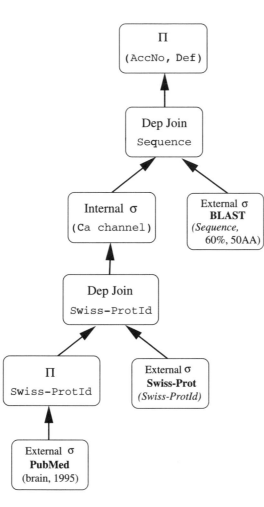

First plan for evaluating query (Q) [75].

related to navigation over linked resources as will be presented in Section 4.5.2. In addition, two semantically equivalent plans may differ dramatically in terms of efficiency, as is explained in the next section.

4.4.3 Query Optimization

Query optimization [79, 80] is the science and the art of applying equivalence rules to rewrite the tree of operators evoked in a query and produce an optimal plan. A plan is optimal if it returns the answer in the least time or using the least space.

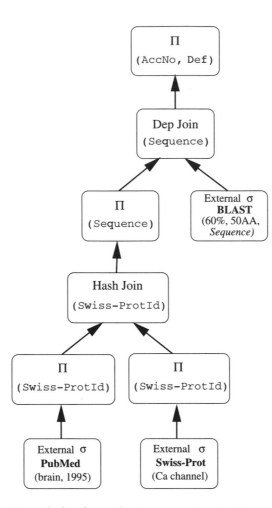

4.2   Second plan for evaluating query (Q) [75].

FIGURE

There are well known syntactic, logical, and semantic equivalence rules used during optimization [79]. These rules can be used to select an optimal plan among semantically equivalent plans by associating a cost with each plan and selecting the lowest overall cost. The cost associated with each plan is generated using accurate metrics such as the cardinality or the number of result tuples in the output of each operator, the cost of accessing a source and obtaining results from that source, and so on. One must also have a cost formula that can calculate the processing cost for each implementation of each operator. The *overall cost* is typically defined as the total time needed to evaluate the query and obtain all of the answers.

The characterization of an optimal, low-cost plan is a difficult task. The complexity of producing an optimal, low-cost plan for a relational query is NP-complete [79–81]. However, many efforts have produced reasonable heuristics to solve this problem. Both dynamic programming and randomized optimization based on simulated annealing provide good solutions [82–84].

A BIS could be improved significantly by exploiting the traditional database technology for optimization extended to capture the complex metrics presented in Section 4.4.1. Many of the systems presented in this book address optimization at different levels. K2 (see Chapter 8 Section 8.1) uses rewriting rules and a cost model. P/FDM (see Chapter 9) combines traditional optimization strategies, such as query rewriting and selection of the best execution plan, with a query-shipping approach. DiscoveryLink (see Chapter 11) performs two types of optimization: query rewriting followed by a cost-based optimization plan. KIND (see Chapter 12) is addressing the use of domain knowledge into executable meta-data. The knowledge of biological resources can be used to identify the best plan with query (Q) defined in Section 4.4.2 as illustrated in the following.

The two possible plans illustrated in Figures 4.1 and 4.2 do not have the same cost. Evaluation costs depend on factors including the number of accesses to each data source, the size (cardinality) of each relation or data source involved in the query, the number of results returned or the selectivity of the query, the number of queries that are submitted to the sources, and the order of accessing sources.

Each access to a data source retrieves many documents that need to be parsed. Each object returned may generate further accesses to (other) sources. Web accesses are costly and should be as limited as possible. A plan that limits the number of accesses is likely to have a lower cost. Early selection is likely to limit the number of accesses. For example, the call to PubMed in the plan illustrated in Figure 4.1 retrieves 81,840 citations, whereas the call to GenBank in the plan in Figure 4.2 retrieves 1616 sequences. (Note that the statistics and results cited in this paper were gathered between April 2001 and April 2002 and may no longer be up to date.) If each of the retrieved documents (from PubMed or GenBank) generated an additional access to the second source, clearly the second plan has the potential to be much less expensive when compared to the first plan.

The size of the data sources involved in the query may also affect the cost of the evaluation plan. As of May 4, 2001, Swiss-Prot contained 95,674 entries whereas PubMed contained more than 11 million citations; these are the values of cardinality for the corresponding relations. A query submitted to PubMed (as used in the first plan) retrieves 727,545 references that mention *brain*, whereas it retrieves 206,317 references that mention *brain* and were published since 1995. This is the selectivity of the query. In contrast, the query submitted to Swiss-Prot in the second plan returns 126 proteins annotated with *calcium channel*.

In addition to the previously mentioned characteristics of the resources, the order of accessing sources and the use of different capabilities of sources also affects the total cost of the plan. The first plan accesses PubMed and extracts values for identifiers of records in Swiss-Prot from the results. It then passes these values to the query on Swiss-Prot via the join operator. To pass each value, the plan may have to send multiple calls to the Swiss-Prot source, one for each value, and this can be expensive. However, by passing these values of identifiers to Swiss-Prot, the Swiss-Prot source has the potential to constrain the query, and this could reduce the number of results returned from Swiss-Prot. On the other hand, the second plan submits queries in parallel to both PubMed and Swiss-Prot. It does not pass values of identifiers of Swiss-Prot records to Swiss-Prot; consequently, more results may be returned from Swiss-Prot. The results from both PubMed and Swiss-Prot have to be processed (joined) locally, and this could be computationally expensive. Recall that for this plan, 206,317 PubMed references and 126 proteins from Swiss-Prot are processed locally. However, the advantage is that a single query has been submitted to Swiss-Prot in the second plan. Also, both sources are accessed in parallel.

Although it has not been described previously, there is a third plan that should be considered for this query. This plan would first retrieve those proteins annotated with *calcium channel* from Swiss-Prot and extract MEDLINE identifiers from these records. It would then pass these identifiers to PubMed and restrict the results to those matching the keyword *brain*. In this particular case, this third plan has the potential to be the least costly. It submits one sub-query to Swiss-Prot, and it will not download 206,317 PubMed references. Finally, it will not join 206,317 PubMed references and 126 proteins from Swiss-Prot locally.

Optimization has an immediate impact in the overall performance of the system. The consequences of the inefficiency of a system to execute users' queries may affect the satisfaction of users as well as the capabilities of the system to return any output to the user. These issues are presented in Chapter 13.

4.5   VISUALIZATION

An important issue when designing a BIS is visualization. Scientific data are available in a variety of media, and life scientists expect to access all these data sets by browsing through correspondences of interest, regardless of the medium or the resource used. The ability to combine and visualize data is critical to scientific discovery. For example, KIND presented in Chapter 12 provides several visual interfaces to allow users to access and annotate the data. For example, the spatial annotation tool displays 2D maps of brain slices when another interface shows the UMLS concept space.

4.5.1   Multimedia Data

Scientific data are multimedia; therefore, a BIS should be designed to manage images, pathways, maps, 3D structures, and so on regardless of their various formats (e.g., raster, bitmap, GIF, TIFF, PCX). An example of the variety of data formats and media generated within a single application is illustrated in Chapter 10. Managing multimedia data is known to be a difficult task. A multimedia management system must provide uniform access transparent to the medium or format. Designing a multimedia BIS raises new challenges because of the complexity and variety of scientific queries. The querying process is an intrinsic part of scientific discovery. A BIS user's interface should enable scientists to visualize the data in an intuitive way and access and query through this representation. Not only do scientists need to retrieve data in different media (e.g., images), but they also need the ability to browse the data with maps, pathways, and hypertext. This means a BIS needs to express a variety of relationships among scientific objects. The difficulties mentioned previously regarding the identification of scientific objects (see Section 4.2.4) are dramatically increased by the need to capture the hierarchy of relationships. Scientific objects such as genes, proteins, and sequences can be seen as classes in an entity-relationship (ER) model; and a map can be seen as the visualization of a complex ER diagram composed of many classes, *isa* relationships, relationships, and attributes. Each class can be populated by data collected from different data sources and the relationships corresponding to different source capabilities. For example, two classes, gene and publication, can be respectively populated with data from GeneCards and PubMed. The relationship from the class gene to the class publication, expressing the publications in which the gene was published, can be implemented by capturing the capability available at GeneCards that lists all publications associated with a gene and provides their PubMed identifiers. The integration data schema is very complex because data and relationships must be integrated at different levels of the nested hierarchy. Geographical information systems address similar issues by representing maps at different granularities, encompassing a variety of information.

Many systems have been developed to manage geographical and spatial data, medical data, and multimedia data. Refer to *Spatial Databases—with Applications to GIS* [85], *Neural Networks and Artifical Intelligence for Biomedical Engineering,* [86] and *Principles of Multimedia Database Systems* [87] for more information. However, very little has been done to develop a system to integrate scientific multimedia systems seamlessly. The following section partially addresses the problem by focusing on relationships between scientific objects.

4.5.2   Browsing Scientific Objects

Scientific entities are related to each other. A gene comprises one or more sequences. A protein is the result of a transcription of dioxyribonucleic acid (DNA) into RNA followed by a translation. Sequences, genes, and proteins are related to reference publications. These relationships are often represented by links (and hyperlinks). For example, there is a relationship between an instance of a gene and instances of the set of sequences that comprise the gene.

The attributes describing an entity, the relationship associating the entity to other entities, and most importantly, the semantics of the relationships, correspond to the complete functional characterization of an entity. Such a characterization, from multiple sources and representing multiple points of view, typically introduces discrepancies. Examples of such discrepancies include dissimilar concepts (GenBank is sequence-centric whereas GeneCards is gene-centric), dissimilar attribute names (the primary GeneCards site has an attribute `protein` whereas a mirror represents the same information as an attribute `product`), and dissimilar values or properties (the gene `TP53` is linked to a single citation in the data source HUGO, 35 citations in the data source GDB, and two citations in the data source GeneCards).

A BIS integrating multiple data sources should allow life scientists to browse the data over the links representing the relationships between scientific objects. A path is a sequence of classes, starting and ending at a class and intertwined with links. Two paths with identical starting and ending classes may be equivalent if they have the same semantics. The resolution of the equivalence of paths is also critical to developing efficient systems as discussed in Section 4.4.3. Semantic equivalence is a difficult problem, as illustrated in the following example. Consider a link from PubMed citations to sequences in GenBank. This link can be physically implemented in two different ways: (1) by extracting GenBank identifiers from the MEDLINE format of the PubMed citation, or (2) by capturing the `Nucleotide Link` as implemented via the Entrez interface. Both implementations expect to capture all the GenBank identifiers relevant to a given PubMed citation. These two links have same starting and ending classes; however, they do not appear to be equivalent. Using the first implementation, the PubMed citation 8552191 refers to four GenBank identifiers. In contrast, the Nucleotide Link representing the second property returns eight GenBank identifiers. Based on the dissimilar cardinality of results (the number of returned sequences), the two properties are not identical. This can also be true for paths (informal sequences of links) between entities. To make the scenario more complex, there could be multiple alternate paths (links) between a start entity and an end entity implemented in completely different sources.

A BIS able to exploit source capabilities and information on the semantics of the relationships between scientific objects would provide users the ability to browse scientific data in a transparent and intuitive way.

4.6   CONCLUSION

Traditional technology often does not meet the needs of life scientists. Each research laboratory uses significant manpower to adjust and customize as much as possible the available technology. Because of the failure of traditional approaches to support scientific discovery, life scientists have proven to be highly creative in developing their own tools and systems to meet their needs. Traditional database systems lack flexibility: Life scientists use flat files instead. Some of the tools and systems developed in scientific laboratories may not meet the expectations of computer scientists, but they perform and support thousands of life scientists. The development of BIS is driven by the needs of a community. But practice shows that the community now needs the development of systems that are more engineered than before, and computer scientists should be involved.

There are good reasons for traditional technology to fail to meet the requirements of BIS. Databases are data-driven and lack flexibility at the level of data representation. XML and other semi-structured approaches may offer this needed flexibility, but the development of native semi-structured systems still is in its infancy. Knowledge bases offer different semantic layers to leverage queries with the exploration process, but they should be coupled with databases to perform traditional data manipulation. On the other hand, agent architectures and grids provide flexible and transparent management of tools. Each of these approaches may and should contribute to the design of BIS.

The systems presented in this book constitute the first generation of BIS. Each system addresses some of the requirements presented in Chapter 2. Each presented system still is successfully used by life scientists; however, the development of each of these systems told a lesson. To be successful, the design of the next generation of BIS should take advantage of these lessons and exploit and combine all existing approaches.

ACKNOWLEDGMENTS

The author wishes to thank Louiqa Raschid and Barbara Eckman for fruitful discussions that contributed to some of the material presented in this chapter.

REFERENCES

[1]   A. Baxevanis. "The Molecular Biology Database Collection: 2002 Update." *Nucleic Acids Research* 30, no. 1 (2002): 1–12.

[2]   C. Burks. "Molecular Biology Database List." *Nucleic Acids Research* 27, no. 1 (1999): 1–9, http://nar.oupjournals.org/cgi/content/full/27/1/1.

[3]   A. Baxevanis. "The Molecular Biology Database Collection: An Online Compilation of Relevant Database Resources." *Nucleic Acids Research* 28, no. 1(2000): 1–7.

[4]   A. Baxevanis. "The Molecular Biology Database Collection: An Updated Compilation of Biological Database Resources." *Nucleic Acids Research* 29, no. 1 (2001): 1–10.

[5]   U. S. Department of Energy Human Genome Program. "Report from the 1999 U. S. DOE Genome Meeting: Informatics." *Human Genome News* 10, no. 3–4 (1999): 8.

[6]   D. Benson, I. Karsch-Mizrachi, D. Lipman, et al. "GenBank." *Nucleic Acids Research* 28, no. 1 (January 2000): 15–18. http://www.ncbi.nlm.nih.gov/Genbank.

[7]   A. Bairoch and R. Apweiler. "The SWISS-PROT Protein Sequence Databank and Its Supplement TrEMBL in 1999." *Nucleic Acids Research* 27, no. 1 (January 1999): 49–54. http://www.expasy.ch/sprot.

[8]   Saccharomyces Genome Database (SGD). http://genome-www.Stanford.edu/saccaromyces/. Department of Genetics, Stanford University.

[9]   M. Rebhan, V. Chalifa-Caspi, J. Prilusky, et al. "GeneCards: A Novel Functional Genomics Compendium with Automated Data Mining And Query Reformulation Support." *Bioinformatics* 14, no. 8 (July 1998): 656–664, http://bioinformatics.weizmann.ac.il/cards/CABIOS_paper.html.

[10]   PubMed. http://www.ncbi.nlm.nih.gov/pubmed/. National Library of Medicine.

[11]   GenBank. "Growth of GenBank." *NCBI GenBank Statistics*, revised March 12, 2002, http://www.ncbi.nlm.nih.gov/Genbank/genbankstats.html.

[12]   International Human Genome Sequencing Consortium. "Initial Sequencing and Analysis of the Human Genome." *Nature* 409 (February 2001): 860–921.

[13]   J. Venter, M. Adams, E. Myers, et al. "The Sequence of the Human Genome." *Science* 291, no. 5507 (February 2001): 1304–1351.

[14] L. Croft, S. Schandorff, F. Clark, et al. "Isis: The Intron Information System Reveals the High Frequency of Alternative Splicing in the Human Genome." *Nature Genetics* 24 (2000): 340–341.

[15] B. Graveley. "Alternative Splicing: Increasing Diversity in the Proteomic World." *Trends in Genetics* 17, no. 2 (2001): 100–107.

[16] S. Misener and S. Krawetz. *Bioinformatics: Methods and Protocols.* Methods in Molecular Biology, no. 132. Totowa, NJ: Humana Press, 1999.

[17] D. Hollingsworth. *The Workflow Reference Model.* Hampshire, UK: Workflow Management Coalition, 1995, http://www.wfmc.org/standards/docs/tc003v11.pdf.

[18] P. Buneman, S. Davidson, G. Hillebrand, et al. "A Query Language and Optimization Techniques for Unstructured Data." In *Proceedings of the 1996 ACM SIGMOD International Conference on Management of Data (Montreal, June 4-6, 1996),* 505–516. New York: ACM Press, 1996.

[19] S. Nestorov, J. Ullman, J. Wiener, et al. "Representative Objects: Concise Representations of Semi-Structured Hierarchical Data. In *Proceedings of the Thirteenth International Conference on Data Engineering (April 7-11, 1997 Birmingham U.K.),* 79–90. Washington, D.C.: IEEE Computer Society, 1997.

[20] W. Fan. "Path Constraints for Databases with or without Schemas." Ph.D. dissertation, University of Pennsylvania, 1999.

[21] D. Fallside. *XML Schema Part 0: Primer: World Wide Web Consortium (W3C) Recommendation,* May 2, 2001, http://www.w3.org/TR/2001/REC-xmlschema-0-20010502/.

[22] H. Thompson, D. Beech, M. Maloney, et al. *XML Schema Part 1: Structures: World Wide Web Consortium (W3C) Recommendation,* May 2, 2001, http://www.w3.org/TR/2001/REC-xmlschema-1-20010502/.

[23] P. Biron and A. Malhotra. *XML Schema Part 2: Datatypes: World Wide Web Consortium (W3C) Recommendation,* May 2, 2001, http://www.w3.org/TR/2001/REC-xmlschema-2-20010502.

[24] S. Abiteboul, P. Buneman, and D. Suciu. *Data on the Web.* San Francisco: Morgan Kaufmann, 2000.

[25] H. Wain, E. Bruford, R. Lovering, et al. "Guidelines for Human Gene Nomenclature (2002)." *Genomics* 79, no. 4 (April 2002): 464–470.

[26] Human Genome Database (GDB). http://www.gdb.org. The Hospital for Sick Children, Baltimore, MD: Johns Hopkins University.

[27] J. Frezal. "GenAtlas Database, Genes and Development Defects." *Comptes Rendus de l'Académie des Sciences–Series III: Sciences de la Vie* 321, no. 10 (October 1998): 805–817.

[28] GenAtlas. Direction des Systèmes d'Information, Université Paris 5, France. http://www.dsi.univ-paris5.fr/genatlas/.

[29] A. Hamosh, A. F. Scott, J. Amberger, et al. "Online Mendelian Inheritance in Man (OMIM)." *Human Mutation* 15, no. 1 (2000): 57–61.

[30] Online Mendelian Inheritance in Man (OMIM). http://www.ncbi.nlm.nih.gov/entrez/query.fcgi?db=omim. World Wide Web interface developed by the National Center for Biotechnology information (NCBI, National Library of Medicine.

[31] K. Pruitt, K. Katz, H. Sciotte, et al. "Introducing Refseq and LocusLink: Curated Human Genome Resources at the NCBI." *Trends in Genetics* 16, no. 1 (2000): 44–47.

[32] LocusLink. http://www.ncbi.nlm.nih.gov/locuslink/. National Center for Biotechnology Information (NCBI), National Library of Medicine.

[33] G. D. Schuler. "Pieces of the Puzzle: Expressed Sequence Tags and the Catalog of Human Genes." *Journal of Molecular Medicine* 75, no. 10 (1997): 694–698.

[34] I. M. A. Chen, V. M. Markowitz. "An Overview of the Object-Protocol Model (OPM) and OPM Data Management Tools." *Information Systems* 20, no. 5 (1995): 393–418.

[35] J. Berlin and A. Motro. "Autoplex: Automated Discovery of Content for Virtual Databases." In *Proceedings of the 9th International Conference on Cooperative Information Systems (Trento, Italy, September 5-7, 2001)*, 108–122. New York: Springer, 2001.

[36] A. Doan, P. Domingos, and A. Levy. "Learning Source Descriptions for Data Integration." In *Proceedings of the Third International Workshop on the Web and Databases (Dallas, May 18-19, 2000)*, 81–86.

[37] W. Pearson and D. Lipman. "Improved Tools for Biological Sequence Comparison." *Proceedings of the National Academy of Science* 85, no. 8 (April 1988): 2444–2448.

[38] S. Altschul, W. Gish, W. Miller, et al. "Basic Local Alignment Search Tool." *Journal of Molecular Biology* 215, no. 3 (October 1990): 403–410. http://www.ncbi.nlm.nih.gov/BLAST.

[39] E. Glemet and J-J. Codani. "LASSAP: A LArge Scale Sequence CompArison Package." *Bioinformatics* 13, no. 2 (1997): 137–143. http://www.gene-it.com.

[40] R. D. Stevens, C. A. Goble, P. Baker, et al. "A Classification of Tasks in Bioinformatics." *Bioinformatics* 17, no. 2 (2001): 180–188.

[41] S. Abiteboul, R. Hull, and V. Vianu. *Foundations of Databases.* Boston: Addison-Wesley, 1995.

[42] National Institute of Standards and Technology. *Database Language SQL*, June 2,1993. http://www.itl.nist.gov/fipspubs/fip127-2.htm.

[43] U. Manber and W. Sun. "GLIMPSE: A Tool to Search Through Entire File Systems." In *Proceedings of USENIX Conference*, 23–32. Berkeley, CA: USENIX Association. 1994.

[44] T. Etzold and P. Argos. "SRS: An Indexing and Retrieval Tool for Flat File Data Libraries." *Computer Applications of Biosciences* 9, no. 1 (1993): 49–57. See also http://srs.ebi.ac.uk.

[45] G. Schuler, J. Epstein, H. Ohkawa, et al. "Entrez: Molecular Biology Database and Retrieval System." *Methods in Enzymology* 266, (1996): 141–162.

[46] W. Wilbur and Y. Yang. "An Analysis of Statistical Term Strength and Its Use in the Indexing and Retrieval of Molecular Biology Texts." *Computers in Biology and Medicine* 26, no. 3 (1996): 209–222.

[47] L. Wong. "Some MEDLINE Queries Powered by Kleisli." *ACCESS* 25 (June 1998): 8–9.

[48] G. Stoesser, W. Baker, A. Van Den Broek, et al. "The EMBL Nucleotide Sequence Database: Major New Developments." *Nucleic Acids Research* 31, no. 1 (2003): 17–22.

[49] J. Clark and S. DeRose. *XML Path Language (XPath): World Wide Web Consortium (W3C) Recommendation*, November 16, 1999. http://www.w3.org/TR/xpath.

[50] D. Chamberlin, D. Florescu, J. Robie, et al. *XQuery: A Query Language for XML: World Wide Web Consortium (W3C) Recommendation*, 2000. http://www.w3.org/TR/xmlquery.

[51] S. Russel and P. Norvig. *Artificial Intelligence: A Modern Approach*. Upper Saddle River, NJ: Prentice Hall, 1995.

[52] S. Kraus, J. Dix, and V. S. Subrahmanian. "Probabilistic Temporal Databases." *Artificial Intelligence Journal* 127, no. 1 (2001): 87–135.

[53] T. Eiter, J. Lu, T. Lucasiewicz, et al. "Probabilistic Object Bases." *ACM Transactions on Database Systems* 26, no. 3 (2001): 264–312.

[54] R. Stevens and A. Brass. "Using CORBA Clients in Bioinformatics Applications." *Collaborative Computational Project 11 Newsletter* 2.1, no. 3 (1998). http://www.hgmp.mrc.ac.uk/CCP11/CCP11newsletters/CCP11NewsletterIssue3.pdf.

[55] L. Wang, P. Rodriguez-Tomé, N. Redaschi, et al. "Accessing and Distributing EMBL Data Using CORBA (Common Object Request Broker Architecture)." *Genome Biology* 1, no. 5 (2000).

[56] Globus. Open grid services architecture (OGSA). http://www. Globus.org/ogsa/.

[57] Cactus: open source problem solving environment. http://www.cactuscode.org/.

[58] NSF funded. Teragrid. http://www.teragrid.org/.

[59] European Union. Datagrid. http://eu-datagrid.web.cern.ch/eu-datagrid/.

[60] T. Berners-Lee, D. Connoly, and R. Swick. "Web Architecture: Describing and Exchanging Data." *W3C Note*, June 1999. http://www.w3.org/1999/04/WebData.

[61] T. Berners-Lee, J. Hendler, and O. Lassila. "The Semantic Web." *Scientific American* (May 2001).

[62] D. Fensel, I. Horrocks, R. van Harmelem, et al. "OIL in a Nutshell." In *Proceedings of the 12th International European Knowledge Acquisition Conference, EKAW 2000,* Juan-les-Pins, France, October 2–6, 2000. LNAI: Springer Verlag, 2000.

[63] T. Critchlow. *Report on XEWA-00: The XML-Enabled Wide-Area Searches for Bioinformatics Workshop.* New York: IEEE Computer Society, 2000.

[64] M. Friedman and D. Weld. "Efficiently Executing Information-Gathering Plans." In *Proceedings of the Fifteenth International Joint Conference on Artificial Intelligence (Nagoya, Japan, August 23-29, 1997),* 785–791. San Francisco: Morgan Kaufmann, 1997.

[65] E. Lambrecht, S. Kambhampati, and S. Gnanaprakasam. "Optimizing Recursive Information-Gathering Plans." In *Proceedings of the Sixteenth International Joint Conference on Artificial Intelligence (Stockholm, July 31-August 6, 1999),* 1204–1211. San Francisco: Morgan Kaufmann, 1999.

[66] D. Florescu, D. Koller, A. Y. Levy. "Using Probabilistic Information in Data Integration." In *Proceedings of 23rd International Conference on Very Large Data Bases (August 25-29, 1997, Athens),* 216–225. San Francisco: Morgan Kaufmann, 1997.

[67] G. Mihaila, L. Raschid, and M-E. Vidal. "Using Quality of Data Metadata for Source Selection and Ranking." In *Proceedings of the Third International Workshop on the Web and Databases (Dallas, May 18-19, 2000),* 93–98. In conjunction with the ACM SIGMOD, 2000.

[68] G. A. Mihaila, L. Raschid, and M.-E. Vidal. "Using Quality of Data Metadata for Source Selection and Ranking." In D. Suciu, G. Vossen (eds.). *Proceedings of the Third International Workshop on the Web and Databases, WebDB 2000.* Dallas, Texas, May 18–19, 2000, in conjunction with ACM PODS/SIGMOD 2000. Informal proceedings.

[69] C. Baru, A. Gupta, B. Ludäscher, et al. "XML-Based Information Mediation with MIX." In *Proceedings ACM SIGMOD International Conference on Management of Data (June 1-3, 1999, Philadelphia),* 597–599. New York: Association for Computing Machinery (ACM) Press, 1999.

[70] A. Levy, A. Rajaraman, and J. Ordille. "Querying Heterogeneous Information Sources Using Source Descriptions." In *Proceedings of 22nd International Conference on Very Large Data Bases (September 3-6, 1996, Mumbai, India),* 251–262. San Francisco: Morgan Kaufmann, 1996.

[71] Y. Papakonstantinou and V. Vassalos. "Query Rewriting for Semistructured Data." In *Proceedings of the ACM SIGMOD International Conference on Management of Data (June 1-3, 1999, Philadelphia),* 455–466. New York: ACM Press, 1999.

[72] M. E. Vidal. "Mediation Techniques for Multiple Autonomous Distributed Information Sources." Ph.D. dissertation, Universidad Simón Bolívar, Caracas, Venezuela, 2000.

[73] V. Vassalos and Y. Papakonstantinou. "Describing and Using Query Capabilities of Heterogeneous Sources." In *Proceedings of 23rd International Conference on Very Large Data Bases (August 25-29, 1997, Athens)*, 256–265. San Francisco: Morgan Kaufmann, 1997.

[74] R. Yerneni, C. Li, H. Garcia-Molina, et al. "Computing Capabilities of Mediators." In *Proceedings ACM SIGMOD International Conference on Management of Data (June 1-3, 1999, Philadelphia)*, 443–454. New York: ACM Press, 1999.

[75] B. Eckman, Z. Lacroix, and L. Raschid. "Optimized Seamless Integration of Biomolecular Data." In *2nd IEEE International Symposium on Bioinformatics and Bioengineering (Bethesda, Maryland, November 4-5, 2001)*, 23–32. Washington D.C.: IEEE Computer Society, D.C., 2001.

[76] Z. Lacroix. "Biological Data Integration: Wrapping Data and Tools." *IEEE Transactions on Information Technology in Biomedicine* 6, no. 2 (June 2002): 123–128.

[77] Z. Nie and S. Kambhampati. "Joint Optimization of Cost and Coverage of Query Plans in Data Integration." In *Proceedings of the 2001 ACM CIKM International Conference on Information and Knowledge Management (November 5-10, 2001, Atlanta)*, 223–230. New York: ACM Press, 2001.

[78] B. Eckman, A. Kosky, and L. Laroco. "Extending Traditional Query-Based Integration Approaches for Functional Characterization of Post-Genomic Data." *Bioinformatics* 17, no. 7 (2001): 587–601.

[79] J. Ullman. *Principles of Database and Knowledge-Base Systems*, volume II. Palo Alto, CA: Computer Science Press, 1989.

[80] J. Ullman. "Information Integration Using Logical Views." In *Proceedings of the Sixth International Conference on Database Theory*, 19–40. Springer, 1997.

[81] K. Morris. "An Algorithm for Ordering Subgoals in NAIL!" In *Proceedings of the Seventh ACM SIGACT-SIGMOD-SIGART Symposium on Principles of Database Systems (March 21-23, 1988, Austin, Texas)*, 82–88. New York: ACM Press, 1988.

[82] Y. Ioanidis and Y. Kang. "Randomized Algorithms for Optimizing Large Join Queries." In *Proceedings of the 1990 ACM SIGMOD International Conference on Management of Data (Atlantic City, NJ, May 23-25, 1990)*, 312–321. New York: ACM Press, 1990.

[83] P. Selinger, M. Astrahan, D. Chamberlin, et al. "Access Path Selection in a Relational Database Management System." In *Proceedings of the 1979 ACM SIGMOD International Conference on Management of Data (Boston, May 30-June1)*, 23–34. New York: ACM Press, 1979.

[84] M. Steinbrunn, G. Moerkotte, and A. Kemper. "Heuristic and Randomized Optimization for the Join Ordering Problem." *VLDB Journal* 6, no. 3 (1997): 191–208.

[85] P. Rigaux, M. Scholl, and A. Voisard. *Spatial Databases—With Applications to GIS*. San Francisco: Morgan Kaufmann, 2001.

[86] D. Hudson and M. Cohen. *Neural Networks and Artificial Intelligences for Biomedical Engineering*. New York: IEEE Press, 2000.

[87] V. S. Subrahmanian. *Principles of Multimedia Database Systems*. San Francisco: Morgan Kaufmann, 1998.

SRS: An Integration Platform for Databanks and Analysis Tools in Bioinformatics

Thure Etzold, Howard Harris, and Simon Beaulah

The Sequence Retrieval System (SRS) approach to data integration has evolved over many years to address the needs of researchers in the life sciences to query, retrieve, and analyze complex, ever increasing, and changing biological data. SRS follows a federation approach to data integration, leaving the underlying data sources in their original formats. For example, Genbank [1] is used in flat file format; the Genome Ontology (GO) [2] is used in either XML format or as relational tables stored in MySQL [3]. Databanks generated and provided by the major technologies available are integrated through meta-data, which is provided for the majority of the common public data sources. SRS customers use this functionality to integrate their own in-house data, such as gene expression databases, with third-party data such as Incyte LifeSeq Foundation data [4], and public data such as EMBL [5] and Swiss-Prot [6].

Databanks in SRS can be queried and analyzed via a Web interface or through a variety of application programming interfaces (APIs) as described in Section 5.8. Using one of a variety of query forms, the user can search a single or a combination of databanks. Search results can be further analyzed using a suite of tools like BLAST [7] and FASTA [8] for sequence similarity searching. SRS provides support for about 200 tools including a major part of EMBOSS [9]. This is further described in Section 5.7.

Meta-data is at the heart of SRS. Each data source is fully described, including the type and structure of data, relationships to other data sources, how the data should be indexed or presented to users, and how it can be mapped to external object models. SRS uses a *meta-data only* approach, which is based on its internal programming language, Icarus. Administrators can customize SRS by editing

Icarus files or through the use of a graphical user interface. No access programs or wrappers need to be written by programmers as they do with other integration systems like DiscoveryLink (Chapter 11) and Kleisli (Chapter 6). An exception is the set of syntactic and semantic rules that need to be composed for the integration of flat file databanks. The result of the meta-data approach is a flexible and modular system that has adapted to all the changes and developments in bioinformatics over the past 10 years. Many approaches to data integration have been proposed over this time to address the needs described in this book, but SRS has surpassed them all to provide the only proven and widely used flexible data integration environment.

SRS aims to remain independent of the technology used for data storage. Extensible markup language (XML), flat files, and relational databases bring with them a range of benefits and problems that often create a particular mind-set for the people who use and maintain them. Flat file databanks are the "dinosaurs" in this field, albeit very successful in defying extinction. Flat file data are compact and generally very flexible to work with. They are mostly semi-structured and are presented in a vast variety of formats, which in their multitude make parser writing an almost impossible task. This is further described in Section 5.1. XML is an elegant way of representing data and is ideally suited for transferring information between tools and applications (e.g., communicating genomic data to a genome browser). XML offers great flexibility, which can present formidable challenges for integration. How SRS meets these challenges is described in Section 5.2. Relational databases create a world of tables, columns, and relationships, providing a structured and maintainable data store. However, the Structured Query Language (SQL) is not a common skill of most researchers, and the scientific concepts researchers wish to analyze are often lost somewhere in the ever-growing database schema. For almost all relational databases in molecular biology, a bespoke interface had to be built. Section 5.3 covers this technology. SRS supports these three technologies (flat file, XML, and relational databases) and can map all data into flexible and extendable object models as described in Section 5.6. Using the object loader, users can define their own views of the data to display, for example, gene expression data with information from GenBank and from InterPro [10]. This type of view combines XML, relational, and flat file data seamlessly and is completely in the control of the user. Section 5.6 also describes how data can be exported as XML to other applications in a standard or customized way.

Providing access to all data sources is only the first step of data integration. The relationships between the different data sources are represented in SRS and are used to form an interconnected set of data sources referred to as the *SRS Universe*. Mapping of attributes and approaches to semantic integration is addressed

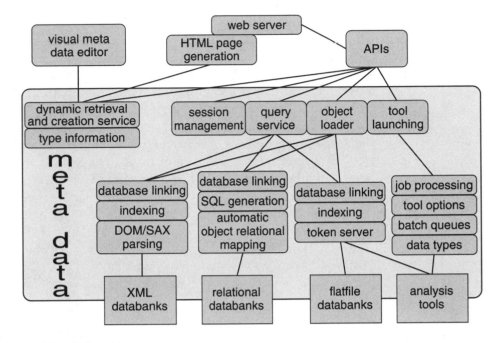

5.1   The SRS architecture.

FIGURE

in Section 5.5. When a new data source is added to the SRS Universe, relationships to already integrated resources can be defined. This allows the SRS administrator and a domain expert to combine their knowledge to implement an SRS Universe that reflects the intricacies of their own data alongside a tried and tested public data SRS Universe. In this way, SRS provides the flexibility and extensibility that is required in a changing environment.

Figure 5.1 gives an overview of the SRS architecture. It shows the three data source types: XML, relational, and flat file databanks. The output of analysis tools is treated in the same way as flat file databanks. SRS provides specific technology to deal with each data source type. For example, flat file databanks use the token server, whereas relational databanks are integrated through object relational mapping and SQL generation modules. On top of these technologies are services such as the *query service* and the *object loader*. They can be applied to all data sources in a transparent way. The APIs can be used by programmers to make use of these services. The SRS Web server gives an example of such an application. Meta-data plays an important role in SRS. All data sources and analysis tools are fully described (e.g., file location, ftp source address, format) and all SRS modules are configured through meta-data. A visual editor can be used to access, modify, and create all SRS meta-data. All of the components are described in this book, if only briefly. Unfortunately, the scope of this book does not allow a complete description of them all.

The SRS server at the European Bioinformatics Institute (EBI) [11] has provided genomic and related data to the European bioinformatics community since 1994. It now serves more than 4 million hits per month, returning results in seconds, and supporting thousands of researchers. The EBI SRS server has approximately 200 data sources integrated with many analysis tools. It also links to an access page with other major academic SRS servers and gives access to the freely available SRS meta-definition files for (currently) more than 700 public databases (see also Krell and Etzold's article "Data banks" [12]). SRS is extensively used in large pharmaceutical and biotech companies and is the basis of the Celera Discovery System [13], Incyte LifeSeq Foundation distribution, Affymetrix NetAffyx portal [14], and Thomson Derwent Geneseq portal [15]. The SRS community of academic and commercial companies makes SRS the most widely used life science integration product.

Note: Throughout this text the words *databank, database,* and *library* are used in a seemingly interchangeable way. To clarify, this *database* is used to refer to the sum of data and the actual system within which it is stored, *databank* to refer to the data only, and *library* to refer to the representation of a databank or database within SRS.

5.1   INTEGRATING FLAT FILE DATABANKS

Before the advent of XML, almost all data collections in molecular biology were available as sets of text files, also called *flat file databanks*. New databanks are now generally available in XML and, increasingly, flat file databanks can be obtained in an alternative XML format. However, flat files continue to be the only available form for many data collections and will stay an important source of information for years to come.

Overall, flat file databanks have a simple structure and usually consist of a single stream of entries represented in a text format with a special syntax. The entries can be very rich, containing comprehensive information about a protein, a DNA sequence, or a tertiary 3D structure. The formats for these flat file databanks vary greatly and are only rarely shared. Once defined, individual formats will change continuously to reflect the growth of complexity in the associated content. Hundreds of these formats have been created, which makes parsing data in molecular biology a highly daunting task. SRS meets this challenge by providing tools that make writing parsers easy and very maintainable. Parsers to disseminate flat file entries, or token servers, are written in Icarus, the internal programming language for SRS.

5.1.1   The SRS Token Server

SRS has a unique approach for parsing data sources that has proved effective for supporting many hundreds of different formats. With traditional approaches, a parser would be written as a program. This program would then be run over the data source and would return with a structure, such as a parse tree that contains the data items to be extracted from the source. In the context of structured data retrieval, the problem with that approach is that depending on the task (e.g., indexing or displaying) different information must be extracted from the input stream. For instance, for data display, the entire description field must be extracted, but to index the description field needs to be broken up into separate words.

A new approach was devised called *token server*. A token server can be associated with a single entity of the input stream, such as a databank entry, and it responds to requests for individual tokens. Each token type is associated with a name (e.g., *descriptionLine*) that can be used within this request. The token server parses only upon request (lazy parsing), but it keeps all parsed tokens in a cache so repeated requests can be answered by a quick look-up into the cache.

A token server must be fed with a list of syntactic and optionally semantic rules. The syntactic rules are organized in a hierarchic manner. For a given databank there is usually a rule to parse out the entire entry from the databank, rules to extract the data fields within that entry, and rules to process individual data fields. The parsed information can be extracted on each level as tokens (e.g., the entire entry, the data fields, and individual words within an individual data field). Semantic rules can transform the information in the flat file. For example, amino acid names in three-letter code can be translated to one-letter code or a particular deoxyribonucleic acid (DNA) mutation can be classified as a missense mutation.

Figure 5.2 shows an example of how two different identifiers for protein mutations can be transformed into four different tokens using a combination of syntactic and semantic rules.

The following example of Icarus code defines rules for the tokens AaChange, ProteinChangePos, and AaMutType for the first variation of the mutation key Leu39Arg.

```
Key:        ~ {$Out  $In:Entry} ln {$Wrt} ~
AaChange:   ~ {$In:Key $Out
              $code={ala:A arg:R asn:N asp:D
```

```
                        cys:C glu:E gln:Q gly:G
                        his:H ile:I leu:L lys:K
                        met:M phe:F pro:P ser:S
                        thr:T try:W tyr:Y val:v}
            }
            aa {$aaSave = $Ct} num aa
            {
              $aa1 = $code.($aaSave.lower)
              $aa2 = $code.($Ct.lower)
              $Wrt:[s:"($aa1)>($aa2)"]
            }
                      ~
ProteinChangePos:  ~ {$In:Key $Out} aa num {$Wrt} ~
AaMutType:         ~ {$In:AaChange $Out}
                     /[A-Z]>[A-Z]/{$Wrt:[s:substitution]}
                     |/[A-Z]>\*/{$Wrt:[s:termination]}
                      ~
ln:                ~ /[^\n]*\n/ ~
aa:                ~ /[a-Z]+/ ~
num:               ~ /[0-9]+/ ~
```

The rules are specified in Icarus, the internal programming language of SRS. Icarus is in many respects similar to Perl [16]. It is interpreted and object-oriented with a rich set of functionality. Icarus extends Perl with its ability to define formal rule sets for parsing. Within SRS it is also used extensively to define and manipulate the SRS meta-data.

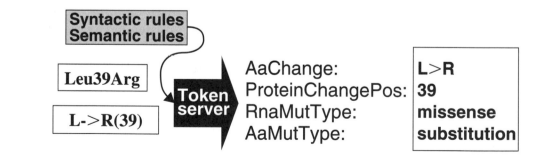

5.2   The SRS token server applied on mutation identifiers.
FIGURE

The previous code example contains seven rule definitions. Each starts with a name, followed by a colon and then the actual rule, which is enclosed within ~ characters. These rules are specified in a variant of the Extended Backus Naur Form (EBNF) [17] and contain symbols such as literals, regular expressions (delimited by / characters), and references to other rules. In addition they can have commands (delimited by { and }), which are applied either after the match of the entire rule (command at the beginning of the rule) or after matching a symbol (command directly after the symbol). In the example three different functions are called within commands. $In specifies the input tokens to which the rule can be applied. For example, AaChange specifies as input the token table Key, which is produced by the rule Key. $Out specifies that the rule will create a token table with the current rule. The command $Wrt writes a string into the token table opened by the current rule. For example, the $Wrt command following the reference to the ln rule in the rule Key will write the line matched by ln into the token table Key. Using commands $In and $Out, rules can be chained by feeding each other with the output token tables they produce. For example, rule AaChange processes the tokens in token table Key and provides the input for rule AaMutType. Lazy parsing means that only the rules necessary to produce a token table will be activated. To retrieve the Key tokens, the rules Key and Entry need to be processed. To obtain AaMutType, the rules AaMutType, AaChange, Key, and Entry are invoked. Production AaChange uses an associative list to convert three-letter amino acid codes to their one-letter equivalents. AaMutType is a semantic rule that uses standard mutation descriptions provided by AaChange to determine whether a mutation is a simple substitution or leads to termination of the translation frame by introducing a stop codon.

Advantages of the token server approach are:

+ It is easy to write a parser where the overall complexity can be divided into layers: entry, fields, and individual field contents.

+ The parser is very robust; a problem parsing a particular data field will not break the overall parsing process.

+ A rule set consists of simple rules that can be easily maintained.

+ Lazy parsing allows adding rules that will only be used in special circumstances or by only a few individuals.

+ Lazy parsing allows alternative ways of parsing to be specified (e.g., retrieval of author names as encoded in the databank or converted to a standard format).

+ The parser can perform reformatting tasks on the output (e.g., insertion of hypertext links).

5.1.2   Subentry Libraries

Flat file databanks are often described as being semi-structured. This stems from the lack of a formal description of the contents, which may just be mentioned briefly in a readme file or user manual. While individual entries in a flat file databank describe a real world object such as a gene or protein, it is often possible to discover entities within these entries that are worth querying and retrieving as independent entities.

Consider a nucleotide sequence entry in EMBL or GenBank that describes an entire genome, or a large part of it, encoding hundreds of genes. It contains for every gene, or coding sequence, a sub-entity, or sub-entry, which can look like the one shown in Figure 5.3.

With SRS these sub-entries can be parsed, indexed, and retrieved as separate entities. There is still a tight association to the parent entry, but a separate databank of sub-entries is created effectively next to the databank of parent entries. Sequence features have a special property in that they are contained within the sequence of the parent entry. The exact location of that sequence can be specified in the sub-entry as shown in Figure 5.3 as a complex *join* statement following the CDS keyword. SRS uses this information to retrieve the sub-sequence of the sequence feature as part of the sub-entry.

Many other flat file libraries have sub-entries (e.g., literature citations and comments). In addition, sequence feature tables of the sequence databanks are parsed to produce a new sub-entry type counter, which is a list of counts of each feature type within an entry. Indexing these allows the scientist to make highly specific queries such as *"all Swiss-Prot entries with exactly seven trans-membrane segments."*

5.2   INTEGRATION OF XML DATABASES

XML is becoming increasingly important within the bioinformatics community. There are several good reasons for using XML as a medium for the storage and transmission of bioinformatics data.

✦ Because XML has a universally recognized format built on a stable foundation [18], it has become the primary means of exchanging information over the Internet.

✦ A variety of tools make it relatively easy to manage XML data and transform it into other formats (e.g., an extensible stylesheet language transformation

```
FT   CDS              join(12151..12199,12319..12483,26154..26312,26771..27004,
FT                    28068..28415,29142..29342,30433..30554,30859..30926,
FT                    31311..31341)
FT                    /codon_start=1
FT                    /db_xref="SWISS-PROT:P01730"
FT                    /note="major receptor for HIV-1; member of immunoglobulin
FT                    supergene family; T cell surface glycoprotein T4"
FT                    /gene="CD4"
FT                    /function="T-cell coreceptor; involved in antigen
FT                    recognition; participant in signal transduction pathway"
FT                    /product="surface antigen CD4"
FT                    /protein_id="AAB51309.1"
FT                    /translation="MNRGVPFRHLLLVLQLALLPAATQGKKVVLGKKGDTVELTCTASQ
FT                    KKSIQFHWKNSNQIKILGNQGSFLTKGPSKLNDRADSRRSLWDQGNFPLIIKNLKIEDS
FT                    DTYICEVEDQKEEVQLLVFGLTANSDTHLLQGQSLTLTLESPPGSSPSVQCRSPRGKNI
FT                    QGGKTLSVSQLELQDSGTWTCTVLQNQKKVEFKIDIVVLAFQKASSIVYKKEGEQVEFS
FT                    FPLAFTVEKLTGSGELWWQAERASSSKSWITFDLKNKEVSVKRVTQDPKLQMGKKLPLH
FT                    LTLPQALPQYAGSGNLTLALEAKTGKLHQEVNLVVMRATQLQKNLTCEVWGPTSPKLML
FT                    SLKLENKEAKVSKREKAVWVLNPEAGMWQCLLSDSGQVLLESNIKVLPTWSTPVQPMAL
FT                    IVLGGVAGLLLFIGLGIFFCVRCRHRRRQAERMSQIKRLLSEKKTCQCPHRFQKTCSPI
FT                    "
```

5.3

FIGURE

A Protein Coding Sequence (CDS) Feature in EMBL.

(XSLT) [19] style sheet may be used to transform XML data to hypertext markup language (HTML) format for display in a Web browser).

✦ By allowing users to create their own syntax (element and attribute names) and structure (hierarchical parent-child relationships between elements), XML gives database designers great freedom to transform their mental models of an information system into a concrete form.

However, people conceptualize information in very different ways, particularly in a complex field like bioinformatics. This makes it difficult, if not impossible, to create widely accepted XML standards for bioinformatics data. Furthermore, different organizations are interested in different aspects and constellations of the bioinformatics data universe, which includes DNA sequences, proteins, structures, expressed sequence tags (ESTs), transcripts, metabolic pathways, patents, mutations, publications, and so forth. If all of these data types were incorporated into a single format, it would be extremely complex and unwieldy.

For these reasons, many companies and organizations have given up the quest for a universal XML standard for bioinformatics data. Instead, they have created their own standards, which are often customized versions of existing standards, optimized for use in internal applications. SRS has remained neutral in the standards war by striving to develop flexible tools that support all the existing and emerging bioinformatics XML formats.

5.2.1   What Makes XML Unique?

Data formatting in XML is similar to data formatting in flat files. Figure 5.4 shows how the EMBL flat file data in Figure 5.3 might appear if rendered in XML format. The key features that make XML formats different from flat file formats are as follows. Figure 5.4 illustrates both types of data encapsulation (the only piece of data expressed as an attribute value is the `feature ID`, CDS).

1. XML uses two distinct kinds of tags for *wrapping* data: elements and attributes. There is no hard-and-fast rule for what kinds of data should be encapsulated in attributes rather than in elements. In general, attributes tend to be used for short pieces of data that have a one-to-one relationship with the data in the parent element, such as IDs and classifications.

2. There are two types of syntax that can be used for XML elements.

   a. *Normal* syntax encloses the data belonging to an element between a start tag (e.g.,`<join>`) and an end tag (e.g., `</join>`).

   b. *Empty* syntax may be used for elements that either have no data content or have content that may be stored efficiently in attribute values. The InterPro format created by the EBI uses empty `db_xref` elements for specifying references to external databases:

      `<db_xref db="EC" dbkey="2.7.4.9"/>`

3. Some XML elements (e.g.,`feature_list`) are used as structural components that define hierarchical relationships between other elements but contain no data of their own.

```
<feature_list>

  <feature id="CDS">

    <join>(12151..12199,12319..12483,26154..26312,26771..27004,28068..28415,

          29142..29342,30433..30554,30859..30926,31311..31341)</join>

    <codon_start>1</codon_start>

    <db_xref>SWISS-PROT:P01730</db_xref>

    <note>major receptor for HIV-1; member of immunoglobulin supergene family;

          T cell surface glycoprotein T4</note>

    <gene>CD4</gene>

    <function>T-cell coreceptor; involved in antigen recognition;

          participant in signal transduction pathway</function>

    <product>surface antigen CD4</product>

    <protein_id>AAB51309.1</protein_id>

    <translation>MNRGVPFRHLLLVLQLALLPAATQGKKVVLGKKGDTVELTCTASQ

          KKSIQFHWKNSNQIKILGNQGSFLTKGPSKLNDRADSRRSLWDQGNFPLIIKNLKIEDS

          IVLGGVAGLLLFIGLGIFFCVRCRHRRRQAERMSQIKRLLSEKKTCQCPHRFQKTCSPI

    </translation>

  </feature>

</feature_list>
```

5.4

FIGURE

EMBL flat file data rendered as XML.

4. Empty elements may be used as structure-only elements to delimit entries (or sub-entries). To support both normal and empty element entry delimiters, the SRS XML parser must have two different types of behavior.

   a. For entries delimited by start and end tags, entry processing terminates when the end tag is found.

   b. For entries delimited by empty element tags, there is no end tag, so entry processing terminates when the start tag of the next entry is found or when the end of the file is reached.

5. XML allows users to define *shorthand* expressions to represent commonly used strings. These expressions are called *general entities*. For example, the

entity &spdb; could stand for the name of a database (e.g., SwissProt-Release). When an XML parser encounters a general entity reference like &spdb; in an attribute value or element content, it must replace the reference with the replacement text.

6. Some commonly used characters have special meaning in XML.

   a. Less thans [<] and greater thans [>] are used in markup tags.

   b. Apostrophes ['] and quotation marks ["] are used to delimit attribute values.

   c. Ampersands [&] are used to specify general entity references.
      If these characters occur within XML attribute values or element content, they can create ambiguities for an XML parser, so they must be handled with care.

7. XML data may also be encapsulated in CDATA sections that may appear wherever character data may appear. Inside CDATA sections, less thans and ampersands are treated as literals (i.e., they do not need to be replaced with entity references).

5.2.2   How Are XML Databanks Integrated into SRS?

XML is fully integrated into the SRS universe of databanks, and it is relatively easy to incorporate XML libraries into an SRS installation. The only prerequisite is a document type definition (DTD) that accurately describes the structure of the XML. If a DTD does not exist, a utility such as Michael Kay's DTDGenerator [20] can be used to create one.

The first step in the configuration process is to run an SRS utility, which analyzes the DTD and creates templates for all the meta-data objects needed to define the new library. The user must then edit the resulting object definitions. Initially, the user must supply all of the extra information needed to perform the basic indexing and loading tasks. The next step is to register the new databank with SRS and index the library. Once the library has been indexed, all the standard library operations become available.

If the new XML library contains sub-entry libraries or takes advantage of any special indexing or loading features, the administrator must perform additional editing to define the sub-entry libraries or to activate these features. Integrating an XML library into SRS is easier than integrating a flat file library because SRS does most of the work of creating the library meta-information. Also, the use of a built-in generic XML parser eliminates the need for writing a library-specific parser.

5.2.3 Overview of XML Support Features

Support for Complex DTDs

A DTD is a set of declarations that defines the syntax and structure of a particular class of XML documents. DTDs may consist of an *internal subset* (inside an XML document) and/or any number of *external subsets* in separate files. External subsets may be invoked recursively from within other external subsets. DTDs may also incorporate INCLUDE and IGNORE blocks (*conditional* sections) containing different sets of declarations to be used in different applications, and these blocks may be activated or deactivated using variables called parameter entities. Thus, DTDs can be quite complex.

The SRS utility used to parse DTD files employs a sophisticated algorithm to process external DTDs recursively in accordance with the guidelines laid down in the World Wide Web Consortium's XML Version 1.0 Recommendation [18]. This ensures that if a DTD includes multiple declarations of the same general entity or default attribute value, the correct values are used in generating the SRS meta-information. The utility also supports the use of parameter entities and correctly processes conditional sections.

Support for Indexing and Querying

SRS provides several powerful features to give users control over the way XML data is indexed and queried. *Micro-parsing* allows users to pre-process data before it is written to an index field. For example, suppose the data contains an author element that uses initials-first formatting (e.g., <author>J. K. Rowling</author>), but the user would like to index this in initials-last format (e.g., Rowling, J. K.). The indexing metaphor for the author element would refer to an Icarus syntax file containing a production to transform the data. Micro-parsing allows users to apply the same types of syntactic and semantic rules used for flat file parsing to the contents of individual XML tags.

Splitting allows users to subdivide input data strings containing lists into their component index values. For example, suppose the data contains an authors element containing a list of author names separated by commas and white space (e.g., <authors>J. K. Rowling, William Shakespeare, Stephen King </authors>), but the user would like to index this list as three separate author names. The indexing metaphor for the authors element would include a split attribute (e.g., split:[,]) specifying a regular expression containing a list of the characters used to separate individual items in the list.

Conditional indexing allows users to process meta-data specified within the XML stream. For example, the Bioinformatic Sequence Markup Language (BSML) format [21] uses an XML element called Attribute as a container for three different types of data: version, source, and organism. The name attribute is a meta-data field that identifies the *type* of data contained in the associated content attribute.

```
<Attribute name="version" content="AB003468.1 GI:2656021"/>
<Attribute name="source" content="Cloning vector pAP3neo DNA."/>
<Attribute name="organism" content="Cloning vector pAP3neo"/>
```

Conditional indexing may be used to channel the data contained in the three content attributes into three separate index fields designed to hold version, source, and organism data.

SRS provides solid support for indexing and querying subentry libraries. In some XML formats, a single type of element is used in more than one sub-entry library, and the element may have a different meaning in each library. To index the data contained in these elements into the correct set of target index fields, SRS allows users to create separate fields and indexing metaphors for each unique instance of the element. The indexing metaphors use a special path attribute to determine which sub-entry library the element currently being processed belongs to so that the data can be indexed into the correct field. Conversely, SRS also allows users to index data from a single element or attribute into multiple index fields.

5.2.4   How Does SRS Meet the Challenges of XML?

Problems with managing XML data can be divided into two main categories: syntactical/semantic (microscopic) and structural (macroscopic). Data formatting varies widely between standard XML formats, and pre-processing is often required before data can be indexed or loaded. Table 5.1 describes several common syntactical problems and explains how SRS solves them.

XML formats, like flat file formats, can be large, complex, and unwieldy, making data access difficult and inefficient. Table 5.2 describes several common structural problems that occur in XML formats used in bioinformatics. SRS provides solutions to some of these problems, but some can only be solved by restructuring the data.

Problem	SRS Solution
Fields may include special characters (e.g., colons, square brackets, dashes, and asterisks) that can interfere with SRS query syntax.	Use micro-parsing to *purify* data fields during indexing.
Fields may contain data whose type depends on the value of another (meta-data) field.	Use conditional indexing to index data into different fields based on the value in a *condition* (meta-data) field.
Fields may contain characters that require special handling in XML (e.g., less thans, greater thans, apostrophes, quotation marks, and ampersands).	Use micro-parsing to replace problematic characters with pre-defined character entity references.
Entity references (both pre-defined and user-defined) must be replaced before data is indexed or loaded. Entity references may include markup.	SRS provides sophisticated entity replacement functionality.
A single element may be used in two or more subentry libraries to contain different types of information.	Users can create separate fields and indexing metaphors for each instance of an element used in a different subentry library. The indexing metaphors use a `path` attribute to index the correct data into the correct fields.
Mixed content elements are difficult to parse because content belonging to the parent element is interspersed with content belonging to child elements.	SRS provides two special loading commands (`xsl:copy-of` and `xsl:value-of`) that emulate useful features found in the extensible stylesheet language transformations (XSLT) language [19].
Fields may contain lists of values that must be separated into individual values.	Indexing metaphors can include a `split` attribute to split a string into sub-strings using a set of separator characters contained in a regular expression.

5.1

TABLE

Syntactical problems and SRS solutions.

Problem	SRS Solution
Some libraries use large numbers of *structure-only* tags. Tags can take up a lot of disk space without providing much useful information.	No solution; inherent in XML.
Excessive numbers of tags make a format difficult to understand and manage.	The SRS utility that generates library definition files uses *intelligent* parsing to eliminate structure-only elements from the set of metadata objects that are included in the library definition file.
Deep nesting and large numbers of sub-entries slow down querying and loading performance.	The SRS loading algorithm builds a document object model (DOM) object for each entry. This approach provides both optimal performance and highly reliable handling of sub-entries. It also provides some special loading commands that improve performance for particular types of loading tasks.
Content belonging to a single entity may be spread across multiple files.	No solution; inherent in certain XML formats. Data should be restructured.
A single entity may appear repeatedly in multiple files.	No solution; inherent in certain XML formats. Data should be restructured.
Excessive data redundancy slows down performance.	No solution; inherent in certain XML formats. Data should be restructured.

5.2

TABLE

Structural problems and SRS solutions.

5.3   INTEGRATING RELATIONAL DATABASES

5.3.1   Whole Schema Integration

For relational databases, a schema organizes the data defining the data entities and their relationships to each other. Because individual entities can only be modeled as flat tables, real world concepts such as genes or metabolic pathways often use many tables to store the information faithfully. Conversely, to make full use of this data, the whole schema needs to be made available to the user. The user must be able to query one or more tables and then collect the necessary data from all related tables. For example, in a relational database storing genes, the user may query the author table for Lee, which will return the set of genes published by the author Lee. Behind the scenes the results from the query in the author table need to be related to other data (e.g., accession code, keyword, references, sequence), which is stored in other tables. The data is represented as a whole and must be assembled from many different tables before it is presented to the user.

The problem of mapping a table structure into a more complex object structure has been addressed before by object relational mapping techniques. Traditional approaches start with a class description of the objects to be stored and then generate the relational schema from the class information. This is in conflict with the SRS approach of integrating existing schemas where often an object model has not yet been defined. The overwhelming majority of the schemas relevant to life science informatics (LSI) have been obtained by more traditional methods, such as entity-relationship (ER) modeling, and not by object-relational modeling.

The SRS approach is to use a semi-automated process to define object-relational mapping on top of an existing schema. This is achieved by selecting a table manually to be the *hub table*, or the table containing values equivalent to an object ID (usually an accession number or unique ID), and other tables that can be defined to *belong* to the object. Using the hub table, the selection of tables, and foreign key relationships, SRS can automatically create an object model, which is introduced to the system as a dynamic type. The resulting object can then be queried and retrieved as a fixed entity, much like an entry in a flat file or XML databank.

When a relational databank is integrated, no indexing on the SRS side needs to be done. SRS will generate SQL statements for querying and retrieval of objects that will emulate the same behavior as users expect when dealing with flat file and XML databanks.

5.3.2 Capturing the Relational Schema

SRS Relational includes a Java program, schemaXML, which uses a standard Java Database Connectivity (JDBC) [22] interface to capture the relational database schema, including all the tables, columns, keys, and foreign key relationships. The program schemaXML passes this information to SRS providing the base information to integrate the relational databank. All further meta-data for customization can be added by editing this schema information using a graphical interface or by direct manipulation of Icarus files. This provides a much simpler solution than would be required by writing individual integration programs for each relational databank to be integrated. If the schema changes the program, schemaXML just needs to be re-run and the edits reapplied. A tool is being developed that reapplies edits to an updated version of the original schema definition.

5.3.3   Selecting a Hub Table

SRS Relational is based on the concept of hub tables, which are used, conceptually, to relate relational database tables to data objects. Hub tables are central points of interest in a relational schema and must contain a unique name (typically a primary key) that can be used as an entry ID (e.g., an accession code in a sequence database). Using foreign key relationships, all data held in surrounding tables can be linked directly or indirectly back to the hub table and entry ID using table joins. All tables that belong to a hub table must be directly or indirectly linked with it. In cases where these links are not apparent from the schema information retrieved by schemaXML, they can be set manually within the visual administration interface.

Figure 5.5 shows a section of the relational schema that is used to maintain the GO databank in the MySQL relational database management system (RDBMS).

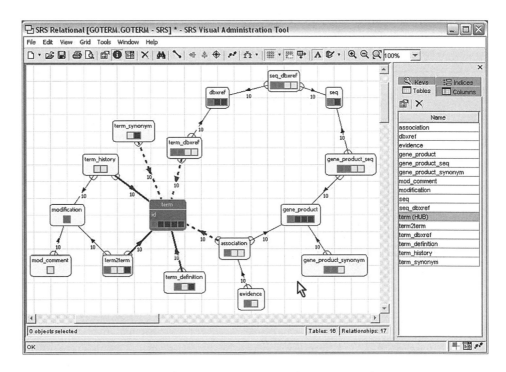

<table>
<tr><td>5.5</td><td>Visual representation of part of the GO term schema within the SRS Visual Admin-</td></tr>
<tr><td>━━━━━</td><td>istration Tool. The table term is selected as a hub table. Individual lines between</td></tr>
<tr><td>FIGURE</td><td>the tables represent foreign key relationships.</td></tr>
</table>

The `term` table is clearly the central point of interest in the schema with related tables surrounding it. It would be selected by the SRS administrator as a hub table for use in SRS. In other databases, the hub table selection may be less clear. For example, a Laboratory Information Management System (LIMS) database has many concepts such as sample, project, and experiment, each with its own collection of related tables. In these cases, multiple hub tables can be selected and associated to separate SRS libraries by the SRS administrator.

5.3.4   Generation of SQL

SRS sees the relational schema as a graph with tables as nodes and foreign key relationships as edges. The hub table is at the center of this graph. An idealized form of such a graph is shown in Figure 5.6. To translate an SRS query into SQL it maps the predicates to the appropriate columns and then to tables in the graph, and a shortest path is derived to relate these predicate queries to rows in the hub table using joins. An example is shown for three predicates in Figure 5.6 (A), which are all linked to the hub table. The SQL query will return a number of rows in the hub table, which are processed to create a list of entry IDs. To retrieve particular entries, a search path is again used, this time radiating out from the hub table and including the required tables using joins. See Figure 5.6 (B).

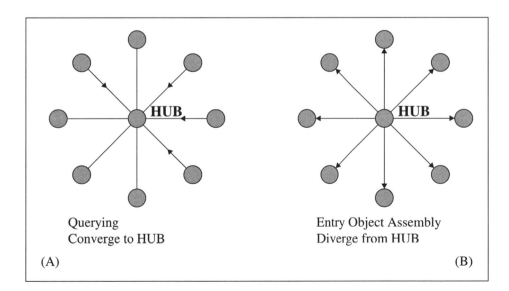

Querying
Converge to HUB

Entry Object Assembly
Diverge from HUB

(A)

(B)

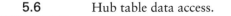

5.6   Hub table data access.

FIGURE

5.3.5   Restricting Access to Parts of the Schema

Once the relational schema held on the SRS side has been configured to define a library, it can be further modified to restrict or allow access to the tables within the schema. Individual tables can be hidden from SRS so general access to the data is not available. In addition, the SRS access permissions can also be used to control access to the whole or sections of the schema (when using multiple hub tables). It is also possible to modify, add, and remove links between tables without altering the original database schema.

5.3.6   Query Performance to Relational Databases

During the development and use of SRS Relational a number of performance optimizations have been added. A few of these are outlined as follows.

- ✦ SRS is case insensitive, and it is well known that case insensitive queries in relational databases can be expensive. Therefore, if all the values in a column are known to be in the same case, this can be indicated within the meta-description of the schema and used to reduce the querying time significantly.

- ✦ When required to do many table joins with One-to-Many (1:N) relationships within a single SQL query, the creation of the result table will suffer from *combinatorial explosion*. The definition of a table link contains a *junction* option, which, if turned on, will generate multiple small queries that can be run simultaneously and joined externally. This provides significant performance improvements for the object assembly process.

- ✦ For text and pattern searches it is possible to make use of text indices produced by the relational database. This results in much faster searches for text-based queries, such as keywords or author name.

- ✦ All query results are cached in a user-owned space. This is inexpensive because the query result is represented as a simple list of entry IDs. To repeat a query, it can be looked up in the cache and retrieved. The cache speeds up reinspection of queries, combining them with other queries, or displaying individual chunks of the result list in a Web interface.

5.3.7   Viewing Entries from a Relational Databank

As mentioned previously, the selection of a hub table and associated tables is used to build an object model automatically. The resulting object can be displayed as an XML stream, which is, however, inconvenient for the user. One option for presenting the object in a human, readable form is to apply XSLT to the XML output. Another more convenient one is to use the mechanism SRS provides to present objects to the Web by a layout description. Section 5.6 will describe how data from relational databanks can be combined with data from flat file or XML databanks into a single data structure.

5.3.8   Summary

Consistent with the SRS philosophy, relational databanks can be added through a *meta-data only* approach. With the exception of defining the HTML representation of the entry data, the entire process of creating and editing the meta-data can be done through mouse clicks in the visual administration interface of SRS.

Not all the options in the configuration have been described here, including setting up sub-entry libraries, automatic handling of binary data (such as images and Microsoft Office documents), and *table cloning* to handle recursive and conditional relationships between tables.

SRS uses a simple interface class to interact with the relational systems. For speed and efficiency the C/C++ interfaces are preferred over JDBC. At present the following Relational Database Management Systems (RDBMS) are supported:

+ Oracle [23]
+ MySQL [24]
+ Microsoft SQLServer [25]
+ DB2 [26]

Relational databanks offer a lot of functionality, which needs to be matched by any system that mediates access to them. The meta-data approach of SRS Relational has proven to provide the flexibility to cope with new user requirements to exploit this functionality.

5.4   THE SRS QUERY LANGUAGE

SRS has its own query language. It supports string comparison including wildcards or regular expressions, numeric range queries, Boolean operators, and the unique link operators (see Section 5.5). Queries always return sets of entries or lists of entry IDs. Sets obtained by querying all databanks can be sorted using various criteria. The query language has no provision to specify sorting. Instead it is invoked using a method of the result set object that has been obtained by evaluating a query. To extract information from entries of result sets, further methods are available.

These methods can retrieve the entire entry as a text or XML stream, retrieve individual token or field values, or load entries into data structures using predefined object loaders (see Section 5.6).

5.4.1   SRS Fields

A query predicate must refer to an SRS field, which has been assigned to the fields in the databank before query time. A query into the Swiss-Prot description field with the the word "*kinase*" looks as follows in the SRS query language:

```
[swissprot-description:kinase]
```

This denotes a string search enclosed in [and]. The databank name swissprot is followed by the field name description. The search term kinase follows the delimiter :. Because the description field is shared with the databank EMBL, the query can be extended to search both Swiss-Prot and EMBL, which then have to be enclosed in curly braces:

```
[{swissprot embl}-description:kinase]
```

Importantly, SRS fields are entities outside a given library definition, which must be mapped onto each field in a library. Whenever possible, the same SRS field is mapped to equivalent fields in different libraries. Through that mechanism each SRS library has a list of associated SRS fields that can be used for searching. Whenever the user selects multiple libraries for searching at the same time, it is possible to find out all the SRS fields that these have in common and represent them in a query form. SRS fields are an important mechanism to integrate heterogeneous databanks with different, but overlapping, content, and they also provide an important simplification because no knowledge of the internal structure of a given databank is required when retrieving and using the list of SRS fields.

A special SRS field exists with the name AllText. It is shared by all databanks and refers to all the text fields in all databanks. Through the use of this field, full-text queries can be specified.

5.5   LINKING DATABANKS

A common theme in databanks in molecular biology is that they all have explicit cross-references to other databanks. Especially now, in the *postgenomic era* in which many known proteins can be linked to a genome location and where results from gene expression and proteomics experiments can be used to understand how these proteins are regulated within the cell, individual data items have very limited value if they are not connected to other databanks. SRS supports and makes use of explicit cross-references in three ways:

+ Hypertext links
+ Indexed links
+ Composite structures (see Section 5.6)

Hypertext links are the simplest mechanism and are ubiquitous on the Web. They are inserted into the appropriate places when displaying information to the user. Linking in this form can be operated on single entries and is very convenient and easy to understand. These links are easy to set up for an SRS Web server. Definitions can be shared among libraries and include options like displaying a link only if it contains a valid reference to an existing entry.

More powerful is the use of indexed links. A simple example of a query using indexed links is "give me all entries in Swiss-Prot that are linked to EMBL." SRS has a general capability to index links based on explicit or even implicit cross-reference information. Link indices are built using information from one side only. All links, once indexed, become bi-directional. An SRS server with many libraries and links can be seen as a graph where nodes are libraries and the edges are the links between them. Figure 5.7 shows such a graph for a comparatively small installation.

In this graph it is possible to link databanks that are not directly connected. For any pair of databanks, the shortest route can be determined and carried out by a multi-step linking process. SRS knows the topology of a given installation and can therefore always determine and execute this shortest path. If the shortest path is not what is desired, this can be specified explicitly within an SRS query language statement.

5.5.1   Constructing Links

Links can be constructed by identifying two SRS fields (see Section 5.4), each from one of the two SRS libraries to be linked that contain identical field values. For instance, to create a link between Swiss-Prot and EMBL, you would select the accession field from EMBL and the data reference (DR) field from Swiss-Prot with explicit cross-references to other databanks. Another example is to link Swiss-Prot and Enzyme [27], which can be defined by the ID field from Enzyme and the description field of Swiss-Prot. The description field of Swiss-Prot carries one or more Enzyme IDs if the protein in question is known to have an enzymatic function. For flat file and XML databanks, link indices

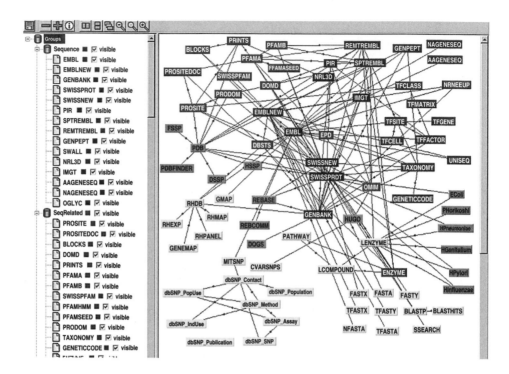

5.7

An SRS library network.

FIGURE

must be built to make the link queryable. Indices can be built by comparing existing indices or by parsing the databank defined to contain the cross-reference information.

Links between and from relational databanks are defined in the same way. However, no indices need to be built. A link query can be executed by querying the information provided in the table structure.

5.5.2　The Link Operators

Link operators are unique to the SRS query language. The two link operators, < and >, allow sets of data from different databanks to be combined. Figure 5.8 shows two databanks, A and B, in which some entries in A have cross-references to entries in B. These cross-references are processed to build link indices, which provide the basis for the link operation. Figure 5.8 also shows the results of two link queries between sets A and B, using the operators < and >.

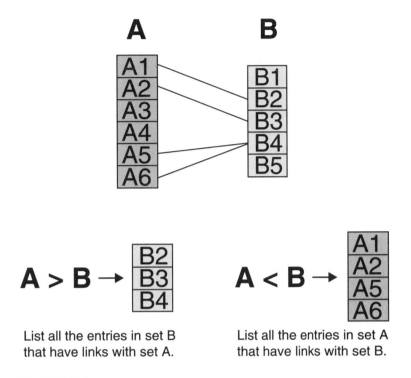

A > B → List all the entries in set B that have links with set A.

A < B → List all the entries in set A that have links with set B.

5.8

FIGURE

The SRS link operators.

By combining predicate queries with link operators it is possible to perform complicated cross-databank queries such as *"retrieve all proteins in Swiss-Prot with calcium binding sites for which their tertiary structure is known with a resolution better than 2 Angstrom."*

Another important use of the link operator is to convert sub-entries (e.g., sequence features) into entries and vice versa. With this link it is possible to search in EMBL all human CDS features (i.e., all sequence features describing coding sequences or all human DNA sequences that *have* CDS features).

5.6   THE OBJECT LOADER

The SRS *object loader* is a technology originally designed to transform semi-structured text data into well-defined data structures that can be accessed in a programmatic way. The object loader processes data according to a loader specification, or a class definition, which, for all of its attributes, specifies how it can obtained from the text file.

The following example shows such a loader for the example of the mutation data in Figure 5.2.

```
$LoadClass:[Mutation attrs:{
    $LoadAttr:[mutation load:$Tok:AaChange]
    $LoadAttr:[position type:int load:$Tok:AaChangePos]
    $LoadAttr:[aaMutType load:$Tok:AaMutType]
    $LoadAttr:[rnaMutType load:$Tok:RnaMutType]
  }
]
```

This definition needs no information on how the required information is to be parsed out of the flat file. Only the name of the token is needed to make the association.

The object loader has been extended to support, in addition to flat file databanks, XML and relational databanks. A variety of ways have been added to specify the origin of the original data to be loaded. This includes using the SRS field abstraction, an XPath-like syntax for XML files (XPath is the XML Path language used for addressing parts of an XML document) [28] or pairs of table and row names for relational databanks. A single loader can be defined for a broad range of databanks. For example, a single *sequence* loader can be specified for all databanks with sequence information, and the original format can be flat file, XML, or relational.

In Section 5.8 an example is described for accessing the loaded objects within a client program.

5.6.1   Creating Complex and Nested Objects

The loader specification supports other useful features, such as class inheritance, and supports various value types like string, integer, and real values and various types of lists.

Using token indices (TINs), a feature of the token server that allows iteration over lists of complex structures inside a text entry, object classes can be nested to an arbitrary degree. Object loaders can build a structure to reflect the entry subentry structure used for indexing, but can have deeper levels of nesting. A good example is an EMBL entry, which contains a list of sequence feature objects, each of which contains a list of qualifier value and name pairs (see Figure 5.3 for an example of an EMBL sequence feature).

5.6.2   Support for Loading from XML Databanks

SRS provides several features to give users control over loading from XML databanks. The utility that generates library definition files from an XML document type definition can generate two types of loaders, a *flat* loader based on the traditional main library/sub-entry library structure or a special *structured* loader, which is a collection of separate loaders for each element that mimics the tree structure of the original XML document. The structured loader makes it easier to load data from libraries that are heavily nested, and it is particularly useful for tasks like writing SRS page layout modules for HTML display.

SRS provides special support for random access of sub-entities within XML files (e.g., individual entries, sub-entries, fields, or collections of fields) or for loading high-level structures without some of the data nested within them (e.g., main entries without sub-entries). In addition, SRS offers two access mechanisms that mimic functionality available in XSLT. Figure 5.9 shows a *mixed content* element called `prolog` that has child elements `child_1` and `child_2` interspersed with its content.

The `xsl:value-of` functionality allows users to extract and concatenate all of the character data content of the `prolog` element and its child elements. A field loaded using the `xsl:value-of` keyword word would contain the following text: "To be, or not to be, that is the question." Note that none of the attribute values are included. The `xsl:copy-of` functionality allows users to extract XML tree fragments, complete with markup. A field loaded using the `xsl:copy-of` keyword word would contain the entire XML fragment shown in Figure 5.9. If neither of these special keywords were used, the `prolog` field would only include the content of the `prolog` element: "To be, to be, the question."

```
<prolog>To be,

    <child_1 attribute_1="this text won't"> or not </child_1>

        to be,

    <child_2 attribute_2="be included"> that is </child_2>

        the question.

</prolog>
```

5.9   Mixed content sample data.

FIGURE

5.6.3　Using Links to Create Composite Structures

An important feature of the object loader is that it can perform links to retrieve single attributes or entire data objects from another linked library. Assuming that the *Mutation* databank is linked to Swiss-Prot, an example is adding an attribute to the *Mutation* loader with the description line from a linked Swiss-Prot entry. The following line instructs the object loader to link to Swiss-Prot using the shortest path and to extract the description token from the linked entry.

```
$LoadAttr:[proteinDescription
  load:$Tok:[description link:swissprot]
]
```

Another possibility would be, rather than extracting a single token, to attach an entire object as defined by an already existing loader class for Swiss-Prot, in this case the loader SeqSimple.

```
$LoadAttr:[protein
  load:$Tok:[link:swissprot loader:$SeqSimple_Loader]
]
```

As information about a certain real-world object is scattered across many databanks, object loaders can provide an extremely valuable foundation for writing programs to display or disseminate these real-world objects. It is possible to define and design these objects freely and in a second step decide where the individual information pieces can be retrieved for their assembly.

5.6.4　Exporting Objects to XML

SRS allows users to export data assembled by the object loader to a generic XML format or to any of the standard XML formats. When converting data to a generic format, SRS creates a well-formed XML document with an accompanying DTD. This functionality can be invoked from the SRS Web interface or from one of the APIs (see Section 5.8). This functionality is provided for every object loader specification by default. If users wish to convert data to a specific format, a public standard, or a format the user invented, they must use a set of XML print metaphor objects that represent and describe the elements and attributes in the target format.

The process for creating XML print metaphors is similar to the process for creating an XML databank definition file. Before a new set of print metaphors can be generated, the user must obtain an accurate DTD for the target XML format. An SRS utility analyzes the DTD and creates print metaphor object templates for all of the XML elements and attributes in the target format. The user must then edit the resulting file to identify data sources for each element and attribute in the target format. The new SRS Visual Administration Tool includes a graphical user interface (GUI) that greatly simplifies the process of editing print metaphors.

Data objects can also be exported to a target XML format using an XSLT style sheet. This process is slightly less convenient than using print metaphors because it involves an extra conversion step. The user must first export the data to the generic XML format, then invoke an XSLT style sheet that converts the generic format to the target format.

XML print metaphors can also be used to transform data from any source into an XML format that is compatible with Microsoft's Office Web Components (OWC) [29]. This technology allows data to be displayed and manipulated using either an Excel spreadsheet or a pivot table embedded in the SRS browser. The pivot table component allows the user to do sophisticated sorting and grouping operations on the data. Both components have an "Export to Excel" button that allows the data to be easily saved to an Excel workbook file.

5.7 SCIENTIFIC ANALYSIS TOOLS

A key feature of SRS is its ability to integrate and use scientific analysis tools that can be applied to user data or to data resulting from database queries. The results generated by these tools can be stored, in turn, in tool-specific databanks, which can then be treated like any other SRS databank. The difference in these databanks is that they are user owned and constitute part of the user session with SRS.

All tools that can be integrated fulfill the following requirements:

✦ It can be launched with a UNIX command line.

✦ It receives input through command line argument or input files.

✦ It writes output to files or to the standard output device.

In bioinformatics, hundreds of tools can be found with these properties. They include BLAST, FASTA for sequence similarity searching, or Clustal [30] for multiple sequence alignment. A selection of these can be combined within an automated annotation pipeline to predict all genes for a genome or, for all proteins derived from these genes, the protein function annotation. Pipelines like this, together with their output, can be integrated as a single tool into SRS. Currently SRS supports about 200 tools, including BLAST, FASTA, stackPACK [31], and the majority of the tools in EMBOSS.

A tool can be added to SRS through meta-data by defining the SRS library with the syntax and data fields of the tool output, information about all tool input options, validation rules to test a parameter set specified by the user, pre-defined parameter sets, association to a data type, and so forth.

SRS has a growing set of pre-defined data types, such as *protein sequence*, which can be extended by the administrator. These data types can be associated with databanks that contain data of this type, tools that take it as input, and tools that produce it as output. This information can be used to build user interfaces that *know* which tools apply to which databases or workflows that feed tools with outputs of other tools.

5.7.1   Processing of Input and Output

Many tools require some pre-processing steps like setting up the run-time environment or conversion of the input sequence to a format they recognize, and post-processing such as cleanup of additional output or preserving input data values. All of these can be specified as part of the tool definition or by using pre-defined *hooks* for shell scripting.

Output can be processed at many levels, depending on the detail required. A simple text view of the output is enough for some applications, but where the results can be parsed for object loaders, this is much preferred. A key decision is the level at which an *entry* in the output should be returned by a later query. The entire output is usually a single entry for simple analysis of a sequence, but for search tools like BLAST it is preferable to represent each *hit* in the sequence databases as a separate entry so these can be linked to the source data. This seriously complicates the task of developing a parser as the entry information is split in several sections of a file, which can be several megabytes in size, but the increased flexibility more than justifies the extra effort.

An important implication of parsing and indexing tool outputs is that the respective tool libraries can become part of the SRS Universe if link information exists. For instance, all outputs from sequence similarity search tools can be linked to the sequence databank searched. Assuming that the search databank is connected to the SRS Universe, questions like *"How many proteins from a certain protein family or metabolic pathway were found?"* can be asked. Links also can be used to compare results obtained by different search tools; for instance, through a single SRS query a list of hits that were found by both FASTA and BLAST can be obtained.

5.7.2   Batch Queues

Batch queues allow the administrator to specify where and when analyses can be run. Once batch queuing is enabled, it is possible to associate a tool with one or more queues with different characteristics. SRS provides support for several popular batch queuing systems such as LSF [32], the Network Queuing System (NQS) [33], the Distributed Queuing System (DQS) [34], or the SUN Grid Engine [35].

If a tool associated with a batch queue is launched, the job is submitted to this batch queue and the Web interface (see Section 5.8, Interfaces to SRS) reports the command line and provides a link to the job status page. This page displays the full list of batch runs. Selecting a completed run will bring up the results. When an application has not been assigned to a batch queue it will be run interactively.

5.8   INTERFACES TO SRS

Several interfaces to SRS exist, which provide full access to all its functions. They include:

✦ Creating and managing a user session

✦ Performing queries over the databanks

✦ Sorting query result sets

✦ Launching analysis tools

✦ Accessing meta-information

The Web interface is implemented as a Common Gateway Interface (CGI), a stateless server that is invoked for every request. However, using the APIs of SRS Objects it is possible to write stateful and multi-threaded servers.

5.8.1   The Web Interface

The most popular access to SRS is through a Web interface. With it the user creates a session that can be temporary or permanent. Within the session the results of many user actions are stored. These include queries, tool launches, and creations of views. The Web interface provides several query forms, one of which is the highly customizable *canned query* form that allows administrators to set up intuitive forms that enable an inexperienced user to launch even complex queries.

5.8.2   SRS Objects

SRS Objects is a package of object-oriented interfaces to SRS. It is designed for software developers who want to access the functionality of SRS from within their own object-oriented application. SRS Objects includes four language-specific APIs, which are:

+ C++

+ Java

+ Perl

+ Python

SRS Objects also includes the SRS Common Object Request Broker Architecture (CORBA) Server, compliant with the CORBA 2.4 specification, which is generally referred to as *SRSCS*.

The C++ API represents the foundation both for the other three APIs (generated automatically from the C++ declarations using the public domain SWIG package) and SRSCS, whose interfaces and operations wrap the C++ API classes and methods. As a consequence, in terms of SRS interaction, the four APIs and SRS CORBA Server are almost identical and provide the same types and method signatures.

The package SRS Objects provides the following major functionalities:

+ Creation of temporary or permanent SRS sessions and interaction with them

+ Access to meta-information about the installed databank groups, databanks, tools, links, etc.

+ Querying of databanks using the SRS query language

+ Accessing databank entries in a variety of ways

+ Launching of analysis tools and managing their results

+ Use and dynamic creation of the SRS object loaders

+ Working with the SRS Objects manager system to create and use dynamic types

In addition, SRS Objects abstracts from the developer tasks such as the initialization of the SRS system, SRS memory management, and SRS error handling.

Central to SRS Objects, as in the Web server, is the session object. It must always be created at the beginning of the program. As in the Web server, the session is associated to a directory where the results of the user actions are stored.

This means the Web client and a program written with SRS Objects can share a session and its contents.

The following program example in Perl illustrates the use of SRS Objects. It starts by creating a session object, then queries all Swiss-Prot entries with *kinase* in the description field, and finally prints a few attributes for each entry in the result.

```perl
$session = new Session;
$set = $session->query("[swissprot-description:kinase]","");
for ($i=0; $i<$set->size(); ++$i) {
  $entry = $set->getEntry($i);
  $obj = $entry->load("SwissEntry");
  print "Accession: ", $obj->attrStr("Accession"), "\n";
  print "Description: ", $obj->attrStr("Description"), "\n";
  print "SeqLength: ", $obj->attrInt("SeqLength"), "\n";
}
```

5.8.3   SOAP and Web Services

Currently SRSCS is the only client server interface to SRS. The others are in-process APIs and require the client application to be run on the same computer as the SRS server. CORBA is well suited for client server applications on the same local area network (LAN), but it is of much more limited use across the Internet or an intranet. The simple object access protocol (SOAP) and the Web Services standard are much better suited for this type of application and are also very compatible with SRS functionality. A Web Services interface, which will provide the same functionality as the existing SRS Objects APIs is currently being built.

5.9   AUTOMATED SERVER MAINTENANACE WITH SRS PRISMA

SRS Prisma is an extension package for SRS that can assist a site administrator with the sometimes onerous task of keeping the flat files, XML files, and indices for installed libraries as up to date as possible. This is done by comparing the status of the local files and indices with the corresponding data files at an appropriate remote FTP site. Any files or indices found to be out of date are replaced by downloading new data and/or by rebuilding the appropriate indices. In addition, SRS Prisma can be used as a more general data management tool, carrying out tasks such as reformatting newly downloaded data files, or creating new data files from existing SRS data files and indices. SRS Prisma can be used on an *ad hoc* basis by the administrator, but it is also ideal for daily scheduling to ensure that all databases are kept as up to date as possible. To assist the administrator in monitoring the completion status of any update processes, Prisma creates a complete archive of Web reports from up to 31 days prior, including easy-to-use graphical views.

In a situation where many databases need to be updated and where a large range of tasks is involved (from downloading, to indexing, to data reformatting), Prisma will determine the minimum number of tasks to be carried out and the dependencies between these tasks. For example, the building of a link index requires the `to` and `from` indices to be up to date. In such a case, the link task would be delayed until any required rebuilding of the `to` and `from` indices was complete. In the event that any of the required tasks fails, the architecture employed by Prisma ensures that any other tasks that do not depend on failed tasks are completed. The Prisma job will finish when all the tasks that can be completed have been done. For instance, if the download phase fails for SWISSNEW, the downloading and indexing of other databases should be unaffected. Other important features of SRS Prisma are as follows:

+ Prisma allows a failed job to be re-run from the point at which it failed, thereby minimizing the repetition of tasks, which can be time-consuming and processor-intensive. For example, if the download phase for a particular databank has failed due to a transient external problem (e.g., a problem accessing the relevant FTP site), the Prisma job can be re-run once this problem has been resolved. In such a case only the failed tasks and those dependent on them will be run.

+ Tasks can be carried out in parallel to optimize performance on multiple processor machines. This type of parallelization includes indexing/merging and downloading. If a databank consists of several files that can be indexed in parallel, Prisma will interleave downloading and indexing of these files.

+ Offline processing of downloads and indexing ensures that during the updating job the SRS server continues to function in an uninterrupted way. The new databanks and indices are only moved online after completion of the entire job.

+ Staged Prisma runs allow controlled and automated decision making to ensure robustness and minimized maintenance. This allows Prisma to bring the update job to an end even if individual tasks fail.

+ An integral part of Prisma is a facility to check the quality of all integrated databanks. Every day every databank is checked for configuration errors, compliance of flat file data to the rules of the token server, the validity of the schema information that SRS holds for relational databanks, and so forth.

Prisma is normally set up to run every night. It provides extensive reporting for the jobs it ran during the night and all the quality checks. Prisma archives the reports within a *sliding window* of 31 days.

Apart from relational databanks that can be accessed through a network connection, flat files and XML sources must be on the same LAN as the SRS server. This is expensive because the storage must be provided, but it guarantees stability and speed. SRS Prisma can keep all local copies of the databanks current in a completely automated fashion, checking every day the integrity of the system and the consistency of each databank and tool.

5.10 CONCLUSION

SRS can integrate the main sources of structured or semi-structured data, flat file databanks, XML files, relational databanks, and analysis tools. It provides technology to access these data, but also to transform them to a common mind-set. Data from the different sources will look and behave in exactly the same way, effectively shielding users from the complexities of the underlying data sources. This is also true for developers who use SRS APIs to write custom programs. SRS forms a truly scalable data and analysis tool integration platform onto which developers can build new databases, analysis tools, user views, and canned queries to tailor the environment to the needs of their company or institution.

Using bi-directional and high-speed links, SRS transforms the multitude of integrated databanks into a network, which paves the way for the full exploration of the relationships between the data sources (e.g., through cross-databank queries). The different sources can be combined using object loaders, which are able to build data objects by extracting data fields from the entire network.

The federated approach to integration, in combination with the use of metadata, means that data can be maintained in its original format. This is important so there is no data loss due to normalization or reformatting. Object loaders can be designed either to provide standardized access to diverse data sources or to extract information transparently from across the entire databank network. SRS, therefore, is both capable of supporting the native structure of databanks and abstractions or unified versions. It supports data in their native format, but it also supports standards derived from them or imposed onto them.

SRS does not improve the data it integrates, nor does it create a super schema over the data, but with its linking capability and object loaders, it provides the perfect framework for the semantic integration of the data sources in bioinformatics. The inherent flexibility and extensibility of SRS means that bioinformaticians can use SRS as a solid foundation for development where they can incorporate their own data and knowledge of the scientific domain to provide a truly comprehensive view of genomic data.

REFERENCES

[1]　D. Benson, I. Karsch-Mizrachi, D. Lipman, et al. "GenBank." *Nucleic Acids Research* 1, no. 28 (2000): 15–18. http://www.ncbi.nlm.nih.gov/Genbank.

[2]　The Gene Ontology consortium. http://www.geneontology.com.

[3]　MySQL open source database. http://www.mysql.com.

[4]　Incyte LifeSeq Foundation database. http://www.incite.com/sequence.foundation.shtml.

[5]　G. Stoesser, W. Baker, A. van den Broek, et al. "The EMBL Nucleotide Sequence Database." *Nucleic Acids Research* 30, no. 1 (2002): 21–26.

[6]　C. O'Donovan, M. J. Martin, A. Gattiker, et al. "High-Quality Protein Knowledge Resources: SWISS-PROT and TrEMBL." *Briefings in Bioinformatics* 3, no. 3 (2002): 275–284.

[7]　S. Altschul, W. Gish, W. Miller, et al. "Basic Local Alignment Search Tool." *Journal of Molecular Biology* 215, no. 3 (October 1990): 403–410. http://www.ncbi.nlm.nih.gov/BLAST.

[8]　W. R. Pearson. "Flexible Sequence Similarity Searching with the FASTA3 Program Package." In *Methods in Molecular Biology* 132 (2000): 185–219.

[9]　P. Rice, I. Longden, and A. Bleasby. "EMBOSS: The European Molecular Biology Open Software Suite." *Trends in Genetics* 16, no. 6 (2000): 276–277. http://www.emboss.org.

[10]　N. J. Mulder, R. Apweiler, T. K. Attwood, et al. "InterPro: An Integrated Documentation Resource for Protein Families, Domains and Functional Sites." *Briefings in Bioinformatics* 3, no. 3 (2002): 225–235. http://www.ebi.ac.uk/interpro/.

[11]　E. M. Zodobnov, R. Lopez, R. Apweiler, et al. "The EBI SRS Server: New Features." *Bioinformatics* 18, no. 8 (2002): 1149–1150. http://srs.ebi.ac.uk.

[12]　D. P. Kreil and T. Ezold. "DATABANKS: A Catalogue Database of Molecular Biology Databases." *Trends in Biochemical Sciences* 24, no. 4 (1999): 155–157.

[13]　Celera Discovery System. http://www.celeradiscoverysystem.com.

[14]　NetAffx Analysis Center from Affymetrix. http://www.netaffx.com.

[15]　Thomson Derwent GENESEQ portal. http://www.derwent.com/geneseqweb/.

[16]　Perl. http://www.perl.org, http://www.perl.com.

[17] K. Jensen and N. Wirth. *Pascal User Manual and Report*, second edition. Heidelberg, Germany: Springer-Verlag, 1974.

[18] T. Bray, J. Paoli, C. M. Sperberg-McQueen, et al. *Extensible Markup Language (XML) 1.0: World Wide Web Consortium (W3C) Recommendation*, 2nd edition, October 6, 2000, http://www.w3.org/TR/REC-xml/html.

[19] J. Clark. XSL Transformations (XSLT) Version 1.0: World Wide Web Consortium (W3C) Recommendation, November 16, 1999. http://www.w3.org/TR/xslt.

[20] Michael Kay's DTDGenerator Utility. http://users.iclway.co.uk/mhkay/saxon/saxon5-5-1/dtdgen.html.

[21] LabBook, Inc. BSML XML format. http://www.labbook.com.

[22] SUN Microsystems. Java Database Connectivity (JDBC). http://java.sun.com/products/jdbc.

[23] Oracle database. http://www.oracle.com.

[24] MySQL. http://www.mysql.com.

[25] Microsoft SQL Server. http://www.Microsoft.com/sql/.

[26] IBM. DB2 Database Software. http://www-3.ibm.com/softward/data/db2/.

[27] A. Bairoch. "The ENZYME Database in 2000." *Nucleic Acids Research* 28, no. 7 (2000): 304–305. http://www.expasy.ch/enzyme/.

[28] J. Clark and S. DeRose. XML Path Language (XPath) Version 1.0: World Wide Web Consortium (W3C) Recommendation, November 16, 1999. http://www.w3.org/TR/xpath.

[29] "The Office Web Components Add Analysis Tools to Your Web Page." Microsoft, 2003. http://office.microsoft.com/Assistance/2000/owebcom1.aspx.

[30] J. D. Thompson, T. J. Gibson, F. Plewniak, et al. "The ClustalX Windows Interface: Flexible Strategies for Multiple Sequence Alignment Aided by Quality Analysis Tools." *Nucleic Acids Research* 24 (1997): 4876–4882.

[31] R. T. Miller, A. G. Christoffels, C. Gopalakrishnan, et al. "A Comprehensive Approach to Clustering of Expressed Human Gene Sequence: The Sequence Tag Alignment and Consensus Knowledge Base." *Genome Research 9*, no. 11 (1999): 1143–1155.

[32] Platform LSF. http://www.platform.com/products/wm/LSF/.

[33] Network Queuing System (NQS). http://umbc7.umbc.edu/nqs/nqsmain.html.

[34] Distributed Queuing System (DQS). http://www.scri.fsu.edu/~pasko/dqs.html.

[35] Sun ONE Grid Engine. http://www.sun.com/software/gridware/.

The Kleisli Query System as a Backbone for Bioinformatics Data Integration and Analysis

Jing Chen, Su Yun Chung, and Limsoon Wong

Biological data is characterized by a wide range of data types from the plain text of laboratory records and literature, to nucleic acid and amino acid sequences, 3D structures of molecules, high-resolution images of cells and tissues, diagrams of biochemical pathways and regulatory networks, to various experimental outputs from technologies as diverse as microarrays, gels, and mass spectrometry. These data are stored in a large number of databases across the Internet. In addition to online interfaces for querying and searching the underlying repository data, many Web sites also provide specific computational tools or programs for analysis of data. In this chapter the term *data sources* is used loosely to refer to both databases and computational analysis tools.

Until recently, data sources were set up as autonomous Web sites by individual institutions or research laboratories. Data sources vary considerably in contents, access methods, capacity, query processing, and services. The major difficulty is that the data elements in various public and private data sources are stored in extremely heterogeneous formats and database management systems that are often *ad hoc*, application-specific, or vendor-specific. For example, scientific literature, patents, images, and other free-text documents are commonly stored in unstructured formats like plain text files, hypertext markup language (HTML) files, and binary files. Genomic, microarray gene expression, or proteomic data are routinely stored in Excel spreadsheets, semi-structured extensible markup language (XML), or structured relational databases like Oracle, Sybase, DB2, and Informix. The National Center for Biotechnology Information (NCBI) in Bethesda, Maryland, which is the largest repository for genetic information, supplies GenBank reports and GenPept reports in HTML format with an underlying highly nested data model based on ASN.1 [1]. The computational analysis tools or applications suffer from a similar scenario: They require specific input and output data formats. The output of one program is not immediately compatible with the input requirement of other programs. For example, the most popular Basic Local Alignment Search Tool (BLAST) database search tool requires a specific format called FASTA for sequence input.

In addition to data format variations, both the data content and data schemas of these databases are constantly changing in response to rapid advances in research and technology. As the amount of data and databases continues to grow on the Internet, it generates another bottleneck in information integration at the semantic level. There is a general lack of standards in controlled vocabulary for consistent naming of biomedical terms, functions, and processes within and between databases. In naming genes and proteins alone, there is much confusion. For example, a simple transcription factor, the CCAAT/enhancer-binding protein beta, is referred to by more than a dozen names in the public databases, including CEBPB, CRP2, and IL6DPB.

For research and discovery, the biologist needs access to up-to-date data and best-of-breed computational tools for data analyses. To achieve this goal, the ability to query across multiple data sources is not enough. It also demands the means to transform and transport data through various computational steps seamlessly. For example, to investigate the structure and function of a new protein, users must integrate information derived from sequence, structure, protein domain prediction, and literature data sources. Should the steps to prepare the data sets between the output of one step to the input of the next step have to be carried out manually, which requires some level of programming work (such as writing Perl scripts), the process would be very inefficient and slow.

In short, many bioinformatics problems require access to data sources that are large, highly heterogeneous and complex, constantly evolving, and geographically dispersed. Solutions to these problems usually involve many steps and require information to be passed smoothly and usually to be transformed between the steps. The Kleisli system is designed to handle these requirements directly by providing a high-level query language, simplified SQL (sSQL), that can be used to express complicated transformations across multiple data sources in a clear and simple way.[1]

The design and implementation of the Kleisli system are heavily influenced by functional programming research, as well as database query language research. Kleisli's high-level query language, sSQL, can be considered a functional

---

1. Earlier versions of the Kleisli system supported only a query language based on comprehension syntax called Collection Programming Language (CPL) [2]. Now, both CPL and sSQL are available.

programming language[2] that has a built-in notion of *bulk* data types[3] suitable for database programming and has many built-in operations required for modern bioinformatics. Kleisli is implemented on top of the functional programming language Standard ML of New Jersey (SML). Even the data format Kleisli uses to exchange information with the external world is derived from ideas in type inference in functional programming languages.

This chapter provides a description of the Kleisli system and a discussion of various aspects of the system, such as data representation, query capability, optimizations, and user interfaces. The materials are organized as follows: Section 6.1 introduces Kleisli with a well-known example. Section 6.2 presents an overview of the Kleisli system. Section 6.3 discusses the data model, data representation, and exchange format of Kleisli. Section 6.4 gives more example queries in Kleisli and comments on the expressive power of its core query language. Section 6.5 illustrates Kleisli's ability to use flat relational databases to store complex objects transparently. Section 6.6 lists the kind of data sources supported by the Kleisli system and shows the ease of implementing wrappers for Kleisli. Section 6.7 gives an overview of the various types of optimizations performed by the Kleisli query optimizer. Section 6.8 describes both the Open Database Connectivity (ODBC)- or Java Database Connectivity (JDBC)-like programming interfaces to Kleisli in Perl and Java, as well as its Discovery Builder graphical user interface. Section 6.9 contains a brief survey of other well-known proposals for bioinformatics data integration.

6.1   MOTIVATING EXAMPLE

Before discussing the guts of the Kleisli system, the very first bioinformatics data integration problem solved using Kleisli is presented in Example 6.1.1. The query was implemented in Kleisli in 1994 [5] and solved one of the so-called "impossible" queries of a U.S. Department of Energy Bioinformatics Summit Report published in 1993 [6].

---

2. Functional programming languages are programming languages that emphasize a particular paradigm of programming technique known as *functional programming* [3, 4]. In this paradigm, all programs are expressed as mathematical functions and are generally free from side effects. Examples of functional programming languages are LISP, Haskell, and SML. Some fundamental ideas in functional programming languages, such as garbage collection, have also been borrowed by other modern programming languages such as Java.
3. *Bulk* data types refer to data types that are collections of objects. Examples of bulk data types are sets, bags, lists, and arrays.

Example 6.1.1   The query was to "*find for each gene located on a particular cytogenetic band of a particular human chromosome as many of its non-human homologs as possible.*" Basically, the query means "*for each gene in a particular position in the human genome, find dioxyribonucleic acid (DNA) sequences from non-human organisms that are similar to it.*"

In 1994, the main database containing cytogenetic band information was the Genome Database (GDB) [7], which was a Sybase relational database. To find homologs, the actual DNA sequences were needed, and the ability to compare them was also needed. Unfortunately, that database did not keep actual DNA sequences. The actual DNA sequences were kept in another database called GenBank [8]. At the time, access to GenBank was provided through the ASN.1 version of Entrez [9], which was at the time an extremely complicated retrieval system. Entrez also kept precomputed homologs of GenBank sequences.

So, the evaluation of this query needed the integration of GDB (a relational database located in Baltimore, Maryland) and Entrez (a non-relational database located in Bethesda, Maryland). The query first extracted the names of genes on the desired cytogenetic band from GDB, then accessed Entrez for homologs of these genes. Finally, these homologs were filtered to retain the non-human ones. This query was considered "impossible" as there was at that time no system that could work across the bioinformatics sources involved due to their heterogeneity, complexity, and geographical locations. Given the complexity of this query, the sSQL solution below is remarkably short.

```
sybase-add (name: "gdb", ...);
create view locus from locus_cyto_location using gdb;
create view eref from object_genbank_eref using gdb;
select accn: g.genbank_ref, nonhuman-homologs: H
from
  locus c, eref g,
  {select  u
   from na-get-homolog-summary(g.genbank_ref) u
   where not(u.title like "%Human%") and not(u.title
   like "%H.sapien%")} H
where
  c.chrom_num = "22" and g.object_id = c.locus_id
  and not (H = { });
```

The first three lines connect to GDB and map two tables in GDB to Kleisli. After that, these two tables could be referenced within Kleisli as if they were two locally defined sets, `locus` and `eref`. The next few lines extract from these tables the accession numbers of genes on Chromosome 22, use the Entrez function na-get-homolog-summary to obtain their homologs, and filter these homologs for non-human ones. Notice that the from-part of the outer select-construct is of the form {select u ...} H. This means that H is the entire set returned by select u ..., thus allowing to manipulate and return all the non-human homologs as a single set H.

Besides the obvious smoothness of integration of the two data sources, this query is also remarkably efficient. On the surface, it seems to fetch the locus table in its entirety once and the eref table in its entirety n times from GDB (a naive evaluation of the comprehension would be two nested loops iterating over these two tables). Fortunately, in reality, the Kleisli optimizer is able to migrate the join, selection, and projections on these two tables into a single efficient access to GDB using the optimizing rules from a later section of this chapter. Furthermore, the accesses to Entrez are also automatically made concurrent.

Since this query, Kleisli and its components have been used in a number of bioinformatics projects such as GAIA at the University of Pennsylvania,[4] Transparent Access to Multiple Bioinformatics Information Sources (TAMBIS) at the University of Manchester [11, 12], and FIMM at Kent Ridge Digital Labs [13]. It has also been used in constructing databases by pharmaceutical/biotechnology companies such as SmithKline Beecham, Schering-Plough, GlaxoWellcome, Genomics Collaborative, and Signature Biosciences. Kleisli is also the backbone of the Discovery Hub product of geneticXchange Inc.[5]

6.2   APPROACH

The approach taken by the Kleisli system is illustrated by Figure 6.1. It is positioned as a mediator system encompassing a complex object data model, a high-level query language, and a powerful query optimizer. It runs on top of a large number of lightweight wrappers for accessing various data sources. There is also a number of application programming interfaces that allow Kleisli to be accessed in an ODBC- or JDBC-like fashion in various programming languages for a various applications.

The Kleisli system is extensible in several ways. It can be used to support several different high-level query languages by replacing its high-level query language module. Currently, Kleisli supports a *comprehension syntax-based* language called CPL [2, 14, 15] and a *nested relationalized* version of SQL called sSQL.

---

4. Information about the GAIA project is available at *http://www.cbil.upenn.edu/gaia2/gaia* and [10].

5. Information about Discovery Hub is available at *http://www.geneticxchange.com*.

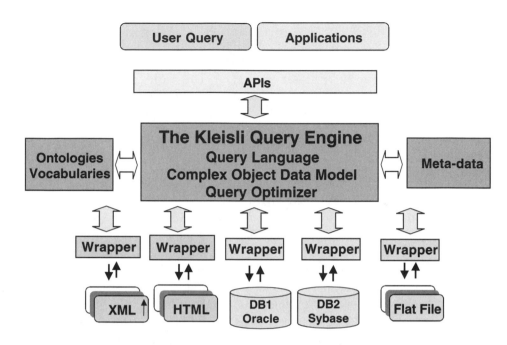

6.1

FIGURE

Kleisli positioned as a mediator.

Only sSQL is used throughout this chapter. The Kleisli system can also be used to support many different types of external data sources by adding new wrappers, which forward Kleisli's requests to these sources and translate their replies into Kleisli's exchange format. These wrappers are lightweight and new wrappers are generally easy to develop and insert into the Kleisli system. The optimizer of the Kleisli system can also be customized by different rules and strategies.

When a query is submitted to Kleisli, it is first processed by the high-level query language module, which translates it into an equivalent expression in the abstract calculus Nested Relational Calculus (NRC). NRC is based on the calculus described in Buneman's "Principles of Programming with Complex Objects and Collection Types" [16] and was chosen as the internal query representation because it is easy to manipulate and amenable to machine analysis. The NRC expression is then analyzed to infer the most general valid type for the expression and is passed to the query optimizer. Once optimized, the NRC expression is then compiled into calls to a library of routines for complex objects underlying the complex object data model. The resulting compiled code is then executed, accessing drivers and external primitives as needed through pipes or shared memory. Each of these components is considered in further detail in the next several sections.

6.3 DATA MODEL AND REPRESENTATION

The data model, data representation, and data exchange format of the Kleisli system are presented in this section. The data model underlying the Kleisli system is a complex object type system that goes beyond the *sets of atomic records* or *flat relations* type systems of relational databases [17]. It allows arbitrarily nested records, sets, lists, bags, and variants. A variant is also called a *tagged union type*, and it represents a type that is *either this or that*. The collection or *bulk* types— sets, bags, and lists—are homogeneous. To mix objects of different types in a set, bag, or list, it is necessary to inject these objects into a variant type.

In a relational database, the sole bulk data type is the set. Furthermore, this set is only allowed to contain records where each field is allowed to contain an atomic object such as a number or a string. Having such a restricted bulk data type presents at least two problems in real-life applications. First, the particular bulk data type may not be a natural model of real data. Second, the particular bulk data type may not be an efficient model of real data. For example, when restricted to the flat relational data model, the GenPept report in Example 6.3.1 must be split into many separate tables to be stored in a relational database without loss. The resulting multi-table representation of the GenPept report is conceptually unnatural and operationally inefficient. A person querying the resulting data must pay the mental overhead of understanding both the original GenPept report and its badly fragmented multi-table representation. The user may also have to pay the performance overhead of having to reassemble the original GenPept report from its fragmented multi-table representation to answer queries. As another example, limited with the set type only, even if nesting of sets is allowed, one may not be able to model MEDLINE reports naturally. A MEDLINE report records information on a published paper, such as its title and its authors. The order in which the authors are listed is important. With only sets, one must pair each author explicitly with a number representing his or her position in order of appearance. Whereas with the data type list, this cumbersome explicit pairing with position becomes unnecessary.

Example 6.3.1 The GenPept report is the format chosen by NCBI to represent information on amino acid sequence. While an amino acid sequence is a string of characters, certain regions and positions of the string, such as binding sites and domains, are of special biological interest. The feature table of a GenPept report is the part of the GenPept report that documents the positions of these regions of special biological interest, as well as annotations or comments on these regions. The following type represents the feature table of a GenPept report from Entrez [9].

```
(#uid:num, #title:string,
   #accession:string, #feature:{(
      #name:string, #start:num, #end:num,
      #anno:[(#anno_name:string, #descr:string)])})
```

It is an interesting type because one of its fields (#feature) is a set of records, one of whose fields (#anno) is in turn a list of records. More precisely, it is a record with four fields #uid, #title, #accession, and #feature. The first three of these store values of types num, string, and string respectively. The #uid field uniquely identifies the GenPept report. The #feature field is a set of records, which together form the feature table of the corresponding GenPept report. Each of these records has four fields: #name, #start, #end, and #anno. The first three of these have types string, num, and num respectively. They represent the name, start position, and end position of a particular feature in the feature table. The #anno field is a list of records. Each of these records has two fields #anno_name and #descr, both of type string. These records together represent all annotations on the corresponding feature.

In general, the types are freely formed by the syntax:

$t ::=$ num $|$ string $|$ bool$|$
$t ::= \{t\} \mid \{\,|t|\,\} \mid [t] \mid (l_1 : t_1, ..., l_n : t_n) \mid <l_1 : t_1, ..., l_n : t_n>$

Here num, string, and bool are the base types. The other types are constructors and build new types from existing types. The types $\{t\}$, $\{\,|t|\,\}$, and $[t]$ respectively construct set, bag, and list types from type t. The type $(l_1 : t_1,..., l_n : t_n)$ constructs record types from types $t_1,..., t_n$. The type $<l_1 : t_1,..., l_n : t_n>$ constructs variant types from types $t_1,..., t_n$. The flat relations of relational databases are basically sets of records, where each field of the record is a base type; in other words, relational databases have no bags, no lists, no variants, no nested sets, and no nested records. Values of these types can be represented explicitly and exchanged as follows, assuming that the instances of e are values of appropriate types: $(l_1 : e_1,..., l_n : e_n)$ for records; $<l : e>$ for variants; $\{e_1,..., e_n\}$ for sets; $\{\,|e_1,..., e_n|\,\}$ for bags; and $[e_1,..., e_n]$ for lists.

Example 6.3.2   Part of the feature table of GenPept report 131470, a tyrosine phosphatase 1C sequence, is shown in the following.

```
(#uid:131470, #accession:"131470",
  #title:"... (PTP-1C)...", #feature:{(
    #name:"source", #start:0, #end:594, #anno:[
      (#anno_name:"organism", #descr:"Mus musculus"),
      (#anno_name:"db_xref", #descr:"taxon:10090")]),
    ...})
```

The particular feature goes from amino acid 0 to amino acid 594, which is actually the entire sequence, and has two annotations: The first annotation indicates that this amino acid sequence is derived from a mouse DNA sequence. The second is a cross reference to the NCBI taxonomy database.

The schemas and structures of all popular bioinformatics databases, flat files, and software are easily mapped into this data model. At the high end of data structure complexity are Entrez [9] and AceDB [18], which contain deeply nested mixtures of sets, bags, lists, records, and variants. At the low end of data structure complexity are the relational database systems [17] such as Sybase and Oracle, which contain flat sets of records. Currently, Kleisli gives access to more than 60 of these and other bioinformatics sources. The reason for this ease of mapping bioinformatics sources to Kleisli's data model is that they are all inherently composed of combinations of sets, bags, lists, records, and variants. Kleisli's data model directly and naturally maps sets to sets, bags to bags, lists to lists, records to records, and variants to variants without having to make any (type) declaration beforehand.

The last point deserves further consideration. In a dynamic, heterogeneous environment such as that of bioinformatics, many different database and software systems are used. They often do not have anything that can be thought of as an explicit database schema. Further compounding the problem is that research biologists demand flexible access and queries in *ad hoc* combinations. Thus, a query system that aims to be a general integration mechanism in such an environment must satisfy four conditions. First, it must not count on the availability of schemas. It must be able to compile any query submitted based solely on the structure of that query. Second, it must have a data model that the external database and software systems can easily translate to without doing a lot of type declarations. Third, it must shield existing queries from evolution of the external sources as much as possible. For example, an extra field appearing in an external database table must not necessitate the recompilation or rewriting of existing queries over that data source. Fourth, it must have a data exchange format that is straightforward to use so it does not demand too much programming effort or contortion to capture the variety of structures of output from external databases and software.

Three of these requirements are addressed by features of sSQL's type system. sSQL has polymorphic record types that allow to express queries such as:

```
create function get-rich-guys (R) as
select x.name from R x where x.salary > 1000;
```

which defines a function that returns names of people in R earning more than $1000. This function is applicable to any R that has at least the name and the salary fields, thus allowing the input source some freedom to evolve.

In addition, sSQL does not require any type to be declared at all. The type and meaning of any sSQL program can always be completely inferred from its structure without the use of any schema or type declaration. This makes it possible to plug in any data source logically without doing any form of schema declaration, at a small acceptable risk of run-time errors if the inferred type and the actual structure are not compatible. This is an important feature because most biological data sources do not have explicit schemas, while a few have extremely large schemas that take many pages to write down—for example, the ASN.1 schema of Entrez [1]—making it impractical to have any form of declaration.

As for the fourth requirement, a data exchange format is an agreement on how to lay out data in a data stream or message when the data is exchanged between two systems. In this context, it is the format for exchanging data between Kleisli and all the bioinformatics sources. The data exchange format of Kleisli corresponds one-to-one with Kleisli's data model. It provides for records, variants, sets, bags, and lists; and it allows these data types to be composed freely. In fact, the data exchange format completely adopts the syntax of the data representation described earlier and illustrated in Example 6.3.2. This representation has the interesting property of not generating ambiguity. For instance, a set symbol { represents a set, whereas a parenthesis (denotes a record. In short, this data exchange format is self describing. The basic specification of the data exchange format of Kleisli is summarized in Figure 6.2. For a more detailed account, please see Wong's paper on Kleisli from the 2000 IEEE Symposium on Bioinformatics and Bio-engineering [19].

A self-describing exchange format is one in which there is no need to define in advance the structure of the objects being exchanged. That is, there is no fixed schema and no type declaration. In a sense, each object being exchanged carries its own description. A self-describing format has the important property that, no matter how complex the object being exchanged is, it can be easily parsed and reconstructed without any schema information. The ISO ASN.1 standard [20] on open systems interconnection explains this advantage. The schema that describes its structure needs to be parsed before ASN.1 objects, making it necessary to write two complicated parsers instead of one simple parser.

Data Type	Data Layout	Remarks
Unit	()	
Booleans	true false	
Numbers	123 123.123 ~123 ~123.123	Positive numbers Negative numbers
Strings	"a string"	A string is put inside double quotes
Records	$(\#l_1 : O_1,$ $\vdots,$ $\#l_n : O_n)$	A record is put inside round brackets. The label-:-value triplets enumerate the fields of the record
Variants	$<\#l : O>$	A variant is put inside angle brackets
Sets	$\{ O_1,$ $\vdots,$ $O_n\}$	A set is put inside curly brackets
Bags	$\{| O_1,$ $\vdots,$ $O_n|\}$	A bag is put inside curly-bar brackets
Lists	$[O_1,$ $\vdots,$ $O_n]$	A list is put inside square brackets
User-defined types	longitude "50E"	A user-defined type is preceded by its name
Errors	error "it goofed"	An error message is preceded by error

6.2

FIGURE

The basic form of the Kleisli Exchange Format. Punctuations and indentations are not significant. The semicolon indicates the end of a complex object. Multiple complex objects can be laid out in the same stream.

6.4   QUERY CAPABILITY

sSQL is the primary query language of Kleisli used in this chapter. It is based on the *de facto* commercial database query language SQL, except for extensions made to cater to the nested relational model and the federated heterogeneous data sources. Rather than giving the complete syntax, sSQL is illustrated with a few examples on a set of feature tables DB.

Example 6.4.1   The query below "*extracts the titles and features of those elements of a data source DB whose titles contain 'tyrosine' as a substring.*"

```
create function get-title-from-featureTable (DB) as
select title: x.title, feature: x.feature
from DB x where x.title like "%tyrosine%";
```

This query is a simple project-select query. A project-select query is a query that operates on one (flat) relation or set. Thus, the transformation such a query can perform is limited to selecting some elements of the relation and extracting or projecting some fields from these elements. Except for the fact that the source data and the result may not be in first normal form, these queries can be expressed in a relational query language. However, sSQL can perform more complex restructurings such as nesting and unnesting not found in SQL, as shown in the following examples.

Example 6.4.2   The following query flattens the source DB completely. l2s is a function that converts a list into a set.

```
create function flatten-featureTable (DB) as
select
  title:x.title, feature:f.name, start:f.start, end:f.end,
  anno-name:a.anno_name, anno-descr:a.descr
from DB x, x.feature f, f.anno.l2s a;
```

The next query demonstrates how to express nesting in sSQL. Notice that the entries field is a complex object having the same type as DB.

```
create function nest-featureTable-by-organism (DB) as
select
  organism: z,
  entries: (select distinct x
```

```
                    from DB x, x.feature f, f.anno a
                    where a.anno_name = "organism" and a.descr = z)
from (select distinct y.anno-descr
      from DB.flatten-featureTable y
      where y.anno-name ="organism") z;
```

The next couple of more substantial queries are inspired by one of the most advanced functionalities of the EnsMart interface of the EnsEMBL system [21].

Example 6.4.3   The feature table of a GenBank report has the type below. The field #position of a feature entry is a list indicating the start and stop positions of that feature. If the feature entry is a CDS, this list corresponds to the list of exons of the CDS. The field #anno is a list of annotations associated with the feature entry.

```
(#uid: num, #title: string, #accession: string,
 #seq: string, #feature: {(
     #name: string,
     #position: [(#start: num, #end: num,
                 #negative: bool, ...)],
     #anno: [(#anno_name: string, #descr: string)],
          ...)}, ...)}
```

Given a set DB of feature tables of GenBank chromosome sequences, one can extract the 500 bases up stream of the translation initiation sites of all disease genes—in the sense that these genes have a cross reference to the Online Mendelian Inheritance of Man database (OMIM)—on the positive strand in DB as below.

```
select
  uid:x.uid, protein:r.descr, flank:string-span(x.seq,
                                     p.start - 500, p.start)
from DB x, x.feature f, {f.position.list-head} p,
                             f.anno.12s a, f.anno.12s r
where not (p.#negative)
and   a.descr like "MIM:%" and a.anno_name  = "db_xref"
and   r.anno_name = "protein_id"
```

Similarly, one can extract the first exons of these same genes as follows:

```
select
  uid:x.uid, protein:r.descr, exon1:string-span
                      (x.seq, p.start, p.end)
```

```
from DB x, x.feature f, {f.position.list-head} p,
                                 f.anno.l2s a, f.anno.l2s r
where not (p.#negative)
and    a.descr like "MIM:%" and a.anno_name = "db_xref"
and    r.anno_name = "protein_id"
```

These two example queries illustrate how a high-level query language makes it possible to extract very specific output in a relatively straightforward manner.

The next query illustrates a more ambitious example of an *in silico* discovery kit (ISDK). Such a kit prescribes experimental steps carried out in computers very much like the experimental protocol carried out in wet-laboratories for a specific scientific investigation. From the perspective of Kleisli, an *in silico* discovery kit is just a script written in sSQL, and it performs a defined information integration task very similar to an integrated electronic circuit. It takes an input data set and parameters from the user, executes and integrates the necessary computational steps of database queries and applications of analysis programs or algorithms, and outputs a set of results for specific scientific inquiry.

Example 6.4.4   The simple *in silico* discovery kit illustrated in Figure 6.3 demonstrates how to use an available ontology data source to get around the problem of inconsistent naming in genes and proteins and to integrate information across multiple data sources. It is implemented in the following sSQL script.[6] With the user input of gene name G, the ISDK performs the following tasks: "*First, it retrieves a list of aliases for* G *from the gene nomenclature database provided by the Human Genome Organization (HUGO). Then it retrieves information for diseases associated with this particular protein in OMIM, and finally it retrieves all relevant references from MEDLINE.*"

```
create function get-info-by-genename (G) as
select
   hugo: w, omim: y, pmid1-abstract: z,
   num-medline-entries: list-sum(lselect ml-get-count-
                                  general(n) from x.Aliases.s2l n)
```

6. s2l denotes a function that converts a set into a list. list-sum is a function to sum a list of numbers. The function ml-get-count-general accesses MEDLINE and computes the number of MEDLINE reports matching a given keyword, whereas ml-get-abstract-by-uid is a function that accesses MEDLINE to retrieve reports given a unique identifier, and webomim-get-id accesses the OMIM database to obtain unique identifiers of OMIM reports matching a keyword. webomim-get-detail is a function that accesses OMIM to retrieve reports given a unique identifier. hugo-get-by-symbol is a function that accesses the HUGO database and returns HUGO reports matching a given gene name.

```
from
  hugo-get-by-symbol(G) w,
  webomim-get-id(searchtime:0, maxhits:0, searchfields:{},
                                          searchterms:G) x,
  webomim-get-detail(x.uid) y,
  ml-get-abstract-by-uid(w.PMID1) z
where
  x.title like ("%" ^ G ^ "%");
```

For instance, this query `get-info-by-genename` can be invoked with the transcription factor `CEBPB` as input to obtain the following result.

```
{(#hugo: (#HGNC: "1834",
    #Symbol: "CEBPB", #PMID1: "1535333", ...
    #Name: "CCAAT/enhancer binding protein (C/EBP), beta",
    #Aliases: {"LAP", "CRP2", "NFIL6", "IL6DBP", "TCF5"}),
```

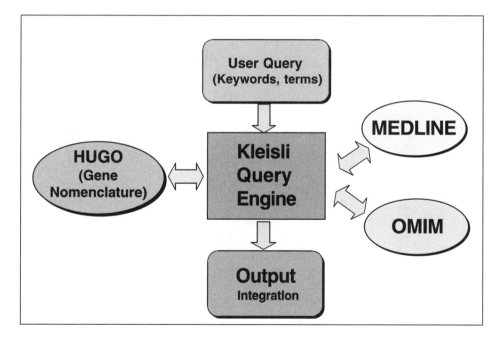

6.3

FIGURE

An "*in silico* discovery kit" that uses an available ontology data source to get around the problem of inconsistent naming in genes and proteins, and integrates information across multiple data sources.

```
  #omim: (#uid: 189965, #gene_map_locus: "20q13.1",
  #allelic_variants: {}, ...),
#pmid1-abstract: (#muid: 1535333,
  #authors: "Szpirer C...", #address: "Departement
                                     de Biologie ...",
  #title: "Chromosomal localization in man and rat of the
                                     genes encoding ...",
  #abstract: "By means of somatic cell hybrids segregating
                                     either human...",
  #journal: "Genomics 1992 Jun; 13(2):292-300"),
  #num-medline-entries: 1936)}
```

Such queries fulfill many of the requirements for efficient *in silico* discovery processes: (1) Their modular nature gives scientists the flexibility to select and combine specific queries for specific research projects; (2) they can be executed automatically by Kleisli in batch mode and can handle large data volumes; (3) their scripts are re-usable to perform repetitive tasks and can be shared among scientific collaborators; (4) they form a base set of templates that can be readily modified and refined to meet different specifications and to make new queries; and (5) new databases and new computational tools can be readily incorporated to existing scripts.

The flexibility and power shown in these sSQL examples can also be experienced in Object-Protocol Model (OPM) [22] and to a lesser extent in Discovery-Link [23]. With good planning, a specialized data integration system can also achieve great flexibility and power within a more narrow context. For example, the EnsMart tool of EnsEMBL [21] is a well-designed interface that helps a non-programmer build complex queries in a simple way. In fact, an equivalent query to the first sSQL query in Example 6.4.3 can be also be specified using EnsMart with a few clicks of the mouse. Nevertheless, there are some unanticipated cases that cannot be expressed in EnsMart, such as the second sSQL query in Example 6.4.3.

While the syntactic basis for sSQL is SQL, its theoretical inspiration came from a paper by Tannen, Buneman, and Nagri [24] where structural recursion was presented as a query language. However, structural recursion presents two difficulties. The first is that not every syntactically correct structural recursion program is logically well defined [25]. The second is that structural recursion has too much expressive power because it can express queries that require exponential time and space.

In the context of databases, which are typically very large, programs (queries) are usually restricted to those that are practical in the sense that they are in a low complexity class such as LOGSPACE, PTIME, or TC^0. In fact, one may even want to prevent any query that has greater than $O(n \times \log n)$ complexity, unless one is confident that the query optimizer has a high probability of optimizing the query to no more than $O(n \times \log n)$ complexity. Database query languages such as SQL, therefore, are designed in such a way that joins are easily recognized because joins are the only operations in a typical database query language that require $O(n^2)$ complexity if evaluated naively.

Thus, Tannen and Buneman suggested a natural restriction on structural recursion to reduce its expressive power and to guarantee it is well defined. Their restriction cuts structural recursion down to homomorphisms on the commutative idempotent monoid of sets, revealing a telling correspondence to monads [15]. A nested relational calculus, which is denoted here by $\mathcal{NRC}$, was then designed around this restriction [16]. $\mathcal{NRC}$ is essentially the simply typed lambda calculus extended by a construct for building records, a construct for decomposing records by field selection, a construct for building sets, and a construct for decomposing sets by means of the restriction on structural recursion. Specifically, the construct for decomposing sets is $\bigcup\{e_1 \mid x \in e_2\}$, which forms a set by taking the big union of $e_1[o/x]$ over each o in the set e_2.

The expressive power of $\mathcal{NRC}$ and its extensions are studied in numerous studies [16, 26–29]. Specifically, the $\mathcal{NRC}$ core has exactly the same power as all the standard nested relational calculi and when restricted to flat tables as input-output, it has exactly the same power as the relational calculus. In the presence of arithmetics and a summation operator, when restricted to flat tables as input-output, it has exactly the power of entry-level SQL. Furthermore, it captures standard nested relational queries in a high-level manner that is easy for automated optimizer analysis. It is also easy to translate a more user-friendly surface syntax, such as the comprehension syntax or the SQL select-from-where syntax, into this core while allowing for full-fledged recursion and other operators to be imported easily as needed into the system.

6.5   WAREHOUSING CAPABILITY

Besides the ability to query, assemble, and transform data from remote heterogeneous sources, it is also important to be able to conveniently warehouse the data locally. The reasons to create local warehouses are several: (1) It increases efficiency; (2) it increases availability; (3) it reduces the risk of unintended *denial of service* attacks on the original sources; and (4) it allows more careful data cleansing that cannot be done on the fly. The warehouse should be efficient to query and easy to update. Equally important in the biology arena, the warehouse should model the data in a conceptually natural form. Although a relational database system is efficient for querying and easy to update, its native data model of flat tables forces an unnatural and unnecessary fragmentation of data to fit third normal form.

Kleisli does not have its own native database management system. Instead, Kleisli has the ability to turn many kinds of database systems into an updatable store conforming to its complex object data model. In particular, Kleisli can use flat relational database management systems such as Sybase, Oracle, and MySQL, to be its updatable complex object store. It can even use these systems simultaneously. This power of Kleisli is illustrated using the example of GenPept reports.

Example 6.5.1   Create a warehouse of GenPept reports and initialize it to reports on protein tyrosine phosphatases. Kleisli provides several functions to access GenPept reports remotely from Entrez [9]. One of them is aa-get-seqfeat-general, which retrieves GenPept reports matching a search string.

```
! connect to our Oracle database system
oracle-cplobj-add (name: "db", ...);
! create a table to store GenPept reports
create table genpept(uid: "NUMBER", detail: "LONG")
                                                using db;
! initialize it with PTP data
select (uid: x.uid, detail: x) into genpept from
                aa-get-seqfeat-general("PTP") x using db;
! index the uid field for fast access
db-mkindex (table: "genpept", index: "genpeptindex",
                                        schema: "uid");
! let's use it now to see the title of report 131470
create view GenPept from genpept using db;
select x.detail.title from GenPept x where x.uid = 131470;
```

In this example, a table genpept is created in the local Oracle database system. This table has two columns, uid for recording the unique identifier and detail for recording the GenPept report. A LONG data type is used for the detail column of this table. However, recall from Example 6.3.2 that each GenPept report is a highly nested complex object. There is therefore a mismatch between LONG (which is essentially a big, uninterpreted string) and the complex structure of a GenPept report. This mismatch is resolved by the Kleisli system, which automatically performs the appropriate encoding and decoding. Thus, as far as the Kleisli user is concerned, x.detail has the type of GenPept report as given in Example 6.3.1. So the user can ask for the title of a report as straightforwardly as x.detail.title. Note that encoding and decoding are performed to map the complex object transparently into the space provided in the `detail` column; that is, the Kleisli system does not fragment the complex object to force it into third normal form.

There are two possible techniques to use a flat relational database system as a nested relational store. This first is to add a layer on top of the underlying flat relational database system to perform automatic normalization of nested relations into the third normal form. This is the approach taken by systems such as OPM [22]. Such an approach may lead to performance problems as the database system may be forced to perform many extra joins under certain situations. The second technique is to add a layer on top of the underlying flat relational database system to perform automatic encoding and decoding of nested components into long strings. This is the technique adopted in Kleisli because it avoids unnecessary joins and because it is a simple extension—without significant additional overhead—to the handling of Kleisli's data exchange format.

6.6 DATA SOURCES

The standard version of the Kleisli system marketed by geneticXchange, Inc. supports more than 60 types of data sources. These include the categories below.

✦ Relational database management systems: All popular relational database management systems are supported, such as Oracle, Sybase, DB2, Informix, and MySQL. The support for these systems is quite sophisticated. For example, the previous section illustrates how the Kleisli system can turn these flat database systems transparently into efficient complex object stores that support both read and write access. A later section also shows that the Kleisli system is able to perform significant query optimization involving these systems.

✦ Bioinformatics analysis packages: Most popular packages for analysis of protein sequences and other biological data are supported. These packages include both Web-based and/or locally installed versions of WU-BLAST [30], Gapped BLAST [31], FASTA, CLUSTAL W [32], HMMER, BLOCKS, Profile Scan (PFSCAN) [33], NNPREDICT, PSORT, and many others.

✦ Biological databases: Many popular data sources of biological information are also supported by the Kleisli system, including AceDB [18], Entrez [9], LocusLink, UniGene, dbSNP, OMIM, PDB, SCOP [34], TIGR, KEGG, and MEDLINE. For each of these sources, Kleisli typically provides many access functions corresponding to different capabilities of the sources. For example,

Kleisli provides about 70 different but systematically organized functions to access and extract information from Entrez.

✦ Patent databases: Currently only access to the United States Patent and Trademark Office (USPTO) is supported.

✦ Interfaces: The Kleisli system also provides means for parsing input and writing output in HTML and XML formats. In addition, programming libraries are provided for Java and Perl to interface directly to Kleisli in a fashion similar to JDBC and ODBC. A graphical user interface called Discovery Builder is also available.

It is generally easy to develop a wrapper for a new data source, or modify an existing one, and insert it into Kleisli. There is no impedance mismatch between the data model supported by Kleisli and the data model necessary to capture the data source. The wrapper is therefore often a very lightweight parser that simply parses records in the data source and prints them out in Kleisli's simple data exchange format.

Example 6.6.1   Let us consider the `webomim-get-detail` function used in Example 6.4.4. It uses an OMIM identifier to access the OMIM database and returns a set of objects matching the identifier. The output is of type:

```
{(#uid: num, #title: string, #gene_map_locus: {string},
  #alternative_titles: {string}, #allelic_variants: {string})}
```

Note that this is a nested relation: It consists of a set of records, and each record has three fields that are also of set types, namely `#gene_map_locus`, `#alternative_titles`, and `#allelic_variants`. This type of output would definitely present a problem if it had to be sent to a system based on the flat relational model, as the information would have to be re-arranged in these three fields to be sent into separate tables.

Fortunately, such a nested structure can be mapped directly into Kleisli's exchange format. The wrapper implementor would only need to parse each matching OMIM record and write it out in a format as illustrated in the following:

```
{(#uid: 189965,
  #title: "CCAAT/ENHANCER-BINDING PROTEIN, BETA; CEBPB",
  #gene_map_locus: "20q13.1",
  #alternative_titles: {"C/EBP-BETA",
                        "INTERLEUKIN 6-DEPENDENT DNA-
                         BINDING PROTEIN; IL6DBP",
                        "LIVER ACTIVATOR PROTEIN; LAP",
```

```
                          "LIVER-ENRICHED TRANSCRIPTIONAL
                          ACTIVATOR PROTEIN",
                          "TRANSCRIPTION FACTOR 5; TCF5"},
     #allelic_variants: {})}
```

Instead of needing to create separate tables to keep the sets nested inside each record, the wrapper simply prints the appropriate set brackets { and } to enclose these sets. Kleisli will automatically deal with them as they were handed over by the wrapper. This kind of parsing and printing is extremely easy to implement. Figure 6.4 shows the relevant chunk of Perl codes in the OMIM wrapper implementing webomim-get-detail.

6.7   OPTIMIZATIONS

A feature that makes Oracle and Sybase much more productive to use than a raw file system is the availability of a high-level query language. Such a query language allows users to express their needs in a declarative, logical way. All low-level details such as opening files, handling disk blocks, using indices, decoding record and field boundaries, and so forth are hidden away and are automatically taken care of. However, there are two prices to pay, one direct and one indirect. The direct one is that if a high-level query is executed naively, the performance may be poor. The same high-level command often can be executed in several logically equivalent ways. However, which of these ways is more efficient often depends on the state of the data. A good optimizer can take the state of the data into consideration and pick the more efficient way to execute the high-level query. The indirect drawback is that because the query language is at a higher level, certain low-level details of programming are no longer expressible, even if these details are important to achieving better efficiency. However, a user who is less skilled in programming is now able to use the system. Such a user is not expected to produce always efficient programs. A good optimizer can transform inefficient programs into more efficient equivalent ones. Thus, a good optimizer is a key ingredient of a decent database system and of a general data integration system that supports *ad hoc* queries.

The Kleisli system has a fairly advanced query optimizer. The optimizations provided by this optimizer include (1) *monadic* optimizations that are derived from the equational theory of monads, such as vertical loop fusion; (2) context-sensitive optimizations, which are those equations that are true only in special contexts and that generally rely on certain long-range relationships between sub-expressions, such as the absorption of sub-expressions in the then-branch of an if-then-else construct that are equivalent to the condition of the construct;

```perl
#!/usr/bin/perl
#
#....stuff for connecting to OMIM omitted...
#....<CMD> is the input stream to be parsed...
#
# default values
$section = "none"; $state = 0; $id = "";
# the main program
print "{\n";
while (<CMD>) {
  chomp;
  if (/dispomim.cgi.cmd=entry.*id=([0-9]+)/) {
    $state = 1; $id = $1; $section = "title"; $line = ""; }
  # look for keywords to being parsing sections
  elsif (($state==1) && (/a href=\"\" name=\"$id\_(.*?)\"/)) {
    $section = "$1"; $line = $_; }
  # parse title
  elsif (($section eq "title") && (/<SPAN CLASS="H3"><font/)) {
    $title = $_; $title =~ s/<.*?>//g; }
  # parse alternative titles
  elsif (($section eq "MIM") && (m-</p></em>(.*)</h4>-)) {
    $tmp = $1; @alts = split /<br>/, $tmp; $alternativeTitles = "";
    foreach $x (@alts) { $alternativeTitles .= "\"$x\", "; }
    $alternativeTitles =~ s/, $//; }
  # parse gene map location
  elsif (($section eq "TEXT") && ($line =~ /^Gene map locus *(.*)/)){
    $geneMapLocus = $1; $geneMapLocus =~ s/<.*?>//g; $line = ""; }
  # parsing for Allelic variants
  # each allelic varient will have it's own section
  # need to group the allelic variants accross sections
  elsif ($section eq "ALLELIC_VARIANTS") { $variantTitle = ""; }
  elsif ($section =~ /AllelicVariant/) {
    $_ = $line; s/<.*?>//g; s/.\d+ *//; $variantTitle .= "\"$_\", "; }
  elsif (($state==1) && ($section eq "CREATION_DATE")) {
    $state = 0; $variantTitle =~ s/, $//; $variantTitle =~ s/\",/\"\n/g;
```

6.4

FIGURE

The Perl code of the wrapper implementing the webomim-get-detail function of Kleisli. It demonstrates the ease of developing wrappers for handling data sources that contain nested objects.

(3) relational optimizations, which are optimizations relating to relational database sources such as the migration of projections, selections, and joins to the external relational database management system; and (4) many other optimizations such as parallelism, code motion, and selective introduction of laziness.

6.7.1   Monadic Optimizations

The restricted form of structural recursion corresponds to the presentation of monads by Kleisli [15, 16] and is expressed by the combinator $\bigcup\{f(x) \mid x \in R\}$ obeying the following three equations:

$$\bigcup\{f(x) \mid x \in \{\}\} = \{\}$$
$$\bigcup\{f(x) \mid x \in \{o\}\} = f(o)$$
$$\bigcup\{f(x) \mid x \in A \cup B) = (\bigcup\{f(x) \mid x \in A\}) \cup (\bigcup\{f(x) \mid x \in B\})$$

This combinator is at the heart of the $\mathcal{NRC}$, the abstract representation of queries in the implementation of sSQL. It earns its central position in the Kleisli system because it offers tremendous practical and theoretical convenience. The direct correspondence in sSQL is: `select y from R x, f(x) y`. This combinator is a key operator in the library of complex object routines in Kleisli. All sSQL queries can be and are first translated into $\mathcal{NRC}$ via Wadler's identities [15, 16].

The practical convenience of the $\bigcup\{f(x) \mid x \in R\}$ combinator is best seen in query optimizations.

A well-known optimization rule is vertical loop fusion [35], which corresponds to the physical notion of getting rid of intermediate data and the logical notion of quantifier elimination. Such an optimization on queries in the comprehension syntax can be expressed informally as $\{e \mid G_1, ..., G_n, x \in \{e' \mid H_1, ..., H_m\}, J_1, ..., J_k\} \rightsquigarrow \{e[e'/x] \mid G_1, ..., G_n, H_1, ..., H_m, J_1[e'/x], ..., J_k[e'/x]\}$. Such a rule in comprehension form is simple to grasp: The intermediate set built by the comprehension $\{e' \mid H_1, ..., H_m\}$ is eliminated in favor of generating the x on the fly. In practice, the rule is quite messy to implement because the informal "..." denotes any number of generator-filters in a comprehension. An immediate implementation would involve a nasty traversal routine to skip over the non-applicable G_i to locate the applicable $x \in \{e' \mid H_1, ..., H_m\}$ and J_i. The effect of the $\bigcup\{f(x) \mid x \in R\}$ combinator on the optimization rule for vertical loop fusion is dramatic. This optimization is now expressed as $\{f(x) \mid x \in \bigcup\{g(y) \mid y \in R\}\}$ $\rightsquigarrow \bigcup\{\bigcup\{f(x) \mid x \in g(y)\} \mid y \in R\}$. The informal and troublesome "..." no longer appears. Such a rule can be coded straightforwardly in almost any implementation language.

To illustrate this point more concretely, it is necessary to introduce some detail from the implementation of the Kleisli system. Recall from the introductory section that Kleisli is implemented on top of SML. The type SYN of SML objects that represent queries in Kleisli is declared as:

```
type VAR = int                      (* Variables, represented
                                                  by int *)
type SVR = int                      (* Server connections,
                                       represented by int *)
type CO = ...                       (* Representation of
                                         complex objects *)
datatype SYN = ...
| EmptySet                          (* { }                    *)
| SngSet of SYN                     (* { E }                  *)
| UnionSet of SYN * SYN             (* E1 U E2                *)
| ExtSet of SYN * VAR * SYN         (* U{ E1 | \x <- E2 }     *)
| IfThenElse of SYN * SYN * SYN     (* if E1 then E2 else E3*)
| Read of SVR * real * SYN          (* process E using S,
        the real is the request priority assigned by
                                                optimizer*)
| Variable VAR                      (* x *)
| Binary (CO * CO -> CO) * SYN * SYN (* Construct for
        caching static objects. This allows the optimizer
        to insert some codes for doing dynamic
                                        optimization *)
```

All SML objects that represent optimization rules in Kleisli are functions and have type RULE:

```
type RULE = SYN -> SYN option
```

If an optimization rule r can be successfully applied to rewrite an expression e to an expression e', then $r(e) =$ SOME(e'). If it cannot be successfully applied, then $r(e) =$ NONE.

Now the vertical loop fusion has a very simple implementation.

Example 6.7.1   Vertical loop fusion.

```
fun Vertfusion(ExtSet(E1,x,ExtSet(E2,y,E3)))
   = SOME(ExtSet (ExtSet(E1,x E2) ,y,E3))
| Vertfusion _ = NONE
```

6.7.2   Context-Sensitive Optimizations

The Kleisli optimizer has an extensible number of phases. Each phase is associated with a rule base and a rule-application strategy. A large number of rule-application strategies are supported, such as BottomUpOnce, which applies rules to rewrite an expression tree from leaves to root in a single pass. By exploiting higher-order functions, these rule-application strategies can be decomposed into a *traversal* component common to all strategies and a simple *control* component special for each strategy. In short, higher-order functions can generate these strategies extremely simply, resulting in a small optimizer core. To give some ideas on how this is done, some SML code fragments from the optimizer module mentioned are presented on the following pages.

The traversal component is a higher-order function shared by all strategies:

```
val Decompose: (SYN -> SYN) -> SYN -> SYN
```

Recall that SYN is the type of SML object that represents query expressions. The Decompose function accepts a rewrite rule r and a query expression Q. Then it applies r to all immediate subtrees of Q to rewrite these immediate subtrees. Note that it does not touch the root of Q and it does not traverse Q—it just non-recursively rewrites immediate subtrees using r. It is, therefore, very straightforward and can be expressed as follows:

```
fun Decompose r (SngSet N) = SngSet(r N)
| Decompose r (UnionSet(N,M)) = UnionSet(r N, r M)
| Decompose r (ExtSet(N,x,M)) = ExtSet(r N, x, r M)
| ...
```

A rule-application strategy S is a function having the following type:

```
val S: RULEDB -> SYN -> SYN
```

The precise definition of the type RULEDB is not important at this point and is deferred until later. Such a function takes in a rule base R and a query expression Q and optimizes it to a new query expression Q' by applying rules in R according to the strategy S.

Assume that Pick: RULEDB -> RULE is an SML function that takes a rule base R and a query expression Q and returns NONE if no rule is applicable, and SOME(Q') if some rule in R can be applied to rewrite Q to Q'. Then the control components of all the strategies mentioned earlier can be generated easily.

Example 6.7.2   The `BottomUpOnce` strategy applies rules in a leaves-to-root pass. It tries to rewrite each node at most once as it moves toward the root of the query expression. Here is its control component:

```
fun BottomUpOnce RDB Qry =
  let fun Pass SubQry =
        let val BetterSubQry = Decompose Pass SubQry
          in case Pick RDB BetterSubQry
            of SOME EvenBetterSubQry => EvenBetterSubQry
             | NONE => BetterSubQry end
    in Pass Qry end
```

The following class of rules requires the use of multiple rule-application strategies. The scope of rules like the vertical loop fusion in the previous section is over the entire query. In contrast, this class of rules has two parts. The inner part is *context sensitive*, and its scope is limited to certain components of the query. The outer part scopes over the entire query to identify contexts where the inner part can be applied. The two parts of the rule can be applied using completely different strategies.

A rule base RDB is represented in the system as an SML record of type:

```
type RULEDB = {
    DoTrace: bool ref,
    Trace: (rulename -> SYN -> SYN -> unit) ref,
    Rules: (rulename * RULE) list ref }
```

The `Rules` field of RDB stores the list of rules in RDB together with their names. The `Trace` field of RDB stores a function f that is to be used for tracing the usage of the rules in RDB. The `DoTrace` field of RDB stores a flag to indicate whether tracing is to be done. If tracing is indicated, then whenever a rule of name N in RDB is applied successfully to transform a query Q to Q', the trace function is invoked as $f\ N\ Q\ Q'$ to record a trace. Normally, this simply means a message like "Q is rewritten to Q' using the rule N" is printed. However, the trace function f is allowed to carry out considerably more complicated activities.

It is possible to exploit trace functions to achieve sophisticated transformations in a simple way. An example is the rule that rewrites `if` e_1 `then` ... e_1 ... `else` e_3 to `if` e_1 `then` ... `true` ... `else` e_3. The inner part of this rule rewrites e_1 to `true`. The outer part of this rule identifies the context and scope of the inner part of this rule: limited to the `then`-branch. This example is very intuitive to a human being. In the `then`-branch of a conditional, all sub-expressions identical to the test predicate of the conditional must eventually evaluate to `true`.

However, such a rule is not so straightforward to express to a machine. The informal "..." are again in the way. Fortunately, rules of this kind are straightforward to implement in Kleisli.

Example 6.7.3 The if-then-else absorption rule is expressed by the AbsorbThen rule below. The rule has three clauses. The first clause says the rule should not be applied to an IfThenElse whose test predicate is already a Boolean constant because it would lead to non-termination otherwise. The second clause says the rule should be applied to all other forms of IfThenElse. The third clause says the rule is not applicable in any other situation.

```
fun AbsorbThen (IfThenElse(Bool _,_,_)) = NONE
| AbsorbThen (IfThenElse(E1,E2,E3)) =
    let fun Then E = if SyntaxTools.Equiv E1 E then
    SOME(Bool true) else NONE
    in case ContextSensitive Then TopDownOnce E2
        of SOME E2' => IfThenElse(E1,E2',E3)
        | NONE => NONE end
| AbsorbThen _ = NONE
```

The second clause is the meat of the implementation. The inner part of the rewrite if e_1 then ... e_1 ... else e_3 to if e_1 then ... true ... else e_3 is captured by the function Then, which rewrites any e identical to e_1 to true. This function is then supplied as the rule to be applied using the TopDownOnce strategy within the scope of the then-branch ... e_1 ... using the ContextSensitive rule generator given as follows.

```
fun ContextSensitive Rule Strategy Qry =
let val Changed = ref false        (* This flag is set if
                                        Rule is applied *)
   val RDB = {                     (* Set up a context-
                                      sensitive rule base *)
    DoTrace = ref true,
    Trace = ref (fn _ => fn _ => fn _ => Changed := true)
                                   (* Changed is true
                                        if Rule is used *)
    Rules = ref [("", Rule)]}
   val OptimizedQry = Strategy RDB Qry
                                   (* Apply Rule using
                                        Strategy. *)
in if !Changed then SOME OptimizedQry else NONE end
```

This `ContextSensitive` rule generator is re-used in many other context-sensitive optimization rules, such as the rule for migrating projections to external relational database systems to be presented shortly.

6.7.3   Relational Optimizations

Relational database systems are the most powerful data sources to which Kleisli interfaces. These database systems are equipped with the ability to perform sophisticated transformations expressed in SQL. A good optimizer should aim to migrate as many operations in Kleisli to these systems as possible. There are four main optimizations that are useful in this context: the migration of projections, selections, and joins on a single database; and the migration of joins across two databases. The Kleisli optimizer has four different rules to exploit these four opportunities.

A special case of the rule for migrating P is to rewrite `select x.name from (process "select * from T" using A) x` to `select x.name from (process "select name from T" using A) x`, where `process Q using A` denotes sending an SQL query `Q` to a relational database `A`. In the original query, the entire table `T` has to be retrieved. In the rewritten query, only one column of that table has to be retrieved. More generally, if `x` is from a relational database system and every use of `x` is in the context of a field projection `x.l`, these projections can be *pushed* to the relational database so that unused fields are not retrieved and transferred.

Example 6.7.4   The rule for migrating projections to a relational database is implemented by `MigrateProj` in this example. The rule requires a function `FullyProjected x N` that traverses an expression `N` to determine whether `x` is always used within `N` in the context of a field projection and to determine what fields are being projected; it returns NONE if `x` is not always used in such a context; otherwise, it returns SOME *L*, where the list *L* contains all the fields being projected. This function is implemented in a simple way using the `ContextSensitive` rule generator from Example 6.7.3.

```
fun FullyProjected x N =
    let val (Count, Projs) = (ref 0, ref [])
        fun FindProjs (Variable y) = (if x = y then inc
                                                Count else (); NONE)
          | FindProjs (Proj (L, Variable y)) =
             (if x = y then Projs := L :: (!Projs) else ();
                                                NONE)
          | FindProjs _ = NONE
```

```
      in ContextSensitive FindProjs BottomUpOnce N;
         if length (!Projs) = !Count then SOME (!Projs) else
                                                         NONE
      end
```

The `MigrateProj` rule is defined below. The function `SQL.PushProj` is one of the many support routines available in the current release of Kleisli that handle manipulation of SQL queries and other `SYN` abstract syntax objects.

```
fun MigrateProj (ExtSet (N, x, Read (S, p, String M))) =
    if Annotations.IsSQL S        (* test if S connects to a
                                             SQL server *)
    then case FullyProjected x N (* test if x is always
                                            in a projection *)
       of SOME Projs => SOME (ExtSet (N, x, Read (S, p,
                             String (SQL.PushProj Projs M))))
        | NONE => NONE
    else NONE
| MigrateProj _ = NONE
```

Besides the four migration rules mentioned previously, Kleisli has various other rules, including reordering joins on two relational databases, parallelizing queries, and large-scale code motion, the description of which is omitted in the chapter due to space constraints.

6.8   USER INTERFACES

Kleisli is equipped with application programming interfaces for use with Java and Perl. It also has a graphical interface for non-programmers. These interfaces are described in this section.

6.8.1   Programming Language Interface

The high-level query language, sSQL, of the Kleisli system was designed to express traditional (nested relational) database-style queries. Not every query in bioinformatics falls into this class. For these non-database-style queries, some other programming languages can be a more convenient or more efficient means of implementation. The Pizzkell suite [19] of interfaces to the Kleisli exchange format was developed for various popular programming languages. Each of these interfaces in the Pizzkell suite is a library package for parsing data in Kleisli's exchange format into an internal object of the corresponding programming language. It also serves as a means for embedding the Kleisli system into that programming language so that the full power of Kleisli is made available within that programming language. The Pizzkell suite currently includes CPL2Perl and CPL2Java, for Perl and Java.

In contrast to sSQL in Kleisli, which is a high-level interface that comes with a sophisticated optimizer and other database-style features, CPL2Perl has a different purpose and is at a lower level. Whereas sSQL is aimed at extraction, integration, and preparation of data for analysis, CPL2Perl is intended to be used for implementing analysis and textual formatting of the prepared data in Perl. Thus, CPL2Perl is a Perl module for parsing data conforming to the data exchange format of Kleisli into native Perl objects.

The main functions in CPL2Perl are divided into three packages:

1. The RECORD package simulates the record data type of the Kleisli exchange format by using a reference of Perl's hash. Some functions are defined in this package:

   ✦ New is the constructor of a record. For example, to create a record such as (#anno_name: "db_xref", #descr: "taxon:10090") in a Perl program, one writes:

   ```
   $rec = RECORD->new ("anno_name",
     "db_xref", "descr","taxon:10090");
   ```

   where $rec becomes the reference of this record in the Perl program.

   ✦ Project gets the value of a specified field in a record. For example,

   ```
   $rec->Project("descr");
   ```

   will return the value of the field #descr in the record referenced by $rec in the Perl program.

2. The LIST package simulates the list, set, and bag data type in the Kleisli data exchange format. These three bulk data types are to be converted as a reference of Perl's list. Its main function is:

   ✦ new is the constructor of bulk data such as a list, a bag, or a set. It works the same way as a list initialization in Perl:

   ```
   $l = LIST->new ( "tom", "jerry" );
   ```

   where $l will be the reference of this list in the Perl program.

3. The CPLIO package provides the interface to read data directly from a Kleisli-formatted data file or pipe. It supports both eager and lazy access methods. Some functions in this package are:

✦ Open1 opens the specified Kleisli-formatted data file and returns the handle of this file in Perl. It supports all the input-related features of the usual open operation of Perl, including the use of pipes. For example, the following expression opens a Kleisli-formatted file "sequences.val":

```
$hd = CPLIO->Open1("sequences.val");
```

✦ Open1a is another version of the Open1 function, which can take a string as input stream. The first parameter specifies the child process to execute and the second parameter is the input string. An example that calls Kleisli from CPL2Perl to extract accession numbers from a sequence file is expressed as follows:

```
$cmd = qq{
create view X from sequences using stdin;
select x.accession from X x; };
$a= CPLIO->Open1a ( "./ssql" , $cmd);
```

✦ Open2 differs from Open1a in that it allows a program to communicate in both directions with Kleisli or other systems. It is parameterized by the Kleisli or other systems to call. It returns a list consisting of a reference of CPLIO object and an input stream that the requests can be sent into. For example:

```
($a, $b) = CLPIO->Open2("./ssql");
print $b $cmd1; flush $b; $res = $a->Parse;
...
print $b $cmd2; flush $b; $res = $a->Parse;
...
```

✦ Parse is a function that reads all the data from an opened file until it can assemble a complete object or a semicolon (;) is found. The return value will be the reference of the parsed object in Perl. For example, printing the values of the field #accession of an opened file may be expressed by:

```
$set = $hd->Parse;
foreach $rec (@{$set} ) {
$n = $rec->Project("accession");
print "$n\n"; }
```

✦ LazyRead is a function that reads data lazily from an opened file. This function is used when the data type in the opened file is a set, a bag, or a list. This function only reads one element into memory at a time. Thus, if the opened file is a very big set, `LazyRead` is just the right function to access the records and print accession numbers:

```
while (1) {
$rec = $hd->LazyRead; last if ($rec eq "");
$n = $rec->Project("accession");
print "$n\n"; }
```

The use of CPL2Perl for interfacing Kleisli to the Graphviz system is demonstrated in the context of the Protein Interaction Extraction System described in a paper by Wong [36]. Graphviz [37] is a system for automatic layout of directed graphs. It accepts a general directed graph specification, which is in essence a list of arcs of the form $x \rightarrow y$, which specifies an arc is to be drawn from the node x to the node y.

Example 6.8.1   Assume that Kleisli produces a file `$SPEC` of type `{ (#actor: string, #interaction: string, #patient: string) }` which describes a protein interaction pathway. The records express that an *actor* inhibits or activates a *patient*. The relevant parts of a Perl implementation of the module `MkGif` that accepts this file, converts it into a directed graph specification, invokes Graphviz to layout, and draws it as a GIF file `$GIF` using CPL2Perl is expressed as follows:

```
use cpl2perl;
$a = CPLIO->Open1 ("$SPEC");
open (DOT, "| ./dot -Tgif  > $GIF");
print DOT "digraph aGraph {\n";
while (1) {
 $rec = $a->LazyRead; last if ($rec eq "");
 $start = $rec->Project ("actor");
 $end = $rec->Project ("patient");
 $type = $rec->Project ("interaction");
 if ($type eq "inhibit") {
   $edgecolor = "red"; }
 else {
   $edgecolor = "green"; }
 $edge = "[color = $edgecolor]";
 print DOT " $start-> $end $edge;\n"; }
print DOT "};\n";
$a->Close;
close (DOT);
```

The first three lines establish the connections to the Kleisli file $SPEC and to the Graphviz program dot. The next few lines use the LazyRead and Project functions of CPL2Perl to extract each interaction record from the file and to format it for Graphviz to process. Upon finishing the layout computation, Graphviz draws the interaction pathway into the file $GIF.

This example, though short, demonstrates how CPL2Perl smoothly integrates the Kleisli exchange format into Perl. This greatly facilitates both the development of data drivers for Kleisli and the development of downstream processing (such as pretty printing) of results produced by Kleisli.

6.8.2 Graphical Interface

The Discovery Builder is a graphical interface to the Kleisli system designed for non-programmers by geneticXchange, Inc. This graphical interface facilitates the visualization of the source data as required to formulate the queries and generates the necessary sSQL codes. It allows users to see all available data sources and their associated meta-data and assists them in navigating and specifying their query on these sources with the following key functions:

+ A graphical interface that can *see* all the relevant biological data sources, including meta-data—tables, columns, descriptions, etc.—and then construct a query as if the data were local

+ Add new wrappers for any public or proprietary data sources, typically within hours, and then have them enjoined in any series of *ad hoc* queries that can be created

+ Execute the queries, which may join many data sources that can be scattered all over the globe, and get fresh result data quickly

The Discovery Builder interface is presented in Figure 6.5.

6.9 OTHER DATA INTEGRATION TECHNOLOGIES

The brief description of several other approaches to bioinformatics data integration problems emphasizes Kleisli's characteristics. The alternatives include Sequence Retrieval System (SRS) [38], DiscoveryLink [23], and OPM [22].

6.9.1 SRS

SRS [38] (also presented in Chapter 5) is marketed by LION Bioscience and is arguably the most widely used database query and navigation system for the life

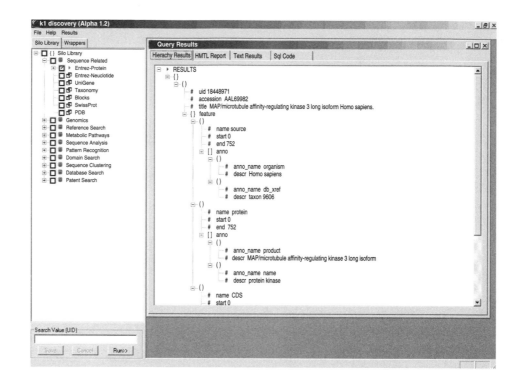

6.5

FIGURE

The Discovery Builder graphical interface to Kleisli.

science community. It provides easy-to-use graphical user interface access to a broad range of scientific databases, including biological sequences, metabolic pathways, and literature abstracts. SRS provides some functionalities to search across public, in-house and in-licensed databases. To add a new data source into SRS, the data source is generally required to be available as a flat file, and a description of the schema or structure of the data source must be available as an Icarus script, which is the special built-in wrapper programming language of SRS. The notable exception to this flat file requirement on the data source is when the data source is a relational database. SRS then indexes this data source on various fields parsed and described by the Icarus script. A biologist then accesses the data by supplying some keywords and constraints on them in the SRS query language, and all records matching those keywords and constraints are returned. The SRS query language is primarily a navigational language. This query language has limited data joining capabilities based on indexed fields and has limited data restructuring capabilities. The results are returned as a simple aggregation of records that match the search constraints. In short, in terms of querying power, SRS is essentially an information retrieval system. It brings back records matching specified keywords and constraints. These records can contain embedded links a user can follow individually to obtain deeper information. However, it does not offer much help in organizing or transforming the retrieved results in a way that might be needed for setting up an analytical pipeline. There is also a Web browser-based interface for formulating SRS queries and viewing results. In fact, this interface of SRS is often used by biologists as a unified front end to access multiple data sources independently, rather than learning the idiosyncrasies of the original search interfaces of these data sources. For this reason, SRS is sometimes considered to serve "more of a user interface integration role rather than as a true data integration tool"[39].

In summary, SRS has two main strengths. First, because of the simplicity of flat file indexing, adding new data sources into the system with the Icarus scripting language is easy. In fact, several hundred data sources have been incorporated into SRS to date. Second, it has a nice user interface that greatly simplifies query formulation, making the system usable by a biologist without the assistance of a programmer. In addition, SRS has an extension known as Prisma designed for automating the process of maintaining an SRS warehouse. Prisma integrates the tasks of monitoring remote data sources for new data sets and downloading and indexing such data sets. On the other hand, SRS also has some weaknesses. First, it is basically a retrieval system that returns entries in a simple aggregation. To perform further operations or transformations on the results, a biologist has to do that by hand or write a separate post-processing program using some external scripting language like C or Perl, which is cumbersome. Second, its principally flat-file based indexing mechanism rules out the use of certain remote data sources—in particular, those that are not relational databases—and does not provide for straightforward integration with dynamic analysis tools. However, this latter shortcoming is mitigated by the Scout suite of applications marketed by LION Bioscience that are specifically designed to interact with SRS.

6.9.2   DiscoveryLink

DiscoveryLink [23] (also presented in Chapter 11) is an IBM product and, in principle, it goes one step beyond SRS as a general data integration system for biomedical data. The first thing that stands out—when DiscoveryLink is compared to SRS and more specialized integration solutions like EnsEMBL and GenoMax—is the presence of an explicit data model. This data model dictates the way DiscoveryLink users view the underlying data, the way they view results, as well as the way they query the data. The data model is the relational data model [17]. The relational data model is the *de facto* data model of most commercial database management systems, including the IBM's DB2 database management system, upon which DiscoveryLink is based. As a result, DiscoveryLink comes with a high-level query language, SQL, that is a standard feature of all relational database management systems. This gives DiscoveryLink several advantages over SRS. First, not only can users easily express SQL queries that go across multiple data sources, which SRS users are able to do, but they can also perform further manipulations on the results, which SRS users are unable to do. Second, not only are the SQL queries more powerful and expressive than those of SRS, the SQL queries are also automatically optimized by DB2. Query optimization allows users to concentrate on getting their queries right without worrying about getting them fast.

However, DiscoveryLink still has to overcome difficulties. The first reason is that DiscoveryLink is tied to the relational data model. This implies that every piece of data it handles must be a table of atomic objects, such as strings and numbers. Unfortunately, most of the data sources in biology are not that simple and are deeply nested. Therefore, there is some impedance mismatch between these sources and DiscoveryLink. Consequently, it is not straightforward to add new data sources or analysis tools into the system. For example, to put the Swiss-Prot [40] database into a relational database in the third normal form would require breaking every Swiss-Prot record into several pieces in a normalization process. Such a normalization process requires a certain amount of skill. Similarly, querying the normalized data in DiscoveryLink requires some mental and performance overhead, as the user needs to figure out which part of Swiss-Prot has gone to which of the pieces and to join some of the pieces back again to reconstruct the entry. The second reason is that DiscoveryLink supports only wrappers written in C++, which is not the most suitable programming language for writing wrappers. In short, it is not straightforward to extend DiscoveryLink with new sources. In addition, DiscoveryLink does not store nested objects in a natural way and is very limited in its capability for handling long documents. It also has limitations as a tool for creating and managing data warehouses for biology.

6.9.3   Object-Protocol Model (OPM)

Developed at Lawrence-Berkeley National Labs, OPM [22] is a general data integration system. OPM was marketed by GeneLogic, but its sales were discontinued some time ago. It goes one step beyond DiscoveryLink in the sense that it has a more powerful data model, which is an enriched form of the entity-relationship data model [41]. This data model can deal with the deeply nested structure of biomedical data in a natural way. Thus, it removes the impedance mismatch. This data model is also supported by an SQL-like query language that allows data to be seen in terms of entities and relationships. Queries across multiple data sources, as well as transformation of results, can be easily and naturally expressed in this query language. Queries are also optimized. Furthermore, OPM comes with a number of data management tools that are useful for designing an integrated data warehouse on top of OPM.

However, OPM has several weaknesses. First, OPM requires the use of a global integrated schema. It requires significant skill and effort to design a global integrated schema well. If a new data source needs to be added, the effort needed to redesign the global integrated schema potentially goes up quadratically with respect to the number of data sources already integrated. If an underlying source evolves, the global integrated schema tends to be affected and significant re-design effort may be needed. Therefore, it may be costly to extend OPM with new sources. Second, OPM stores entities and relationships internally using a relational database management system. It achieves this by automatically converting the entities and relationships into a set of relational tables in the third normal form. This conversion process breaks down entities into many pieces when stored. This process is transparent to OPM users, so they can continue to think and query in terms of entities and relationships. Nevertheless, the underlying fragmentation often causes performance problems, as many queries that do not involve joins at the conceptual level of entities and relations are mapped to queries that evoke many joins on the physical pieces to reconstruct broken entities. Third, OPM does not have a simple format to exchange data with external systems. At one stage, it interfaces to external sources using the Common Object Request Broker Architecture (CORBA). The effort required for developing CORBA-compliant wrappers is generally significant [42]. Furthermore, CORBA is not designed for data-intensive applications.

6.10   CONCLUSIONS

In the era of genome-enabled, large-scale biology, high-throughput technologies from DNA sequencing, microarray gene expression and mass spectroscopy, to combinatory chemistry and high-throughput screening have generated an unprecedented volume and diversity of data. These data are deposited in disparate, specialized, geographically dispersed databases that are heterogeneous in data formats and semantic representations. In parallel, there is a rapid proliferation of computational tools and scientific algorithms for data analysis and knowledge extraction. The challenge to life science today is how to process and integrate this massive amount of data and information for research and discovery. The heterogeneous and dynamic nature of biomedical data sources presents a continuing challenge to accessing, retrieving, and integrating information across multiple sources.

Many features of the Kleisli system [2, 5, 43] are particularly suitable for automating the data integration process. Kleisli employs a distributed and federated approach to access external data sources via the wrapper layer, and thus can access the most up-to-date data on demand. Kleisli provides a complex nested internal data model that encompasses most of the current popular data models including flat files, HTML, XML, and relational databases, and thus serves as a natural data exchanger for different data formats. Kleisli offers a robust query optimizer and a powerful and expressive query language to manipulate and transform data, and thus facilitates data integration. Finally, Kleisli has the capability of converting relational database management systems such as Sybase, MySQL, Oracle, DB2, and Informix into nested relational stores, thus enabling the creation of robust warehouses of complex biomedical data. Leveraging the capabilities of Kleisli leads to the development of the query scripts that give us a high-level abstraction beyond low-level codes to access a combination of the relevant data and the right tools to solve the right problem.

Kleisli embodies many of the advances in database query languages and in functional programming. The first is its use of a complex object data model in which sets, bags, lists, records, and variants can be flexibly combined. The second is its use of a high-level query language that allows these objects to be easily manipulated. The third is its use of a self-describing data exchange format, which serves as a simple conduit to external data sources. The fourth is its query optimizer, which is capable of many powerful optimizations. It has had significant impact on data integration in bioinformatics. Indeed, since the early Kleisli prototype was applied to bioinformatics, it has been used efficiently to solve many bioinformatics data integration problems.

REFERENCES

[1]   National Center for Biotechnology Information (NCBI). *NCBI ASN.1 Specification*, revision 2.0. Bethesda, MD: National Library of Medicine, 1992.

[2]   L. Wong. "Kleisli: A Functional Query System." *Journal of Functional Programming* 10, no. 1 (2000): 19–56.

[3]   J. Backus. "Can Programming Be Liberated from Von Neumann Style? A Functional Style and Its Algebra of Programs." *Communications of the ACM* 21, no. 8 (1978): 613–641.

[4]   J. Darlington. "An Experimental Program Transformation and Synthesis System." *Artificial Intelligence* 16, no. 1 (1981): 1–46.

[5]   S. Davidson, C. Overton, V. Tannen, et al. "BioKleisli: A Digital Library for Biomedical Researchers." *International Journal of Digital Libraries* 1, no. 1 (1997): 36–53.

[6]   R. J. Robbins, ed. *Report of the Invitational DOE Workshop on Genome Informatics*, 26–27. Baltimore, MD: April 1993.

[7]   P. Pearson, N. W. Matheson, D. L. Flescher, et al. "The GDB Human Genome Data Base Anno 1992." *Nucleic Acids Research* 20, supplement (1992): 2201–2206.

[8]   C. Burks, M. J. Cinkosky, and W. M. Fischer. "GenBank." *Nucleic Acids Research* 20, supplement (1992): 2065–2069.

[9]   G. D. Schuler, J. A. Epstein, H. Ohkawa, et al. "Entrez: Molecular Biology Database and Retrieval System." *Methods in Enzymology* 266 (1996): 141–162.

[10]   L. C. Bailey Jr., S. Fischer, J. Schug, et al. "GAIA: Framework Annotation of Genomic Sequence." *Genome Research* 8, no. 3 (1998): 234–250.

[11]   P. G. Baker, A. Brass, and S. Bechhofer. "TAMBIS: Transparent Access to Multiple Bioinformatics Information Sources." *Intelligent Systems for Molecular Biology* 6 (1998): 25–34.

[12]   C. A. Goble, R. Stevens, and G. Ng. "Transparent Access to Multiple Bioinformatics Information Sources." *IBM Systems Journal* 40, no. 2 (2001): 532–552.

[13]   C. Schoenbach, J. Koh, and X. Sheng. "FIMM: A Database of Functional Molecular Immunology." *Nucleic Acids Research* 28, no. 1 (2000): 222–224.

[14]   P. Buneman, L. Libkin, and D. Suciu. "Comprehension Syntax." *SIGMOD Record* 23, no. 1 (1994): 87–96.

[15]   P. Wadler. "Comprehending Monads." *Mathematical Structures in Computer Science* 2, no. 4 (1992): 461–493.

[16]   P. Buneman, S. Naqui, V. Tannen, et al. "Principles of Programming With Complex Objects and Collection Types." *Theoretical Computer Science* 149, no. 1 (1995): 3–48.

[17]   E. F. Codd. "A Relational Model for Large Shared Data Bank." *Communications of the ACM* 13, no. 6 (1970): 377–387.

[18]   S. Walsh, M. Anderson, S. W. Cartinhour, et al. "ACEDB: A Database for Genome Information." *Methods of Biochemical Analysis* 39 (1998): 299–318.

[19]   L. Wong. "Kleisli: Its Exchange Format, Supporting Tools, and an Application in Protein Interaction Extraction." In *Proceedings of the First IEEE International Symposium on Bio-Informatics and Biomedical Engineering*, 21–28. Los Alamitos, CA: IEEE Computer Society, 2000.

[20]   International Standards Organization (ISO). *Standard 8824: Information Processing Systems. Open Systems Interconnection. Specification of Abstraction Syntax Notation One (ASN.1)*. Geneva, Switzerland: ISO, 1987.

[21]   T. Hubbard, D. Barker, and E. Birney. "The ENSEMBL Genome Database Project." *Nucleic Acids Research* 30, no. 1 (2002): 38–41.

[22]   I. M. A. Chen and V. M Markowitz. "An Overview of the Object-Protocol Model (OPM) and OPM Data Management Tools." *Information Systems* 20, no. 5 (1995): 393–418.

[23] L. M. Haas, P. M. Schwarz, and P. Kodali. "DiscoveryLink: A System for Integrated Access to Life Sciences Data Sources." *IBM Systems Journal* 40, no. 2 (2001): 489–511.

[24] V. Tannen, P. Buneman, and S. Nagri. "Structural Recursion as a Query Language." In *Proceedings of the Third International Workshop on Database Programming Languages*, 9–19. San Francisco: Morgan Kaufmann, 1991.

[25] V. Tannen and R. Subrahmanyam. "Logical and Computational Aspects of Programming with Sets/Bags/Lists." In *Proceedings of the 18th International Colloquium on Automata, Languages, and Programming, Lecture Note in Computer Science, vol. 510*, 60–75. Berlin, Germany: Springer-Verlag, 1991.

[26] D. Suciu. "Bounded Fixpoints for Complex Objects." *Theoretical Computer Science* 176, no. 1-2 (1997): 283–328.

[27] G. Dong, L. Libkin, and L. Wong. "Local Properties of Query Languages." *Theoretical Computer Science* 239, no. 2 (2000): 277–308.

[28] L. Libkin and L. Wong. "Query Languages for Bags and Aggregate Functions." *Journal of Computer and System Sciences* 55, no. 2 (1997): 241–272.

[29] D. Suciu and L. Wong. "On Two Forms of Structural Recursion." In *Proceedings of the Fifth International Conference on Database Theory, Lecture Notes in Computer Science, vol. 83*. 111–124. Berlin, Germany: Springer-Verlag, 1995.

[30] S. F. Altschul and W. Gish. "Local Alignment Statistics." *Methods in Enzymology* 266 (1996): 460–480.

[31] S. F. Altschul, T. L. Madden, and A. A. Schaffer. "Gapped BLAST and PSI-BLAST: A New Generation of Protein Database Search Programs." *Nucleic Acids Research* 25, no. 17 (1997): 3389–3402.

[32] J. D. Thompson, D. G. Higgins, and T. J. Gibson. "CLUSTAL W: Improving the Sensitivity of Progressive Multiple Sequence Alignment through Sequence Weighting, Positions-Specific Gap Penalties and Weight Matrix Choice." *Nucleic Acids Research* 22 (1994): 4673–4680.

[33] L. Falquet, M. Pagni, P. Bucher, et al. "The PROSITE Database, Its Status in 2002." In *Nucleic Acids Research* 30, no. 1 (2002): 235–238. http://hits.isb-sib.ch/cgibin/PFSCAN.

[34] A. Murzin, S. E. Brenner, and T. Hubbard, et al. "SCOP: A Structural Classification of Protein Database for the Investigation of Sequences and Structures." *Journal of Molecular Biology* 247, no. 4 (1995): 536–540.

[35] A. Goldberg and R. Paige. "Stream Processing." In *Proceedings of the ACM Symposium on LISP and Functional Programming*, 53–62. New York: ACM, 1984.

[36] L. Wong. "PIES, A Protein Interaction Extraction System." In *Proceedings of the Pacific Symposium in Biocomputing*, 520–531. Singapore: World Scientific, 2001.

[37] E. R. Gansner, E. Koutsofios, S. C. North, et al. "A Technique for Drawing Directed Graphs." *IEEE Transactions on Software Engineering* 19, no. 3 (1993): 214–230.

[38] T. Etzold and P. Argos. SRS: Information Retrieval System for Molecular Biology Data Banks. *Methods in Enzymology* 266 (1996): 114–128.

[39] 3rd Millennium Inc. "Practical Data Integration in Biopharmaceutical R & D: Strategies and Technologies." Cambridge, MA: 3rd Millennium, 2002.

[40] A. Bairoch and R. Apweiler. "The SWISS-PROT Protein Sequence Data Bank and Its Supplement TrEMBL in 1999." *Nucleic Acids Research* 27, no. 1 (1999): 49–54.

[41] P. P. S. Chen. "The Entity-Relationship Model: Toward a Unified View of Data." *ACM Transactions on Database Systems* 1, no. 1 (1976): 9–36.

[42] J. Selletin and B. Mitschang. "Data-Intensive Intra- & Internet Applications: Experiences Using Java and CORBA in the World Wide Web." In *Proceedings of the Fourteenth IEEE International Conference on Data Engineering*, 302–311. Los Alamitos, CA: IEEE Computer Science, 1998.

[43] L. Wong. "The Functional Guts of the Kleisli Query System." In *Proceedings of the Fifth ACM SIGPLAN International Conference on Functional Programming*, 1–10. New York: ACM, 2000.

Complex Query Formulation Over Diverse Information Sources in TAMBIS

**Robert Stevens, Carole Goble, Norman W. Paton,
Sean Bechhofer, Gary Ng, Patricia Baker, and Andy Brass**

Molecular biology is a data-rich discipline that has produced a vast quantity of sequence and other data. Most of the resulting data sets are held in independently developed databanks and are acted upon by separate analysis tools. These information sources and tools are autonomous, distributed, and have differing call interfaces. As such, they manifest classical syntactic and semantic heterogeneity problems [1].

Many bioinformatics tasks are supported by individual sources. However, biologists increasingly wish to ask complex questions that span a range of the available sources [2]. This places barriers between a biologist and the task to be accomplished; the biologist has to know what sources to use, the locations of the sources, how to use the sources (both syntactically and their semantics), and how to transfer data between the sources.

This chapter presents an approach to solving these problems called Transparent Access to Multiple Bioinformatics Information Sources (TAMBIS) [3]. This chapter reports on the first version of the TAMBIS system, which was developed between 1996 and 2000. A second version extends and develops this first version, addressing some of the problems recognized in the approach. This new version is introduced in Section 7.5. The TAMBIS approach attempts to avoid the pitfalls described previously by using an ontology of molecular biology and bioinformatics to manage the presentation and usage of the sources. An *ontology* is a description of the concepts, and the relationships between those concepts, within a domain.

The ontology allows TAMBIS:

+ to provide a homogenizing layer over the numerous databases and analysis tools
+ to manage the heterogeneities between the data sources
+ to provide a common, consistent query-forming user interface that allows queries across sources to be precisely expressed and progressively refined

This ontology is the backbone of the TAMBIS system; it is what the user interacts with to form questions. It allows the same style of query and terms to be used across diverse resources, and it also manages the answering of the query itself.

A *concept* is a description of a set of instances, so a concept or description can also be viewed as a query. The TAMBIS system is used for retrieving instances described by concepts in the model. This contrasts with queries phrased in terms of the structures used to store the data, as are used in conventional database query environments. This approach allows a biologist to ask complex questions that access and combine data from different sources. However, in TAMBIS, the user does not have to choose the sources, identify the location of the sources, express requests in the language of the source, or transfer data items between sources.

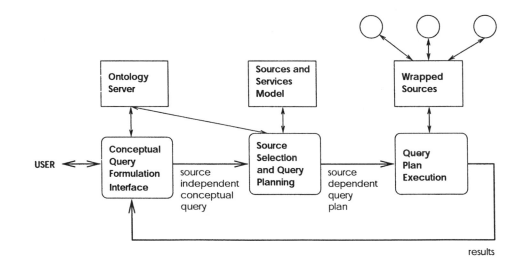

7.1 The flow of information through the TAMBIS architecture.

FIGURE

Figure 7.1 shows how a query is constructed and processed through the TAMBIS system. The steps in processing a TAMBIS query are as follows:

1. A query is formulated in terms of the concepts and relationships in the ontology using the visual *Conceptual Query Formulation Interface*. This interface allows the ontology to be browsed by users and supports the construction of complex concept descriptions that serve as queries. The output of the query formulation process is a *source independent conceptual query*. The query formulation interface makes extensive use of the TAMBIS Ontology Server, which not only stores the ontology but supports various reasoning services over the ontology. These reasoning services serve, for example, to ensure that queries constructed using the query formulation interface are biologically meaningful with respect to the TAMBIS ontology.

2. Given a query, TAMBIS must identify the sources that can be used to answer the query and construct valid and efficient plans for evaluating it given the facilities provided by the relevant sources. The source selection and query planning process makes extensive use of the Sources and Services Model (SSM), which associates concepts and relationships from the Ontology with the services provided by the sources. The output of the source selection and query planning process is a *source dependent query plan* that describes the sources to be used and the order in which calls should be made to the sources.

3. The query plan execution process takes the plan provided by the planner and executes that plan over the *wrapped sources* to yield an answer to the query. Sources are wrapped so they can be accessed in a syntactically consistent manner. In version one of TAMBIS, each source is represented as a collection of function calls, which are evaluated by the collection programming language (CPL) [4]. The sources used in TAMBIS 1.0 were Swiss-Prot, ENZYME, CATH (Classes, Architecture, Topology, Homology), Basic Local Alignment Search Tool (BLAST), and PROSITE.

The remainder of this Chapter is organized as follows. Section 7.1 gives a brief overview of the TAMBIS ontology, describing its scope and the language in which it is implemented. Section 7.2 describes how users interact with TAMBIS, in particular how the ontology is explored and how queries are constructed using the interface from Section 7.2. Section 7.3 describes how queries constructed using the interface from Section 7.2 are evaluated over the individual sources. Section 7.4 describes work in several areas related to TAMBIS and describes how TAMBIS compares to alternative or complementary proposals. Section 7.5 considers issue relating to query construction and source integration raised by experience in TAMBIS and how these are addressed in TAMBIS 2.0.

7.1   THE ONTOLOGY

An *ontology* is a description of the concepts and their relationships within a domain. An ontology is a mechanism by which knowledge about a domain can be captured in computational form and shared within a community [5]. The TAMBIS ontology describes both molecular biology and bioinformatics tasks.

A concept represents a class of individuals within a domain. Concepts such as `Protein` and `Nucleic acid` are part of the world of molecular biology. An `Accession number`, which acts as a unique identifier for an entry in an information source, lies outside this domain but is essential for describing bioinformatics tasks in molecular biology. The TAMBIS ontology contains only concepts and the relationships between those concepts. Individuals that are members of concept classes (P21598 is an individual of the class `Accession number`) do not appear in the TAMBIS ontology. Such individuals are contained within the external resources over which TAMBIS answers queries.

The TAMBIS ontology has been designed to cover the standard range of bioinformatics retrieval and analysis tasks [2]. This means that a broad range of biology has been described. The model is, however, currently quite shallow; although the detail present is sufficient to allow descriptions of most retrieval tasks supportable using the integrated bioinformatics sources. In addition, precision can arise from the ability to combine concepts to create more specialized concepts (see Section 7.2.2).

The model is centered upon the biopolymers `Protein` and `Nucleic acid` and their children, such as `Enzyme`, `DNA`, and `RNA`. Biological functions and processes are also present, so it is possible to describe, for example, the kinds of reactions that are catalyzed by an enzyme. Many tasks in bioinformatics involve comparing or identifying patterns in sequences. As a result, sequence components such as protein motifs and structure classifications are described. For example, a *motif* is a pattern within a sequence that is generally associated with some biological function. The ontology thus supports the description of motifs and various different kinds of motifs. Such descriptions are facilitated by the presence of a rich collection of relationships between concepts in the ontology. These basic concepts are present in the *is a* hierarchy. Other relationships add richness to the model, so that a wide range of biological features can be described. For example, `Motif` (and its children) can be components (parts of) `Protein` or `Nucleic acid`. Other relationships capture associations to functions, processes, sub-cellular locations, similarities, and labels such as species name, gene names, protein names, and accession numbers. The model is described in more detail in an article in *Bioinformatics* [6] and can be browsed via an applet on the TAMBIS Web site.

The TAMBIS ontology is expressed in a description logic (DL) [7], a type of knowledge representation language for describing ontologies [8]. DLs are considered an important formalism for giving a logical underpinning to knowledge representation systems, but they also provide practical reasoning facilities for inferring properties of and relationships between concepts [9]. TAMBIS makes extensive use of these reasoning services.

As well as the traditional *isa* relationships (e.g., a Motif isa Sequence-Component), there are partitive (describing parts), locative (describing location), and nominative (describing names or labels) relationships. This means that the TAMBIS ontology can describe relationships such as: "Motifs are parts of proteins" and "Organelles are located inside cells." The ontology initially holds only asserted concepts, but these can be combined dynamically via relationships to form new, compositional concepts. These compositional concepts are automatically classified using the reasoning services of the ontology. Such compositional concepts can be made in a post-coordinated manner: That is, the ontology is not a static artifact; users can interact with the ontology to build new concepts, composed of those already in the ontology, and have them checked for consistency and placed at the correct position in the ontology's lattice of concepts. For example, Motif can be combined with Protein using the relationship isComponentOf to form a new concept Protein motif, which is placed as a kind of Motif.

The ontology is a dynamic model in that what is present in the model is the description of potential concepts that can be formed in the domain of molecular biology and bioinformatics. As these new, compositional concepts are described, they are placed automatically within the lattice of existing concepts by the DL reasoning services. For example, the compositional concept Protein motif (see above) is automatically classified as a kind of Motif. This new concept is then available to be re-used in further compositional concepts. Most of the other biological ontologies are static; the TAMBIS ontology is dynamic, built around a collection of concept descriptions and constraints on how they can be composed.

The TAMBIS ontology is described using the DL called Galen Representation and Integration Language (GRAIL) [10]. In GRAIL, a new concept can be defined as follows:

```
Base which r₁f₁ ... rₙfₙ
```

where each r_i is a role name and each f_i a filler concept. Each $r_i f_i$ pair is also known as a *criterion*. A *role* is a property of a concept, and the *filler* of a *role* is the name or description of the concept that can play the given role. For example, Motif which isComponentOf Protein is a description of a protein motif. Motif and Protein are names of existing concepts, which are acting here as the *base* concept and a *role filler* respectively. The construct isComponentOf Protein is the criterion of Motif in this case.

Description logic ontologies are organized within a subsumption lattice, which captures the *isa* relationship between two concepts. The fact that one concept is a kind of another can either be asserted as part of the model, or inferred by the reasoning system on the basis of the concept descriptions. Figure 7.2 illustrates both forms of subsumption relationship. For example, Motif has been asserted to be a kind of SequenceComponent, and PhosphorylationSite has been asserted to be a kind of Motif. By contrast, with the asserted hierarchy, the notion of a Motif that can be found within a protein (Motif which isComponentOf Protein) is inferred to be a kind of Motif, as are the other concepts in the three boxes on the bottom in Figure 7.2. In these cases, the criteria describing the concept are used to infer the classification of these concepts. Wherever C_2 is subsumed by C_1, every instance of C_2 is guaranteed to be an instance of C_1 (e.g., every Motif is a SequenceComponent, and every Motif which isComponentOf Protein is a Motif).

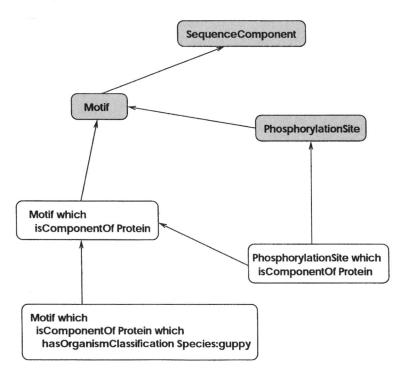

7.2

FIGURE

Example of subsumption relationship within the ontology. The concepts that have been inserted into the lattice are shaded in the three boxes at the top of the figure. The locations of the unshaded concepts in the lattice have been inferred.

In spite of its inexpressiveness compared with some other DLs [11], the GRAIL representation has a useful property in its ability to describe constraints about when relationships are allowed to be formed. For example, it is true that a `Motif` is a component of a `Biopolymer`, but not all motifs are components of all biopolymers. For example, a `PhosphorylationSite` can be a component of a `Protein`, but not a component of a `Nucleic acid`, both of which are `Biopolymers`. The constraint mechanism allows the TAMBIS model to capture this distinction and thus only allow the description of concepts that are described as being biologically meaningful in terms of the model from which they are built. This allows general queries, such as *"find all protein motifs,"* to be expressed as well as specific queries such as *"find phosphorylation motifs upon this protein."*

The TAMBIS ontology is supplied as a software component that acts as a server. Other components can ask questions of the knowledge in the ontology component. It is the backbone of the architecture, and other components either directly or indirectly use the ontology. These other components ask questions such as: "is this a concept;" "what are the parents, children, or siblings of this concept;" "which relationships are held by this concept;" and "what is the natural language version of this concept."

7.2   THE USER INTERFACE

This section describes the user interface to TAMBIS. The interface supports users in carrying out two principal tasks: exploring the ontology and constructing queries, which are described in Sections 7.2.1 and 7.2.2, respectively.

7.2.1   Exploring the Ontology

Although the full TAMBIS ontology contains approximately 1800 concepts, the version of the ontology used in the online system for querying contains approximately 250 concepts. This model concentrates on proteins and enzymes; it describes features such as functions, processes, motifs, and structure. In this and following sections, examples are based on this smaller ontology.

The main window of the TAMBIS system is shown in Figure 7.3. The main window is used to launch exploration or query building tasks. A concept name is either typed into the *find* field directly or obtained from the list of `Bookmarks`. This concept can be used either as the starting point for model exploration or query building by selecting `New query` or `Explore`. If `Explore` is selected in Figure 7.3, the explorer window depicted in Figure 7.4 is launched.

7.3

FIGURE

The TAMBIS main window.

7.4

FIGURE

The explorer window showing motif with all types of relations it has with other concepts.

Relationship	Concept
hasAccessionNumber	accession number
isComponentOf	protein
indicatesFunction	biological function
isAssociatedWithProcess	biological process
hasModification	molecular modification

7.1   The relationships from `motif` to other concepts in the TAMBIS model.

TABLE

The window in Figure 7.4 shows the basic concept description facilities of the model browser. Concepts are shown as buttons; the buttons usually have a title that describes the relationship to the central or *focus* concept, which has no title itself. The button color also indicates the relationship the button has to the central concept, although this is not evident in the monochrome screenshot.

Figure 7.4 shows all the relationships of `motif`. The parent and children concepts are `sequence component` and `site`, respectively. The relationships other than *is-a-kind-of* are shown in the lighter area of the figure. The name of the relationship appears as the button title, and the name of the concept to which the relationship links is the button label. For example, a relationship button title is `hasAccessionNumber` and a concept button label is `accession number`. Table 7.1 shows these relationships for `motif`. The user can explore the *is-a-kind-of* hierarchy or the other relationships by clicking on the buttons representing the concepts to which `motif` is related.

The explorer uses a pie-chart view of the ontology, with different sectors showing the parents, children, definitions, and other relationships. In the TAMBIS ontology, some concepts have a large number of members in one sector, far more than can be shown at any one time. Rather than cramping the view of related concepts, the sectors are scrollable, allowing controlled viewing of the ontology's contents.

Clicking on a concept button that is not the focus causes that button to become the new focus. Thus, a user can move up and down the taxonomy and across the taxonomies by following other relationships. Larger jumps may be made within the model by using a *go to* function.

7.2.2   Constructing Queries

Queries in TAMBIS are essentially concept descriptions. Thus, the task of query formulation involves the user in constructing a concept that describes the information of interest. An example query is illustrated in Figure 7.5, which is a screen

A query builder window containing the concept describing motifs in guppy proteins.

shot of the query builder window containing a request for the motifs that are components of guppy proteins. The equivalent GRAIL concept is:

```
motif which isComponentOf
    protein which hasOrganismClassification species:guppy
```

As its name indicates, the query builder window is used for building descriptions of biological concepts that act as queries. One of the buttons along the bottom of the window, Submit, is used to ask TAMBIS to process the query and collect the results. Part of a results page for this query is given in Figure 7.6. The results

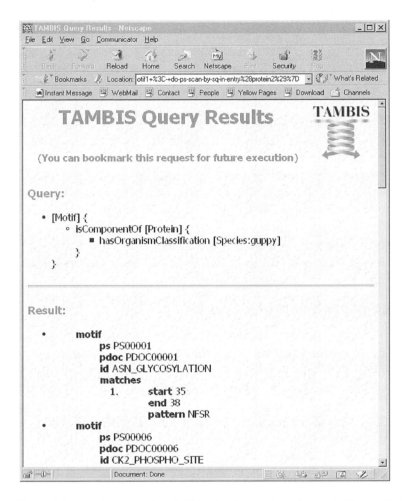

7.6

FIGURE Part of the results page that fulfills the description shown in Figure 7.5.

shown in this figure are the values for the base concept of the query (i.e., the properties of the `motifs` that are components of guppy proteins). This set of results contains only the base concept (`Motif`); other concepts may be included in the results and the relationships maintained between the different instances via the query builder. The pop-up menu on a concept button contains an option *include in results*. Selecting this option causes the concept button to be highlighted in the query builder.

Given that a query is of the form:

```
Base which r₁f₁ ... rₙfₙ
```

where each r_i is a role name and each f_i a filler concept, the query builder essentially supports:

1. The specialization or generalization of the base or filler concepts
2. The addition or removal of criteria associated with a composite concept

This incremental concept construction and modification is possible because of the dynamic model supported by the ontology. In fact, the query interface is driven directly from the model, and the reasoning services are used extensively during query construction to present appropriate options to users and for validating the concepts constructed. The knowledge held in the ontology is used to guide the user through the query building process by offering only appropriate possibilities for modifying a query at each stage [12]. As new concepts are formed and classified, new criteria become available and others are lost as potential additions to the growing concept. Support for the previous query construction operations is illustrated in the following subsections.

Replacing Part of a Query

The query builder can be used either to construct a query from scratch or for modifying previous or bookmarked queries. One way of modifying the query is to *replace* the concept motif with one that is more specific. An example is replacing the concept motif with the more specialized concept phosphorylation site.

Figure 7.7 shows a menu associated with concepts in the query builder. Selecting *replace with-a-kind-of-this* causes a new window to appear, as shown in Figure 7.8. This window is the *replacer* window, which allows a version of the *explorer* to be used to identify a concept that can be used in the query in place of motif.

When launched, the replacer is focused on the motif concept, on which the query builder had focus. Moving down the hierarchy, through site to modified site yields the window shown in Figure 7.9. Selecting phosphorylation site and pressing *replace it* updates the query in the query builder so that the query has the same structure as that in Figure 7.5, but with motif replaced with phosphorylation site (Figure 7.10). The *replacer* limits the user to the *is a* hierarchy during replacement. This helps ensure that only valid concepts are created.

Restricting a Concept

When one concept is joined to another in the query builder with a relationship other than *is-a-kind-of*, the description of the original concept is restricted. In the example query, motifs are restricted to those that occur in proteins, rather than Motifs that can occur in other kinds of molecules. This restriction was added to motif using the *restrict by a relationship* option illustrated in

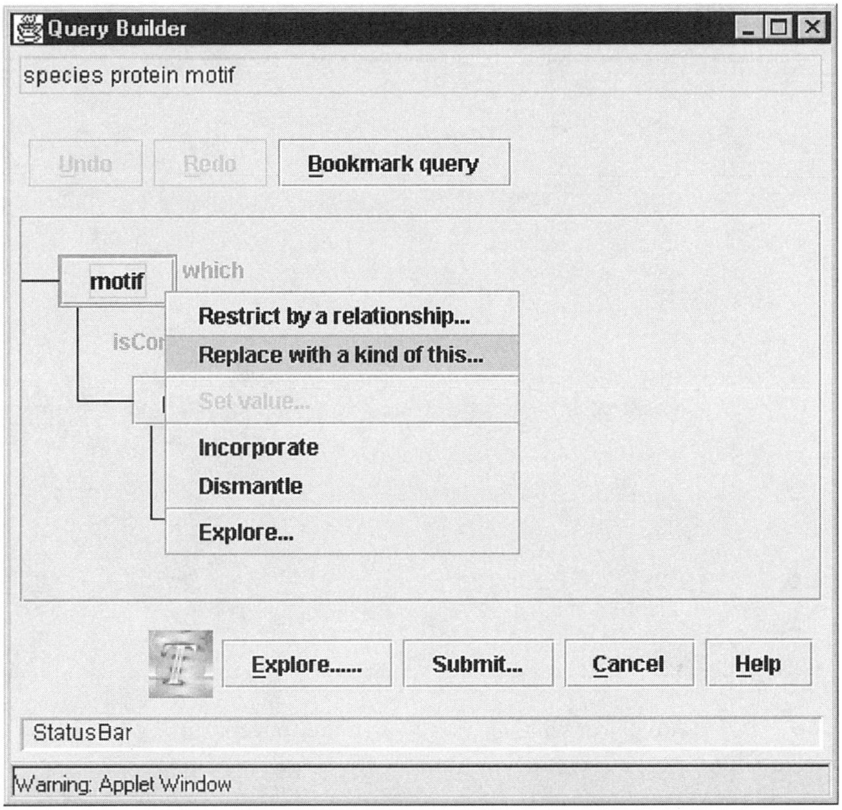

7.7

FIGURE

A query builder window showing the pop-up menu invoked by clicking on the topic concept `motif`.

Figure 7.7. The type of `motif` to be retrieved can be further restricted by adding another concept to the description of `motif`. For example, selecting the *restrict by a relationship* option leads to the user being offered the restrict window shown in Figure 7.11. If the user then selects the `hasModification post transla-tional modification` check box and *accept*, the query in the query builder is replaced with that in Figure 7.12. The query is now "*retrieve all motifs that bring about post translational modifications in guppy proteins.*"

Nonsensical Questions

The TAMBIS model only allows biologically sensible questions to be constructed. By only allowing *is-a-kind-of* relationships to be seen in the replacer, the tendency is to have only biologically sensible queries constructed. It is, however, possible to

FIGURE

A replacer window centered on the concept motif.

replace a valid concept with one that is biologically nonsense. However, the query builder detects this by consulting the ontology and informs the user of the error.

For example, in the previously modified query it would be possible to replace the concept protein with the concept nucleic acid. However, if this replacement is made, TAMBIS notices that in the ontology nucleic acids cannot have phosphorylation sites and changes the color of the offending concept button (nucleic acid in this case) to yellow, indicating that the query is not consistent with the constraints in the ontology.

7.2.3　The Role of Reasoning in Query Formulation

GRAIL, like other DL implementations, provides a classification or reasoning service, which allows the organization of concept descriptions into subsumption (*isa*) hierarchies. In the case of DLs, this is most interesting when applied to composite descriptions. In standard taxonomies, the position of each concept is explicitly stated by the modeler. Within TAMBIS, through the use of the ontology

7.9

FIGURE

A replacer window centered on the concept `modified site` with the pointer about to select the concept `phosphorylation site`.

server, the position of composite concept descriptions can be determined by the reasoner. This is of particular importance when new, previously unseen, descriptions are introduced into the model—particularly when a user forms a new concept to ask a query.

The basic classification hierarchy can be used to navigate through the existing descriptions in the model (e.g., using the explorer). More interesting, however, is TAMBIS's ability to support the formation of new, composite query expressions (through the use of the query builder).

TAMBIS uses a constraint mechanism known as *sanctioning* to drive the query builder user interface [10]. Information included in the ontology specifies the compositions that may be formed, and this in turn determines the specialization options that may be applied to a query. This type of constraint mechanism is peculiar to DLs, as such constraints naturally form part of many frame-based knowledge representation languages. These constraints are, however, important in describing what concepts are allowed to be formed within the ontology.

Query Builder   _ □ ✕

species protein phosphorylation site

| Undo | Redo | Bookmark query |

phosphorylation site   which

      isComponentOf

      protein   which

            hasOrganismClassification

              species: guppy

| Explore...... | Submit... | Cancel | Help |

StatusBar

Warning: Applet Window

7.10

FIGURE

A query builder window showing the query with `motif` replaced by `phosphorylation site`.

For example, the concept `Motif` may be restricted or specialized through a number of relationships including `hasOrganismClassification` or `indicatesFunction`. For each of these relationships, the allowable values are constrained by the values of the sanctions in the model. For example, `hasOrganismClassification` can only be filled with the concept `kingdom` (or one of its subclasses).

It would be an onerous task to specify explicitly the potential values for any combination, so to minimize the information required, sanctions are inherited down the classification hierarchy in the model. Thus, the sanctioning information can be added sparsely. As a query is gradually built up, its position in the classification hierarchy will change, leading to changes in the restriction options offered. The reasoner is key to this process because it is used to determine the appropriate

7.11

FIGURE

The restrict window for `motif`, showing the relationships to other concepts that can be used to *restrict* the description of motif. The cursor lies on the `hasModification post translational modification` check box.

position of a query description and thus, the potential restrictions. As the query is constructed, the interface communicates with the ontology server, updating the restrictions offered to the user. The constraints or sanctions can also be viewed through the explorer; the relationships shown are exactly those that can be used for specialization or restriction of the concept in a query.

7.3 THE QUERY PROCESSOR

The query processor converts a source independent declarative GRAIL query into a source specific execution plan expressed in CPL [4]. CPL allows the concise expression of retrieval requests over collections of data, with data types for representing arbitrarily nested sets, bags, lists, records, and variants. The principal components of the query processor are the wrappers, the sources and services model (SSM), and the planner.

7.12

FIGURE

The query builder containing the example query with an extra restriction on the topic concept motif. Note the lines indicating the relationship between the concepts.

7.3.1   The Sources and Services Model

The SSM stores the relationships between the concepts and roles in the ontology and the functions used to wrap sources in CPL. In the SSM, the ontology is used to index the CPL functions used to evaluate queries written in terms of the ontology.

The SSM contains descriptions of three broad categories of information: iterators that retrieve instances of concepts from sources, role evaluators that retrieve or compute values for the roles of instances, and filters that are used to discard instances not relevant to the query.

Each such description of a CPL function in the SSM includes its *name*, the types of its *arguments*, the type of its *result*, some information on the *cost* of computing the function, and the *source* accessed by the function.

There are seven categories of mapping information supported within the SSM, which are described in detail in a paper of the *Proceedings of 11th International Conference on Scientific and Statistical Data Management* [13]. Four of these categories are described here to illustrate how the query processor works:

1. *Iteration:* Iteration allows the instances of a concept to be retrieved from a source. For example, the fact that the instances or individuals of `Protein` can be obtained from Swiss-Prot is represented by associating the concept `Protein` with the function `get-all-sp-entries`, which has no input arguments and which returns results of type `protein_record`. Given a query in which the instances of protein are required, this SSM entry could be used to retrieve proteins from Swiss-Prot using a function call such as:

   ```
   \p <- get-all-sp-entries()
   ```

   If an alternative, more specialized source of protein information is available, for example, from a database of enzymes (any protein that acts as a catalyst is an enzyme), then an additional SSM entry can be created to indicate this. In fact, there is a source called ENZYME that stores descriptions of enzymes, and thus, there is an SSM entry associating the concept `Protein which hasFunction catalysis` with a function `get-all-enzyme-entries` that supports iteration over the entries in the ENZYME database. During query processing, the planner uses the most specialized source of information available to answer a query. If there are several sources of the same information (e.g., if there is more than one protein source), this must be handled within the wrappers. This restriction within the first version of TAMBIS is to be relieved in future versions of TAMBIS (see Section 7.5).

2. *Roles:* Roles allow the evaluation of a role in an instance to obtain a value for its filler. For example, it is possible to obtain the `AccessionNumber` of a protein given the `Protein`. This is represented in the SSM by the association of the concept `Protein which hasAccessionNumber AccessionNumber` with the function `get-ac-from-sp-entry`, which takes as argument a value of type `protein_record` and returns a value of type `accession_number`. This does not itself directly access a source, but rather it

accesses a data structure retrieved from a source by some other function (such as `get-all-sp-entries` described previously). This SSM entry could be used to retrieve the accession number from a Swiss-Prot entry using a function call such as:

```
\accno <- get-ac-from-sp-entry(p)
```

where p is a variable previously bound to a `protein_record`.

3. *Mapped Roles:* Mapped roles are roles in which the concept provided as the role filler can be used to select instances of the base concept from a source. For example, instances of the concept `Protein which hasOrganism-Classification Species:guppy` can be retrieved from Swiss-Prot by retrieving entries with *guppy* in their *organism species* field. This SSM entry could be used to retrieve Swiss-Prot entries using a function call such as:

```
\p <- get-sp-entries-by-os("guppy")
```

4. *Filters:* When instances of a concept have been retrieved, for example, by iteration, other criteria in the query may be used to discard some of the instances. For example, given an instance of `Protein` in the query `Protein which hasFunction Hydrolase`, the instance of `Protein` must be checked to see if it `hasFunction Hydrolase`. The relevant SSM entry could be used to generate code that tests a Swiss-Prot record for the function hydrolase using a function call such as:

```
check-sp-entry-for-hydrolase(p)
```

where p is a variable previously bound to a `protein_record`.

The filters entries in the SSM are used to select values with the required characteristics at the client (i.e., values are retrieved from sources and then checked to see if they meet the needs of the query). In general, it is desirable to have the sources retrieve only values that are relevant. Mapped roles provide one way of sending filters to the sources to be applied as early as possible in the retrieval process. Unfortunately, at the time of writing, many sources did not offer query interfaces that allowed all filtering to be carried out early in the query process. This left much client-side filtering to take place.

7.3.2   The Query Planner

GRAIL queries are declarative, in that the meaning of a query is not dependent on the order of evaluation of its components. As a result, the TAMBIS system, and not the user, must take responsibility for identifying an efficient evaluation order for the components of a GRAIL query. This section describes how GRAIL queries are represented internally for the purposes of optimization and how this internal representation is generated.

GRAIL queries are intrinsically nested structures. The query internal form (QIF) used in TAMBIS can be seen as an un-nested representation of the original GRAIL query. This representation has been developed to allow easier reordering of the components of a query in the planner. The QIF is a list of query components, an example of which is given in Figure 7.13 for the running example GRAIL query:

```
Motif which isComponentOf
    Protein which hasOrganismClassification Species:guppy
```

The query is represented by two query components, one representing the Motif and the other representing the Protein. Each of the components stores the name of the base concept, a list of the criteria from the query, the name of the CPL variable used to hold values retrieved from sources, and details of the technique identified by the planner for retrieving instances of the concept and of the

$<$ *name* : Motif
  theCriteria :
    $<$ *theCriterion* : isComponentOf
      relatedComponent : component of protein-1
      userValue : "" $>$
  theVariable : motif-1
  theTechnique : ""
  theFetchCriterion : null $>$

$<$ *name* : Protein
  theCriteria :
    $<$ *theCriterion* : hasOrganismClassification Species
      relatedComponent : null
      userValue : "'guppy' $>$
  theVariable : Protein-1
  theTechnique : ""
  theFetchCriterion : null $>$

7.13
FIGURE
    QIF for example query.

```
input: query: List of QueryComponent

finalPlan: List of QueryComponent
while query <> [] do
    bestQC := findBest(query)
    finalPlan := finalPlan ++ bestQC
    query := query -- bestQC
end
return finalPlan
```

7.14   The optimization algorithm.

FIGURE

criteria used during retrieval. The values for *theTechnique* and *theFetchCriterion* are identified during planning. Generation of the QIF from a GRAIL query is straightforward and is carried out in a single pass over the query.

Given the QIF for a query, a search algorithm seeks to identify efficient ways to evaluate the query given the functions available in the SSM. The search algorithm exploits the augmentation heuristic [14], which was selected because it is straightforward to implement and provides a reasonable tradeoff between cost of optimization and quality of plan generated. The algorithm is given in Figure 7.14. The basic strategy is to generate a plan as an ordered list of query components in which the first component in the list is predicted to be the least costly component to evaluate from scratch, and the subsequent components are the least costly to evaluate given what has previously been evaluated.

The optimization algorithm in Figure 7.14 depends heavily on the definition of the *findBest* function. This function, given a query component, considers a variety of ways in which instances of the component can be retrieved from sources. Thus *findBest* considers the alternative ways of implementing the components of a QIF onto CPL functions, using the entries in the SSM.

For example, the CPL generated for the example query is:

```
{motif-1 |
  \protein-1<-get-sp-entry-by-os("guppy"),
  \motif-1<-do-prosite-scan-by-entry-rec(protein-1)}
```

This query contains two query components, one for Protein and the other for Motif, as illustrated in Figure 7.13. The query component for Protein is chosen for evaluation first, and the SSM entry used to obtain instances of Protein is the role that is the inverse of `hasOrganismClassification`. Thus, the query processor accesses the ontology to find the inverse of `hasOrganismClassification`, which is the role `hasProteins` on `Species`. This role has a roles entry in the SSM, which is associated with the function `get-sp-entry-by-os`. This function, given the name of a `Species`, consults Swiss-Prot to find the proteins from the species. The second query component is evaluated in a similar manner, using the inverse of the role `isComponentOf`.

The output from the planner is a QIF annotated with details of how to retrieve its components. Generating the corresponding CPL program involves a single pass through the QIF. For each QIF component, the code generator writes out the CPL functions identified by the planner and iterates over the component's other criteria, writing out function calls associated with roles and filters as required.

7.3.3 The Wrappers

The distribution and heterogeneity within bioinformatics resources means that many applications need to employ *wrappers*. Wrappers include external resources into a system that enable the resource to adopt the same operating paradigms as the host system, as well as transform the resource to common syntactic and semantic conventions. Many applications perform this wrapping on an *ad hoc* basis, using the resources available within many programming languages. Kleisli (presented in Chapter 6) is one of the few systems to offer wrapper services together with a query language that is flexible enough to cope with bioinformatics resources.

The output from the TAMBIS system is a query plan written in CPL using a modified version of the BioKleisli library of biological database wrappers [15]. An example CPL query, which *"retrieves all motifs in guppy proteins,"* is as follows:

```
{m |
  \p<-get-sp-entry-by-os("guppy"),
  \m<-do-prosite-scan-by-entry-rec(p) }
```

In the query, the part before the | is the projection expression, which, in this case, indicates that only the motifs m are of interest. The two function calls in the body of the query to the right of the | are generators, which retrieve values from distinct, wrapped sources. The first line in the query body indicates that the new variable p is to be bound to each of the values that result from the evaluation of the function `get-sp-entry-by-os` with the parameter guppy. The function name can be read as *get Swiss-Prot entry by organism species*, although this is just a name—the structure of the name is not significant in itself. The second function call binds the variable m to each of the motifs of the proteins bound to p. The function name can be read as *scan the prosite database for motifs in the given protein record.*

The CPL system is supplied with function libraries that provide access to a range of bioinformatics sources of different types (e.g., databases, analysis tools [15]). TAMBIS uses these libraries and a number developed to provide a function-based view of the sources. The public release of TAMBIS 1.0 accessed five sources and used a total of approximately 300 CPL functions.

CPL can be seen as providing syntactically consistent, but not source transparent, access to the sources, and thus, CPL can be viewed as a wrapping mechanism tightly coupled with convenient language facilities for accumulating and transmitting results from different sources.

Remarks on Handling Syntactic and Semantic Heterogeneity

The heterogeneity in the bioinformatics resources is handled within this wrapper layer and the SSM. The wrapper layer irons out much of the structural or syntactic heterogeneity, providing a consistent call interface in terms of level of abstraction and services to each of the resources. For example, all CPL functions return sets of data, regardless of the number of instances returned. This means only one operator ever needs to be used to manipulate the results of a query. Any heterogeneity in encoding, such as representation of amino acid sequences, can also be dealt with at this level.

The wrapper layer also gives an opportunity for standardization of naming conventions for services available in the resources, though this is of no consequence to users, except that they may find the CPL query plan useful as a quality check on the task TAMBIS is performing.

The SSM affords the main opportunity for the reconciliation of semantic heterogeneity. The ontology gives the user a global schema against which to form queries. The SSM allows terms seen in this global schema to be mapped to the values used in the various resources. For instance, the concept PhosphorylationSite corresponds to the motif entry PS00001 in the PROSITE databank. Similarly, the concept Kinase maps to node 2.7.*.* in the ENZYME databank, but to the term kinase in the Swiss-Prot databank. The SSM can match filler mappings to the appropriate function mapping via the databank attribute in SSM objects. In this manner, the terms in the ontology may be mapped to different terms appearing in the resources.

7.4 RELATED WORK

7.4.1 Information Integration in Bioinformatics

The difficulties associated with obtaining effective access to multiple biological information resources have long been recognized, and several different approaches have been proposed, making use of widely varying underlying technologies.

Probably the most widely used source integration environment for bioinformatics resources is the Sequence Retrieval Service (SRS) [16] (presented in Chapter 5). SRS is a system designed to integrate flat file databanks, which are the most common data storage form used for bioinformatics resources. SRS has its own proprietary data description and processing language. This is used to parse the flat file entries and create indices over fields and their contents. SRS has a query language for selecting entries or part of entries via Boolean combinations of indexed fields and their values. The language contains operators that can take advantage of the heavy cross-linking between different databanks. SRS is usually accessed via a Web-based interface behind which the construction of queries is hidden. The Web interface also offers supplementary analyses such as similarity and pattern scans over protein or nucleic acid sequences. SRS makes no attempt to reconcile any semantic heterogeneity between the different resources during query execution. Once results have been retrieved, the user can follow hyperlinks between entries and much use is made of this query by navigation style.

Although SRS is successful at providing navigational access between diverse resources, it provides limited facilities to support querying or programming over diverse sources. Several proposals have been made in these directions. In terms of query-oriented access, Kleisli [15] provides both a query language for ranging over data types described using a rich hierarchical data model and a collection of wrappers (known in Kleisli as *drivers*) for accessing biological resources. However, Kleisli has no global schema providing a model of the available data and thus can be seen as providing lower-level access to biological resources than TAMBIS. In fact, as already described, TAMBIS generates Kleisli programs as output. Another query-oriented approach is provided by the Object Protocol Model (OPM) [17], in which queries can be written over an object-oriented global model using an object query language, and tools have been developed to assist in the creation of OPM views over heterogeneous sources. The main factor that differentiates TAMBIS from OPM, from a users' point of view, is that in TAMBIS queries are constructed over an ontology rather than over an object model. The impact of the ontology and its reasoning services on query building in TAMBIS has been discussed in Section 7.2. The ontology shields the user from the query language used, the heterogeneity of the resources, and any demand for knowledge of the resources. Such transparency may not be to all users' tastes and more intricate queries or programs could be hand-crafted in systems such as Kleisli or OPM. Other proposals describing query-based access from object models to biological data include ISYS [18], DiscoveryLink [19] and P/FDM [20]. Kleisli, DiscoveryLink, and P/FDM are respectively presented in Chapters 6, 11, and 9.

Considerable attention has been given in bioinformatics to wrapping sources, thereby providing syntactically consistent access from programming languages to diverse resources. The bioPerl initiative[1] offers a collection of Perl modules that provide access to computational techniques and data commonly found within bioinformatics resources. In the early stages of the first TAMBIS version, however, there was much interest in using the Common Object Request Broker Architecture (CORBA) to wrap bioinformatics resources [21]. CORBA allows development of object views of heterogeneous and distributed resources, regardless of their host platform, operating system, or storage paradigm. The use of CORBA within bioinformatics is promoted by the Life Sciences Research (LSR) group of the Object Management Group.[2] The LSR aims to promote standard descriptions of object interfaces that enable interoperation between distributed bioinformatics resources. Among others, the European Bioinformatics Institute has provided CORBA servers for some of their databases [22]. Recently, access to SRS [16] has been provided through CORBA [23]. This service allows objects representing databank entries to be retrieved through the SRS query language. This should allow remote access to a large number of databanks and analysis programs, along with a rudimentary query facility. TAMBIS has a very different emphasis from the middleware approaches in that interactive user access is the main emphasis and in that individual sources are essentially hidden from the user in TAMBIS.

Unfortunately, the required large number of consistent, CORBA wrapped sources did not arrive to be taken advantage of by TAMBIS. The ability to download a description of a service's interface and automatically generate a client that could act as a wrapper was desirable, but not delivered. Many providers balked at the effort needed to provide a CORBA solution to delivering services. Simple Object Protocol Servers (SOAP) servers and Web services[3] offer a lighter weight solution to delivering bioinformatics services. A SOAP server for a resource is relatively cheap to set up because an object model does not have to be designed

---

1. Information about the bioPerl initiative is available at *http://bioperl.org.*

2. Go to *http://www.omg.org/lsr* for information on the Life Sciences Research effort of the Object Management Group.

3. The Simple Object Protocol Servers (SOAP) and Web service protocols are available at *http://www.w3c.org/soap.*

and implemented, as in CORBA. The operations available through that server can be described in the Web services description language (WSDL),[4] and this description can be compiled into a client for the SOAP server. The idea is much the same as that for CORBA, but as a lighter weight solution it relies on simple message passing, not on a heavyweight object approach. These services transfer their data in extensible markup language (XML) and thus can take advantage of widely adopted XML data formats such as the biopolymer markup language [24] and the Bioinformatics Sequence Markup Language (BSML).[5] XML is also seen as the data format of choice by the Interoperable Informatics Infrastructure Consortium (I3C),[6] which aims to promote standards for protocols and exchange formats. The distributed annotation system (DAS)[7] uses many of these ideas to manage sequence annotations distributed around the network, and delivered by SOAP servers providing an XML description of sequence annotations that allows many annotators to form an integrated, yet varied, view on the biological sequence. These technologies offer a middleware solution to the integration of bioinformatics resources. Vital though such technologies undoubtedly are, they can be seen as plumbing resources together. Choice of resources, locating those resources, knowing how to reconcile their view of the data, and the order in which to use them is still left up to the user of these technologies. TAMBIS, on the other hand, sits upon these middleware technologies and uses the ontology to offer full transparency in query management to the user.

7.4.2   Knowledge Based Information Integration

TAMBIS is one of several systems that uses a knowledge base as a central component in information integration; although, it is the first such system to be used in bioinformatics. A survey of knowledge-based information integration is given in Paton et al.'s article in Information and Software Technology [25].

In common with single interface to multiple sources (SIMS) [26], Information Manifold [27] and Observer [28], TAMBIS uses a description logic [7] to describe the concepts over which queries are to be expressed. A description logic is a modeling notation that supports reasoning over descriptions of concepts and their relationships. Two principal approaches are used in information integration systems to relate concepts in a global schema to the schemas of individual sources, namely *global as view* and *local as view* [29]. In the former, the global schema is defined as

---

4. For more information about the WSDL, refer to *http://www.w3.org/TR/wsdl*.
5. The BSML is available at *http://www.bsml.org*.
6. Refer to *http://www.i3c.org* for information about I3C.
7. Go to *http://www.biodas.org* for information on biological DAS.

a view over the constructs in the schemas of the individual sources; in the latter the constructs in the schemas of the individual sources are defined as a view of those in the global schema. SIMS and Observer essentially use *global as view* techniques for processing queries, whereas Information Manifold is *local as view*. TAMBIS follows the *global as view* approach, but it generally differs from other such approaches in that very few assumptions are made of the query processing capabilities of the individual sources. In fact, as is generally true in bioinformatics, TAMBIS assumes that individual sources lack declarative query interfaces and instead provide rather limited call interfaces, supporting tasks such as iterating through the data items of a particular type or retrieving all data items with a given value for a particular attribute.

A further important feature of TAMBIS is that it supports a distinctive user interface driven from the ontology, which guides the user through the query formulation process in a way that makes it difficult to construct biologically meaningless queries. Other knowledge-based information integration systems lack such sophisticated query formulation interfaces.

7.4.3   Biological Ontologies

The number of ontologies used in bioinformatics applications is still quite small, but it is growing. However, where ontologies have been used, they span a wide range of purposes, subject areas, and representation styles [5]. The uses of bio-ontologies fall into two distinct areas: database schema definition (e.g., EcoCyc and RiboWeb) and annotation and communication (e.g., GO and OMB). The TAMBIS ontology adds a third use, ontology-based search and query formulation, to this list. The version of TAMBIS described in this chapter was the first ontology solution based on Description Logic of its type in the bioinformatics arena.

RiboWeb [30] is an ontology of ribosome structure, components, and experimental methods used to drive a Web interface that supports the analysis of ribosomal data. The ontology acts as a schema, driving the acquisition of instances that create the knowledge base. The knowledge held in the ontology also drives the analysis of new data, guiding the user as to which analysis methods are appropriate for the data in hand and indicating results that contradict current knowledge.

EcoCyc [31] uses an ontology to create an encyclopedia of *E .coli* metabolism, regulation, and signal transduction. As with RiboWeb, this ontology acts as a schema for the knowledge base, capturing the domain knowledge with high fidelity. Both these systems use a frame-based knowledge representation language in which a frame represents a concept and slots within frames represent attributes or roles and their fillers. Such representations can be expressive, hence the richness of the models.

The Ontology for Molecular Biology (OMB) [32] provides a framework for describing computational methods, database representations, and core molecular biological concepts. The OMB is aimed at providing a reference ontology to improve community-wide communication. Data resources would use the OMB to define their classes, relationships, and terms. The OMB uses an object-like structure with an *is a kind of* hierarchy and large use of other relationship types.

The Gene Ontology (GO) [33] is a structured, controlled vocabulary used to annotate gene products for their function, ultimate cellular location, and the processes in which they take part. GO is used in several genomic databases and thus adds consistency across these resources. As a consequence, querying these resources becomes more reliable. The ontology has a simple structure, relying on an *is-a-kind-of* hierarchy and a sparse partonomy to relate natural language phrases. The ImMunoGeneTics ontology holds terminology on the areas of immunoglobulins and their genetics. Again, this acts as a controlled vocabulary, but it has a less well-defined structure than GO and appears more like a glossary with inter-related entries.

Although all these resources can be termed *ontologies*, they fall into a spectrum of expressivity and formality. The frame-based systems are relatively rich, expressive, and formal, whereas the phrase-based terminologies are simpler and less expressive. The first TAMBIS ontology was the first bio-ontology to use a description logic as its representation and, as a consequence, has a more well-defined semantics than the other representations used in bio-ontologies. In contrast to the narrow range of ontology use, however, the scope and detail of the content of these ontologies varies enormously. The ImMunoGeneTics, RiboWeb, and EcoCyc ontologies are highly detailed but highly specialized to one subject area, leaving only some commonality for core areas such as *gene* and *protein*. The OMB is wide ranging and high level, whereas GO lacks any high-level conceptualization but becomes very detailed, starting its conceptualization where the OMB finishes. As has been seen, the TAMBIS ontology is broad in its conceptualization, using an upper-level ontology in which to place these concepts. The ontology is relatively shallow, but detail may be added as the user dynamically creates new concepts as compositions of pre-existing concepts and has them automatically checked and classified by the ontology's reasoning service.

7.5 CURRENT AND FUTURE DEVELOPMENTS IN TAMBIS

The first version of TAMBIS was successful. It is possible to use an ontology describing a complex domain, such as molecular biology and bioinformatics, and use it to give the illusion of a common query interface to multiple, diverse, and heterogeneous information sources. The ontology drove a query formulation interface that allowed users to create complex queries over those multiple sources—queries that would usually need a program written by a trained bioinformatician. Usage of TAMBIS did, however, reveal some issues that needed to be addressed in further work.

Total transparency is not always desirable. The level of transparency offered by the first version of TAMBIS was appreciated by less skilled users who were happy to have decisions on which resources to use taken out of their hands. However, some users, usually those well versed in using bioinformatics resources, wished to express preferences about which resources to use, given that some sources may be more trusted than others. As the number of resources available within TAMBIS increases, such preferences will be able to be expressed. In addition, users may wish to record when and where data they retrieved arose [34]. In version one, the CPL query plan implicitly recorded some such information in the names of functions used in the query plan. It is more desirable, however, to record query provenance directly and explicitly.

The TAMBIS user survey [2] revealed that user intervention during query execution and inspection of intermediate results was desirable. Users often wish to monitor the progress of a complex, multi-source query, inspecting results to evaluate validity of the query so far and to edit data before it proceeds into subsequent parts of the query. Code for managing the execution of a query will have to be included into the code generated by the updated query processor in TAMBIS.

In the first version of TAMBIS, both the SSM and wrappers were hand-crafted. It is an aim of future versions of TAMBIS to build tools to support this process. Concepts in the ontology have to be related to methods or functions in the wrappers and information about argument and return types, and costs recorded. The ontology itself, through the ontology server, could drive such a tool and also help to check that the content of the ontology is covered within the SSM.

In the new version of TAMBIS, the TAMBIS ontology has been remodeled using the DL language DAML+OIL[8] and classified using the FaCT reasoner [35], which is considerably more powerful than GRAIL used in the original TAMBIS ontology.[9] This allows the biological domain to be described more precisely in the ontology and allows more precise questions to be asked by the users. In addition, the reasoning services of the DL are used extensively during query processing to support semantic query optimization based on axioms within the ontology.

---

8. Information on DAML+OIL can be found at *http://www.daml.org*.

9. Versions of this ontology represented in DAML+OIL may be found at *http://img.cs.man.ac.uk/stevens/tambis-oil.html*.

The query processing in version one of TAMBIS was somewhat limited, and these limitations include:

1. The ontology is represented using a relatively inexpressive DL in which certain features of the biological domain are difficult to express.

2. The mapping between concepts in the ontology and collections in the sources is quite restrictive. For example, TAMBIS did not allow multiple sources for the same kind of data (e.g., both Swiss-Prot and the Protein Information Reserve (PIR) as protein sources.

3. Although queries are optimized [13], there is no semantic query optimization making use of axioms from the ontology.

In the second version of TAMBIS, an object-oriented wrapper layer has been adopted to replace that provided by CPL and BioKleisli. Instead of a CPL query plan, a Java program is written. The use of an object-oriented wrapper layer will make TAMBIS compatible with mainstream middleware proposals such as that standardized by the Object Management Group (OMG), which in turn is associated with an important standardization activity in bioinformatics.[10]

7.5.1 Summary

This chapter has provided an overview of the first TAMBIS system for querying distributed bioinformatics sources. The key contributions of TAMBIS are:

1. It is the first ontology-based information integration system to be used in bioinformatics. Although ontologies are becoming important in bioinformatics for annotating databases [33] and for managing complex information resources [30], TAMBIS is the first project to use ontologies to support the important task of integrating bioinformatics resources.

2. TAMBIS is centered on the first description logic-based ontology in bioinformatics. Other ontologies in bioinformatics have made use of frame-based representations or structured terminologies, but they are not amenable to subsumption reasoning as in TAMBIS.

3. The user interface in TAMBIS is driven directly from the ontology, and as such, it both guides the user in constructing well formed requests and detects when biologically nonsensical questions have been asked. Other knowledge-based

---

10. Information about the effort of standardization of bioinformatics by the OMG is available at *http://www.omg.org/homepages/lsr/*.

information integration systems have paid less attention to user interaction issues.

4. The TAMBIS query processor has been integrated with existing wrapping software, allowing re-use of established middleware techniques and existing wrappers. The query processor makes minimal assumptions on the query interfaces made available by sources, reflecting the limited public query interfaces generally available in bioinformatics.

TAMBIS seeks to provide correct answers to precisely formed queries. Queries can be expressed precisely, at a level of detail corresponding to that of the underlying resources, by using the ontology to constrain what it is valid to ask. Answers should be correct because the sources and services model makes explicit how queries expressed over the ontology can be answered using the available sources. However, such quality of service is achieved at some cost; the development of ontologies that describe a domain is a skilled and time-consuming process (the 1800-concept TAMBIS ontology took 2 person-years to write), and incorporating a wrapped source into the SSM is itself a manual and time-consuming task. However, these two tasks involve (1) describing what it is valid to ask of a collection of bioinformatics sources and (2) describing how to obtain answers from a collection of sources. Although the developers and maintainers of a TAMBIS installation must undertake these tasks, the users of the TAMBIS system need not, and thus they can benefit from the knowledge encoded in the ontology and in the SSM.

ACKNOWLEDGMENTS

This work is funded by AstraZeneca pharmaceuticals, the BBSRC/EPSRC Bioinformatics Initiative (grant number BIF/05344), and the EPSRC Distributed Information Management Initiative (grant number GR/M76607), whose support we are pleased to acknowledge. We are also grateful to Alex Jacoby and Martin Peim for their contributions to the implementation of the TAMBIS system.

REFERENCES

[1] V. M. Markowitz and O. Ritter. "Characterizing Heterogeneous Molecular Biology Database Systems." *Journal of Computational Biology* 2, no. 4 (1995): 547–556.

[2] R. D. Stevens, C. A. Goble, P. Baker, et al. "A Classification of Tasks in Bioinformatics." *Bioinformatics* 17, no. 2 (2001): 180–188.

[3]   C. A. Goble, R. Stevens, G. Ng, et al. "Transparent Access to Multiple Bioinformatics Information Sources." *IBM Systems Journal* 40, no. 2 (2001): 532–552.

[4]   P. Buneman, S. B. Davidson, K. Hart, et al. "A Data Transformation System for Biological Data Sources." In *Proceedings of the 21st International Conference on Very Large Data Bases (VLDB)*, 158–169. San Francisco: Morgan Kaufmann, 1995.

[5]   R. Stevens, C. A. Goble, and S. Bechhofer. "Ontology-Based Knowledge Representation for Bioinformatics." *Briefings in Bioinformatics* 1, no. 4 (November 2000): 398–416.

[6]   P. G. Baker, C. A. Goble, S. Bechhofer, et al. "An Ontology for Bioinformatics Applications." *Bioinformatics* 15, no. 6 (1999): 510–520.

[7]   A. Borgida. "Description Logics in Data Management." *IEEE Transactions on Knowledge and Data Engineering* 7, no. 5 (1995): 785–798.

[8]   G. A. Ringland and D. A. Duce. *Approaches to Knowledge Representation: An Introduction*. New York: John Wiley, 1988.

[9]   F. Baader, D. McGuinness, D. Nardi, et al, eds. *The Description Logic Handbook: Theory, Implementation and Applications*. Cambridge, UK: Cambridge University Press, 2003.

[10] A. L. Rector, S. K. Bechhofer, C. A. Goble, et al. "The GRAIL Concept Modelling Language for Medical Terminology." *Artificial Intelligence in Medicine* 9, no. 2 (1997): 139–171.

[11] F. M. Donini, M. Lenzerini, D. Nardi, et al. "Reasoning in Description Logics." In *Foundations of Knowledge Representation*, 191–236. Stanford, CA: Center for the Study of Language and Information (CSLI) Publications, 1996.

[12] S. Bechhofer, R. Stevens, G. Ng, et al. "Guiding the User: An Ontology Driven Interface." In *Proceedings of User Interfaces to Data Intensive Systems (UIDIS99)*, edited by N. W. Paton and T. Griffiths, 158–161. New York: IEEE Press, 1999.

[13] N. W. Paton, R. Stevens, P. Baker, et al. "Query Processing in the TAMBIS Bioinformatics Source Integration System." In *Proceedings of the 11th International Conference on Scientific and Statistical Database Management (SSDBM)*, 138–147. New York: IEEE Press, 1999.

[14] A. N. Swami. "Optimization of Large Join Queries." In *Proceedings of the 1989 ACM SIGMOD International Conference on Managing Error*, 367–376. New York: ACM Press, 1989.

[15] S. B. Davidson, C. Overton, V. Tannen, et al. "BioKleisli: A Digital Library for Biomedical Researchers." *Journal of Digital Libraries* 1, no. 1 (November 1997): 36–53.

[16] T. Ezold, A. Ulyanov, D. Nardi, et al. "SRS: Information Retrieval System for Molecular Biology Data Banks." *Methods in Enzymology* 266 (1996): 114–128.

[17] I-M. A. Chen, A. S. Kosky, V. M. Markowitz, et al. "Constructing and Maintaining Scientific Database Views in the Framework of the Object Protocol Model." In *Proceedings of the 9th International Conference on SSDBM*, 237–248. New York: IEEE Press, 1997.

[18] A. C. Siepel, A. N. Tolopko, A. D. Farmer, et al. "An Integration Platform for Heterogeneous Bioinformatics Software components." *IBM Systems Journal* 40, no. 2 (2001): 570–591.

[19] L. M. Haas, P. Kodali, J. E. Rice, et al. "Integrating Life Sciences Data with a Little Garlic." In *Proceedings of the International Symposium on Bio-Informatics and Biomedical Engineering (BIBE)*, 5–12. New York: IEEE Press, 2000.

[20] G. J. L. Kemp, N. Angelopoulog, and P. M. D. Gray. "A Schema-Based Approach to Building a Bioinformatics Database Federation." In *Proceedings of the International Symposium on Bio-Informatics and Biomedical Engineering (BIBE)*, 13–20. New York: IEEE Press, 2000.

[21] R. Stevens and C. Miller. "Wrapping and Interoperating Bioinformatics Resources Using CORBA." *Briefings in Bioinformatics* 1, no. 1 (2000): 9–21.

[22] P. Rodriguez-Tomé, C. Helgesen, P. Lijnzaad, et al. "A CORBA Server for the Radiation Hybrid DataBase." In *Proceedings of the Fifth International Conference on Intelligent systems for Molecular Biology*, 250–253. Menlo Park, CA: AAAI Press, 1997.

[23] T. Coupaye. "Wrapping SRS With CORBA: From Textual Data to Distributed Objects." *Bioinformatics* 15, no. 4 (1999): 333–338.

[24] D. Fenyo. "The Biopolymer Markup Language." *Bioinformatics* 15, no. 4 (1999): 339–340.

[25] N. W. Paton, C. A. Goble, and S. Bechhofer. "Knowledge Based Information Integration Systems." *Information and Software Technology* 42, no. 5 (2000): 299–312.

[26] Y. Arens, C. A. Knoblock, and W-M. Shen. "Query Reformulation for Dynamic Information Integration." *Journal of Intelligent Information Systems* 6, no. 2-3 (1996): 99–130.

[27] A. Y. Levy, D. Srivastava, and T. Kirk. "Data Model and Query Evaluation in Global Information Systems." *Journal of Intelligent Information Systems* 5, no. 2 (1995): 121–143.

[28] E. Mena, A. Illarramendi, V. Kashyap, et al. "Observer: An Approach for Query Processing in Global Information Systems Based on Interoperation Across Pre-Existing Ontologies." *Distributed and Parallel Databases* 8, no. 2 (2000): 223–271.

[29] J. D. Ullman. "Information Integration Using Logical Views." In *Proceedings of ICDT '97: 6th International Conference on Database Theory*, 19–40. Heidelberg, Germany: Springer-Verlag, 1997.

[30] R. Altman, M. Bada, X. J. Chai, et al. "RiboWeb: An Ontology-Based System for Collaborative Molecular Biology." *IEEE Intelligent Systems* 14, no. 5 (1999): 68–76.

[31] P. Karp, M. Riley, S. Paley, et al. "EcoCyc: Electronic Encyclopedia of *E. coli* Genes and Metabolism." *Nucleic Acids Research* 27, no. 1 (1999): 55–58.

[32] S. Schulze-Kremer. "Ontologies for Molecular Biology." In *Proceedings of the Third Pacific Symposium on Biocomputing*, 693–704. Singapore: World Scientific, 1998.

[33] M. Ashburner, C. A. Ball, J. A. Blake, et al. "Gene Ontology: Tool for the Unification of Biology." *Nature Genetics* 25, no. 1 (2000): 25–29.

[34] P. Buneman, S. Khanna, and W-C. Tan. "Data Provenance: Some Basic Issues." *TSTTCS 2000: 29th Conference on Foundations of Software Technology and Theoretical Computer Science, New Dehli, India. Lecture Notes in Computer Science, vol. 1974*, 87–93. Heidelberg, Germany: Springer-Verlag, 2000.

[35] I. Horrocks. "Using an Expressive Description Logic: Fact or Fiction." In *Principles of Knowledge Representation and Reasoning: Proceedings of the Sixth International Conference (KR '98)*, edited by A. G. Cohn, L. K. Schubert, and S. C. Shapiro, 636–647. San Francisco: Morgan Kaufmann, 1998.

The Information Integration System K2

Val Tannen, Susan B. Davidson, and Scott Harker

In 1993, the invitational Department of Energy (DOE) workshop on genome informatics published a report that claimed that until all sequence data is gathered in a standard relational database, none of the queries in the appendix to the report could be answered (see Figure 8.1 for a listing of the queries) [1]. While the motivation for the statement was largely political, the gauntlet had been laid in plain view for database researchers: The data to answer the queries in the appendix were (by and large) available, but they were stored in a number of physically distributed databases. The databases represented their data in a variety of formats using different query interfaces. The challenge was, therefore, one of integrating heterogeneous, distributed databases and software programs in which the type of data was complex, extending well beyond the capabilities of relational technology.

As an example of the type of genomic data that is available online, consider the EMBL-format Swiss-Prot entry shown in Figure 8.2. Each line begins with a two-character code, which indicates the type of data contained in the line. For example, each entry is identified by an accession number (AC) and is timestamped by up to three dates (DT). The create date is mandatory, while the sequence update and annotation update dates only appear if the sequence or annotation has been modified since the entry was created. The sequence (SQ), a list of amino acids, appears at the end of the entry; the rest of the core data includes citation information (bibliographical references, lines beginning with R), taxonomic data (OC), a description of the biological source of the protein, and database references (DR), explicit links to entries in other databases: EMBL (annotated nucleotide sequence database); HSSP (homology derived secondary structure of proteins); Wormpep (predicted proteins from the *Caenorhabditis elegans* genome sequencing project); InterPro, Pfam, PRINTS, PROSITE (databases of protein families and domains, among other things). Annotation information, which is obtained by publications reporting new sequence data, review articles, and external experts, is mainly found in the feature table (FT), keyword lines (KW), and comment lines (CC), which do not appear in this example due to lack of space. Note that the bibliographical

The following "unanswerable queries" were taken from Appendix 1 of the 1993 Invitational DOE Workshop on Genome Informatics report, available at *http://www.ornl.gov/hgmis/publicat/miscpubs/bioinfo/contents.html*. Rather than saying that all sequence databases must be relationalized, the wording of the report has now been modified to say "until a fully atomized sequence database is available (i.e., no data stored in ASCII text fields), none of the queries in this appendix can be answered."

1. Return all sequences that map "close" to marker M on human chromosome 19, are putative members of the olfactory receptor family, and have been mapped on a contig map of the region; return also the contig descriptions. (This is nominally a link between GenBank, GDB, and LLNL's databases.)

2. Return all genomic sequences for which alu elements are located internal to a gene domain.

3. Return the map location, where known, of all alu elements having homology greater than "h" with the alu sequence "S".

4. Return all human gene sequences, with annotation information, for which a putative functional homologue has been identified in a nonvertebrate organism; return also the GenBank accession number of the homologue sequence where available.

5. Return all mammalian gene sequences for proteins identified as being involved in intracellular signal transduction; return annotation information and literature citations.

6. Return any annotation added to my sequence number #### since I last updated it.

7. Return the genes for zinc-finger proteins on chromosome 19 that have been sequenced. (Note that answering this requires either query by sequence similarity or uniformity of nomenclature.)

8. Return the number and a list of the distinct human genes that have been sequenced.

9. Return all the human contigs greater than 150 kb.

10. Return all sequences, for which at least two sequence variants are known, from regions of the genome within +/- one chromosome band of DS14###.

11. Return all publications from the last 2 years about my favorite gene, accession number ####.

12. Return all G1/S serine/threonine kinase genes (and their translated proteins) that are known (experimentally) or are thought (by similarity) also to exhibit tyrosine phosphorylation activity. Keep clear the distinction in the output.

8.1

FIGURE

The 1993 DOE Report's "unanswerable" queries.

references are *nested* structures; there are two references, and the RP (Reference Position), RC (Reference Comment), RA (Reference Author), and RL (Reference Location) fields are specific to each reference. Similarly, the FT (Feature Table) is a nested structure in which each line contains a start and end position (e.g., 14 to 21), a type of feature (e.g., NP_BIND), and a description. The entry is designed to be read easily by a human being and structured enough to be machine parsed. However, several lines still contain a certain amount of structure that could be separated out during parsing. For example, the author list is a string, which could be parsed into a list of strings so as to be able to index into the individual authors. Similarly, the taxonomic data is also a string spread over several lines and could again be parsed into a list.

```
ID   EF1A_CAEEL      STANDARD;      PRT;    463 AA.
AC   P53013;
DT   01-OCT-1996 (Rel. 34, Created)
DT   01-OCT-1996 (Rel. 34, Last sequence update)
DT   15-DEC-1998 (Rel. 37, Last annotation update)
DE   ELONGATION FACTOR 1-ALPHA (EF-1-ALPHA).
GN   (EFT-3 OR F31E3.5) AND R03G5.1.
OS   Caenorhabditis elegans.
OC   Eukaryota; Metazoa; Nematoda; Chromadorea; Rhabditida; Rhabditoidea;
OC   Rhabditidae; Peloderinae; Caenorhabditis.
RN   [1]
RP   SEQUENCE FROM N.A.(EFT-3).
RC   STRAIN=BRISTOL N2;
RA   Favello A.;
RL   Submitted (NOV-1995) to the EMBL/GenBank/DDBJ databases.
RN   [2]
RP   SEQUENCE FROM N.A. (R03G5.1).
RC   STRAIN=BRISTOL N2;
RA   Waterston R.;
RL   Submitted (MAR-1996) to the EMBL/GenBank/DDBJ databases.
DR   EMBL; U51994; AAA96068.1; -.
DR   EMBL; U40935; AAA81688.1; -.
DR   HSSP; P07157; 1AIP.
DR   WORMPEP; F31E3.5; CE01270.
DR   WORMPEP; R03G5.1; CE01270.
DR   INTERPRO; IPR000795; -.
DR   PFAM; PF00009; GTP_EFTU; 1.
DR   PRINTS; PR00315; ELONGATNFCT.
DR   PROSITE; PS00301; EFACTOR_GTP; 1.
KW   Elongation factor; Protein biosynthesis; GTP-binding;
KW   Multigene family.
FT   NP_BIND       14      21       GTP (BY SIMILARITY).
FT   NP_BIND       91      95       GTP (BY SIMILARITY).
FT   NP_BIND      153     156       GTP (BY SIMILARITY).
SQ   SEQUENCE   463 AA;  50668 MW;  12544AF1F17E15B7 CRC64;
             MGKEKVHINI VVIGHVDSGK STTTGHLIYK CGGIDKRTIE KFEKEAQEMG KGSFKYAWVL
             DKLKAERERG ITIDIALWKF ETAKYYITII DAPGHRDFIK NMITGTSQAD CAVLVVACGT
             GEFEAGISKN GQTREHALLA QTLGVKQLIV ACNKMDSTEP PFSEARFTEI TNEVSGFIKK
             IGYNPKAVPF VPISGFNGDN MLEVSSNMPW FKGWAVERKE GNASGKTLLE ALDSIIPPQR
             PTDRPLRLPL QDVYKIGGIG TVPVGRVETG IIKPGMVVTF APQNVTTEVK SVEMHHESLP
             EAVPGDNVGF NVKNVSVKDI RRGSVCSDSK QDPAKEARTF HAQVIIMNHP GQISNGYTPV
             LDCHTAHIAC KFNELKEKVD RRTGKKVEDF PKFLKSGDAG IVELIPTKPL CVESFTDYAP
             LGRFAVRDMR QTVAVGVIKS VEKSDGSSGK VTKSAQKAAP KKK
```

8.2   Sample Swiss-Prot entry.

FIGURE

As shown in Figure 8.2, the type system for genomic data naturally goes beyond the sets of records of relational databases and include sequential data (lists), deeply nested record structures, and union types (variants). As an example of a union type, the format of the RL line depends on the type of publication: An unpublished entry contains a brief comment; a journal citation includes the journal abbreviation, the volume number, the page range, and the year; the format of a book citation includes the set of editor names, the name of the book, an optional volume, the page range, the publisher, city, and year. The structure of this Swiss-Prot entry can be described precisely in a data definition language with sufficiently rich types. Such a description will be shown later on in Figure 8.5.

The database group at the University of Pennsylvania responded to the DOE challenge by developing a *view integration* (or integration on-the-fly) environment. In such an environment, the schemas of a collection of underlying data sources are merged to form a global schema in some common model (e.g., relational, complex value, or object-oriented). Users query this global schema using a high-level query language, such as Structured Query Language (SQL) [2], Object Query Language (OQL) [3], or Collection Programming Language (CPL) [4]; the system then determines what portion of the global query can be answered by which underlying data source, ships local queries off to the underlying data sources, and then combines answers from the underlying data sources to produce an answer to the global query. The initial view-integration environment developed by our group was called Kleisli, and it was designed and implemented by Limsoon Wong. Wong later re-designed and re-implemented the Kleisli system at Singapore's Institute of Systems Science; this new version of Kleisli is described in Chapter 6 of this book. About the same time, other information integration projects were also developed [5, 6, 7], including the system based on the Object Protocol Model (OPM) [8].

K2 vs. Kleisli

K2 is a successor system to Kleisli that was designed and implemented at the University of Pennsylvania by Jonathan Crabtree, Scott Harker, and Val Tannen. Like Kleisli, K2 uses a complex value model of data and is based on the so-called *monad* approach (see Section 8.4). However, the design of K2 also contains a number of new ideas and redirections: First, the model incorporates a notion of *dictionaries*, which allows a natural representation of object-oriented classes [9] as well as Web-based data. Second, the internal language features a new approach to aggregate and collection conversion operations [10]. Third, the syntax of the language follows a mainstream query language for object-oriented databases called OQL [3] rather than the elegant but less familiar comprehension-style syntax originally used in CPL [4] (Kleisli now uses an adapted SQL syntax). Fourth, a separation is made between the mediator (global schema) level and the query level by introducing a mediator definition language for K2, K2MDL. K2MDL combines an Object Definition Language (ODL) specification of the global schema with OQL statements that describe the data mapping. The ability to specify intermediate mediators allows a large integration environment to be created in layers and componentized.

Finally, to improve its portability, K2 is implemented in Java and makes use of several of the standard protocols and application programming interfaces (APIs) that are part of the Java platform,[1] including Remote Method Invocation (RMI)[2] and Java Data Base Connectivity (JDBC).[3] Thus, K2 is *not an extension* of Kleisli, but rather a system implemented from scratch that shares with Kleisli some of its design principles, while featuring a number of distinct developments just outlined.

Overall, the goal of K2 is to provide a generic and flexible view integration environment appropriate for the complex data sources and software systems found throughout genomics, which is portable and appeals to common practices and standards.

8.1   APPROACH

A number of other techniques have also been developed over the past 11 years in response to the DOE challenge, including link-driven federations and warehouses. In a *link-driven federation*, users start by extracting entries of interest at one data source and then hop to other related data sources via Web links that have been explicitly created by the developers of the system. The Sequence Retrieval System (SRS) [11] presented in Chapter 5, LinkDB [12], and GeneCards [13] are examples of this approach. While the federation approach is easy to use, especially for novices, it does not scale well: When a new data source is added to the federation, connections between its entries and entries of all existing federation data sources must be added; this is commonly referred to as the N^2 *problem*. Furthermore, if users are interested in a join between two data sources in the federation, they must manually perform the join by clicking on each entry in the first data source and following all connections to the second data source.[4] In contrast, a join can be expressed in a single high-level query in a view or warehouse integration strategy. In general, the query languages supporting view or warehouse integration approaches are much more powerful and allow arbitrary restructuring of the retrieved data.

A *warehouse* strategy creates a central repository of information and annotations. One such example is the Genomics Unified Schema (GUS) [14], which integrates and adds value to data obtained from GenBank/EMBL/DDBJ, dbEST,

---

1. See *http://www.javasoft.com/j2se/*.

2. See *http://www.javasoft.com/products/rmi-iiop/*.

3. See *http://java.sun.com/products/jdbc/*.

4. A counterexample to this is SRS, in which a linking operator is provided to retrieve linked entries to a set of entries.

and Swiss-Prot (and others) and contains annotated nucleotide (dioxyribonucleic acid [DNA], ribonucleic acid [RNA]), and amino acid (protein) sequences. Note that view integration systems can also be used to create warehouses, which are instantiations of the global schema. The advantage of a warehouse approach over a view integration is one of speed and reliability; because all data are local, delays and failures associated with networks can be avoided. Furthermore, there is greater control over the data. However, a warehouse is not dynamic: Not only must it be kept up-to-date with respect to the underlying data sources, but including a new data source or algorithm is time-consuming. A more extended discussion of the problems and benefits of the link, warehouse, and view integration approaches can be found in articles in the *IBM Systems Journal* and the *Journal of Digital Libraries* [14, 15]. K2 is a system for generating *mediators*. Mediators are middleware components that integrate domain-specific data from multiple sources, reducing and restructuring data to an appropriate *virtual view* [16]. A major benefit of mediation is the scalability and long-term maintenance of the integration systems structure.

Figure 8.3 is an example of how mediators can help the data integration task. In this example, each of the boxes represents a machine on which a copy of K2 is used to provide a mediator for some local, as well as external, data sources

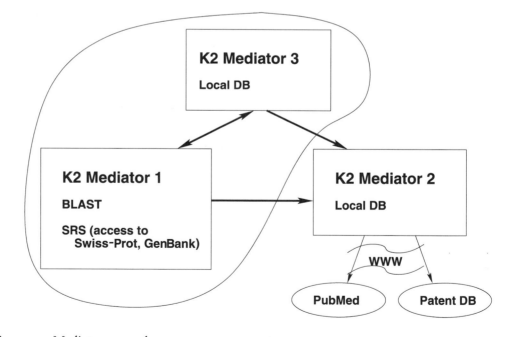

| 8.3 | Mediator example. |

FIGURE

and views. Mediator 1 resides behind a company firewall and was built to integrate data from a local database and local copies of Swiss-Prot and GenBank, accessed through SRS, with the application program BLAST. Mediator 2 was then built outside the firewall to integrate data from some external data sources—PubMed and a patent database, which can only be accessed through a Web interface. An external copy of PubMed was used due to its size and the fact that the most recent version was always needed. Mediator 3 was then built within the company firewall to integrate data from Mediators 1 and 2, and Mediator 1 was enlarged to integrate data from Mediator 3.

K2 (together with Tsimmis [17]) distinguishes itself among approaches based on mediation in that it *generates* mediators starting from a concise, high-level description. This makes K2 especially appropriate for configurations in which many mediators are needed or in which mediators must be frequently changed due to instability in the data sources or in the client needs. Some of the salient features of K2's mediation environment are:

✦ K2 has a *universal* internal data model with an external data exchange format for interoperation with similar components.

✦ It has interfaces based on the Object Data Management Group (ODMG) standard [3] for both data definition and queries.

✦ It integrates nested data, while offering a Java-based interface (JDBC) to relational database systems and an ODMG interface to object-oriented database systems.

✦ It offers a new way to program integration/transformation/mediation in a very high-level declarative language (K2MDL) that extends ODMG.

✦ It has an extensible rule-based and cost-based optimizer.

✦ External or internal decision-support systems can easily be included.

✦ It is written entirely in Java, with corresponding consequences about portability.

The basic functionality of a K2-generated mediator is to implement a data transformation from one or more data sources to one data target. The component contains a high-level (in ODMG/ODL) description of the schemas (for sources and the target) and of the transformation (in K2MDL). From the target's perspective, the mediator offers a view that, in turn, can become a data source for another mediator.

An overview of the K2 architecture is given in Figure 8.4. In this diagram, clients can issue OQL queries or other commands against an integration schema constructed using K2MDL. The queries are then translated to the K2 internal

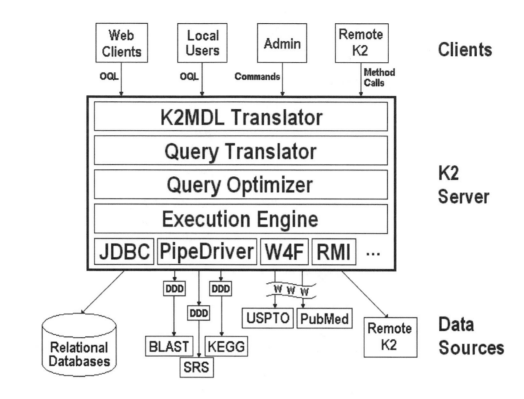

8.4   K2 system architecture.

FIGURE

language using K2MDL and query translators. This internal language expression is then optimized and executed using data drivers to ship sub-queries to external data sources and return results.

The remainder of this chapter walks through the architecture by describing the data model, illustrating K2MDL and OQL, and briefly discussing the internal language, data drivers, query optimization, and user interfaces. The chapter closes with a discussion of scalability and impact.

8.2   DATA MODEL AND LANGUAGES

ODMG was founded by vendors of object-oriented database management systems and is affiliated with the Object Management Group (OMG), who created the Common Object Request Broker Architecture (CORBA). The ODMG standard has two main components: The first is ODL, a data definition language that is used to define data elements. ODL is an extension of CORBA's Interface Definition Language (IDL). The second is OQL, an enhanced SQL92-like language that is used for querying. By building on these standards, K2 leverages the following features:

+ Rich modeling capabilities

+ Seamless interoperability with relational, object-oriented, information retrieval (dictionaries), and electronic data interchange (EDI) formats (e.g., ASN.1)

+ Compatibility with the Universal Modeling Language (UML)

+ Integration with extensible markup language (XML) documents with a given Document Type Definition (DTD)

+ Official bindings to Java, C++, and Smalltalk

+ Industrial support from ODMG members (Ardent, Poet, Object Design)

+ Increasing use in building *ontologies*

K2 uses ODMG's ODL to represent the data sources to be integrated. It turns out that many biological data sources can be described as dictionaries whose keys are simple strings and whose entries are complex values.

A *dictionary* is simply a finite function. Therefore, it has a domain that is a finite set and associates to each element of the domain (called a *key*) a value (called an *entry*). The type of dictionary in ODL is denoted by `dictionary<T1,T2>` where `T1` is the type of the keys and `T2` is the type of the entries. In OQL, the entry in the dictionary, `L`, corresponding to the key, `k`, is denoted by `L[k]`. Because OQL has no syntax for the domain of a dictionary `L`, `dom(L)` is an addition to OQL for this purpose. Note that if `L` has type `dictionary<T1,T2>` then `dom(L)` has type `set<T1>`.

Complex value data are built by arbitrarily nesting records (tuples); collections—such as sets, bags (multisets), and lists; and variants. *Variants* are pieces of data representing tagged alternatives (also known as *tagged unions*). To illustrate complex values (including variants), Figure 8.5 presents an ODL declaration for a class whose objects correspond to (parts of) Swiss-Prot entries. The attribute `Ref` returns complex values obtained by nesting sets, lists, records (`struct` in ODL), and variants, the latter identified by the keyword `choice`.

K2's approach to data integration consists of two stages. In the first stage users specify data transformations between multiple sources and a single target. The target is virtual (unmaterialized) and is, in effect, a new view. The sources

```
class Entry
   (extent Entries)
{
  attribute string ID;
  attribute string AC;
  attribute struct Dates { date Create; date SeqUpdate; date AnnotUpdate; } DT;
  ...
  attribute list<string> OC;
  attribute list<struct {
                        string RP;
                        list<string> RA;
                        string RC;
                        choice { string present; bool absent; } RT;
                        choice {
                           string Unpublished;
                           struct { string JAbbrev;
                                   short Volume;
                                   struct { short from; short to; } Pages;
                                   short Year; } Journal;
                           struct { set<string> Editors;
                                   string Title;
                                   short Volume;
                                   struct { short from; short to; } Pages;
                                   string Publisher;
                                   string City;
                                   short Year; } Book;
                           ... } RL;
                        ... }> Ref;
  ...
  attribute string KW;
  attribute struct { string KeyName; long From; long To; string Desc; } FT;
  attribute string SQ;
}
```

8.5

FIGURE

ODL description of a class of Swiss-Prot entries (partial).

may consist of materialized data or virtual views that have been defined previously through similar data transformations.

In the second stage users formulate queries against the virtual views. OQL is an excellent vehicle for the second stage, but because it does not construct new classes as output, it is not expressive enough for the first stage. Hence, for defining sources-target transformations, K2 uses a new language, (K2MDL), which combines the syntax of ODL and OQL to express high-level specifications of middleware components called mediators, as explained previously. Some examples of K2MDL syntax are in the next section.

The rich type system used in K2 has allowed us to model a large range of practical data sources in a transparent and friendly manner.

8.3   AN EXAMPLE

The K2 approach is illustrated with an example where the target data could be called an *ontology*, that is, a schema agreed upon by a class of users. It shows how a mediator generated by K2 could implement it in terms of standard data sources. Consider the following data description, which is given in ODL syntax:[5]

TARGET DATA DESCRIPTION

```
class Protein
    (extent proteins)
{
  attribute string SwissprotAccession;
  attribute string recommendedName;
  attribute set<string> alternateNames;
  attribute string sequence;
  attribute int seqLength;
  relationship set<Gene> hasSource
      inverse Gene::hasProduct;
}

class Gene
    (extent genes)
{
  struct Range {long from; long to;};
  attribute string name;
  attribute string organism;
  attribute Range location;
  relationship Protein hasProduct
      inverse Protein::hasSource;
}
```

---

5. This is, of course, a very simplified model of proteins and genes, but the intent of this example is to demonstrate K2MDL, not to develop a scientifically viable model.

Data about proteins is recorded using a Swiss-Prot accession number, a recommended name, a set of alternate names, the protein sequence and its length, and a set of references to the genes that code for the protein. Data about genes consists of the name of the gene, the name of the organism from which it comes, the location of the gene in the genome, and a reference to the protein for which it codes. While most of our attributes have simple values, strings, and integers, `alternateNames` is a set of strings and `location` is a record. The schema also specifies that `hasSource` in `Protein` and `hasProduct` in `Gene` are more than just attributes; they form a relationship between the extents of the two classes and are inverses. This means that the two following statements are validated:

+ Given a protein P, for each of the genes G in the set `P.hasSource` it is the case that `G.hasProduct = P`.

+ Given a gene G, it is the case that G belongs to the set `(G.hasProduct).hasSource`.

Now assume that the data about proteins and genes reside in (for illustration purposes) four materialized data sources: Swiss-Prot, Orgs, Genes, and Protein-Synonyms. Swiss-Prot contains some protein data, which can be accessed through an SRS driver that presents an object-oriented schema (i.e., a class). Orgs and Genes contain organism and gene data, respectively, in two relations (in the same or in separate relational databases). The SQL data description is given here for these relations, but in fact K2 uses an equivalent description in ODL syntax, based on the observation that relations are simply sets of records. Finally, a Web-based data source contains protein name synonyms and is modeled as a dictionary.

SOURCE DATA DESCRIPTION

```
class Swissprot
    (extent swissprots key Accession)
{
  attribute string ID;
  attribute string Accession;
  attribute string Description;
  attribute list<string> GeneNames;
  attribute string Sequence;
  attribute int Sequence_Length;
}

CREATE  TABLE  Orgs
      (name   string,
```

```
      orgid string
    );

CREATE   TABLE   Genes
    (name     string,
     geneid   string,
     orgid    string,
     startpos long,
     length   long
    );

ProteinSynonyms : dictionary< string,
                   set<struct{syn string, lang string}> >;
```

The K2MDL description of the integration and transformation that is performed when the sources Swiss-Prot, Orgs, Genes, and ProteinSynonyms are mapped into the ontology view. K2MDL descriptions look like the ODL definition in the ontology, enhanced with OQL expressions that compute the class extents, the attribute values, and the relationship connections (a related idea appears in a paper from the *1991 International Conference on Management of Data* [18]).

The definition of the class `Protein` as a K2MDL `classdef` starts with an OQL statement that shows how to compute the extent `proteins` of this class by collecting the accession numbers from SwissProt. The elements of the extent are used as *object identifiers* (OIDs) for the objects in the class. The rest of the definition shows how to compute the value of each attribute for a generic object identified by the OID (denoted by the keyword `self`). The OQL function `element(c)` extracts the unique element of the collection, `c`, when `c` is a singleton, and raises an exception otherwise.

MEDIATOR DESCRIPTION I

```
classdef Protein
   (extent proteins { select distinct s.Accession
                      from swissprots s;})
{
  attribute string SwissprotAccession  { self; };
  attribute string recommendedName {
   element(select s.Description
          from Swissprot s where s.Accession=self);
  };
```

```
    attribute set<string> alternateNames {
      select distinct ps.syn
      from swissprots s, ProteinSynonyms[s.Description] ps
      where s.Accession=self and ps.lang="eng";
    };
    attribute string sequence {
      element(select s.Sequence
              from swissprots s where s.Accession=self);
    };
    attribute int seqLength {
      element(select s.Sequence_Length
              from swissprots s where s.Accession=self);
    };
    relationship set<Gene> hasSource {
      select distinct gn
      from swissprots s, s.GeneNames gn
      where s.Accession=self;
    } inverse Gene::hasProduct;
}
```

Note, for example, the computation of the value of the attribute `alternate-Names`. For an object identified by `self`, find the Swiss-Prot entry, `s`, whose accession number is `self`, then use the description of `s` as a key in the dictionary `ProteinSynonyms`. The entry retrieved from the dictionary `Protein-Synonyms[s.Description]` is a set of records. Select from this set the records with names in English and collect those names into the answer. The value of the attribute is a set of strings. A further query posed against the class `Protein` may, for example, select objects whose `alternateNames` attribute contains a given synonym.

MEDIATOR DESCRIPTION II

```
classdef Gene
   (extent genes { select distinct g.geneid from Genes g; })
{
 struct Range
 {
   long from;
   long to;
 };
```

```
attribute string name {
  element(select distinct g.name
          from Genes g
          where g.geneid=self);
};
attribute string organism {
  element(select o.name
          from Orgs o, Genes g
          where g.geneid=self and o.orgid=g.orgid);
};
attribute Range location {
  element(select struct
          (from: g.startpos, to: g.startpos+g.length-1)
           from Genes g
           where g.geneid=self);
};
relationship Protein hasProduct {
  element(select distinct s.Accession
          from swissprots s, s.GeneNames gn, Genes g
          where g.geneid=self and gn=g.name);
} inverse Protein::hasSource;
}
```

This example illustrates that relatively complex integrations and transformations can be expressed concisely and clearly, and can be easily modified.

8.4   INTERNAL LANGUAGE

The key to making ODL, OQL, and K2MDL work well together is the expressiveness of the internal framework of K2, which is based on complex values and dictionaries. ODL classes with extents are represented internally as dictionaries with abstract keys (the object identities). This framework opens the door to interesting optimizations that make the approach feasible.

The K2 internal language is organized by its type structure. There are base types such as string and number; record and variant types; collection types, namely sets, bags, and lists; and dictionary types. For each type construction there are two classes of operations: *constructors*, such as empty set and set union, and *deconstructors*, such as record field selection. The operations for collection types are inspired from the theory of *monads* [19] and are outlined in an article in *Theoretical Computer Science* [20]. For details specific to aggregates and collection conversions see *Proceedings of the 7th International Conference on Category Theory and Computer Science* [10]; the operations on dictionaries are described in the *1999 Proceedings of the International Conference on Database Theory* [9].

The internal language derives its expressiveness from its flexibility. The primitives are chosen according to the principle of *orthogonality*, which says that their meaning should not overlap and that one should not be able to simulate one primitive through a combination of the others. This produces a language with fewer, but better understood, primitives. As an example, consider the following basic query statement:

```
select E(x)
from   x in R
where P(x)
```

This translates internally into

```
SetU(x in R)if P(x) then sngset(E(x)) else emptyset
```

where `SetU(x in S)T(x)` is the set deconstructor suggested by the theory of monads and `sngset(e)` is a *singleton* set (i.e., the set with just one element, e).

The semantics of the set deconstructor is that of the union of a family of sets. For example, if $S = \{a_1, \ldots, a_n\}$ then

$$\texttt{SetU(x in S)T(x)} = \texttt{T}(a_1) \cup \cdots \cup \texttt{T}(a_n)$$

This approach increases the overall language expressiveness by allowing any type-correct combination of the primitives (the language is fully *compositional*). At the same time it provides a systematic approach to the identification of optimization transformations by yielding an *equational theory* for the internal language. The formulas of such a theory are equalities between equivalent parts of queries, and they are used for rewriting queries in several of the stages of the optimizer. Finally, K2 exploits known efficient physical algorithms for operations such as joins by automatically identifying within queries the groups of primitives that compute these operations.

8.5   DATA SOURCES

K2 maps data from external sources into its internal language, as described previously. K2 also has a notion of functions, which are used to provide access both to stand-alone applications such as BLAST (a sequence similarity package) and to pre-defined and user-defined data conversion routines. This flexibility allows K2 to represent most data sources faithfully and usefully.

K2 accesses external information through data drivers. This is an intermediate layer between the K2 system proper and the actual data sources. There are two kinds of drivers in K2: those that are tightly integrated with the server and those that are more loosely connected.

Integrated data drivers (IDDs) are created by extending the two abstract Java classes that form K2's driver API. The IDDs export a set of entry points to K2, connect to the data source, send queries to it, receive results from it, and package the results for use in the rest of the K2 system. The tight coupling of IDDs with the K2 system minimizes the overhead associated with connecting to the data source and allows for additional optimizations. K2 can also cache results of queries sent to the IDDs to improve overall speed.

K2 comes with an IDD that can connect to any relational database system that implements Sun's JDBC API. A Sybase-specific IDD is also available, which takes advantage of some features of Sybase that are not available through JDBC. Oracle- and MySQL-specific versions are currently under development. To connect to a new relational database that supports JDBC, one merely needs to add the connection information to a configuration file, and K2 will automatically expose the underlying schema for querying. When the underlying schema changes, the K2 administrator must restart the IDD so it can rediscover the new schema. A procedure for automatically detecting and rediscovering schema changes is planned as a future enhancement.

Another IDD provided with K2 makes use of the World Wide Web Wrapper Factory (W4F), also developed at the University of Pennsylvania.[6] W4F is a toolkit for the generation of wrappers for Web sources. New wrappers can easily be generated using W4F's interface and can then be converted automatically to K2 drivers. Changes to the format of Web sources are partly handled by W4F's declarative wrapper specification language. Large format changes require human intervention to re-define and re-generate wrappers.

A very powerful feature of K2 is its ability to distribute query execution using its IDD for Java RMI. This IDD can make an RMI connection to a remote K2 server and send it part of the local query for processing. Therefore, all that is required to connect to the remote K2 server and start distributing queries is a change to the local K2 configuration file.

Sometimes it is difficult to develop an IDD for a new type of data source. For example, to treat a group of flat files as a data source it is often easier to write a

---

6. Information about W4F is available at *http://db.cis.upenn.edu/Research/w4f.html*.

Perl script to handle the string manipulations involved rather than implementing them in Java, as would be required in an IDD. In fact, some data sources cannot be accessed at all from Java but only through APIs in other languages. To handle this, K2 has an IDD called the PipeDriver that does not connect to the data source directly, but to a decoupled data driver (DDD).

A DDD is a simple, stand-alone application, written in any language at all, that accepts queries through its standard input stream and writes results to its standard output. The PipeDriver takes care of sending queries to the DDD and converting the results into K2's internal representation. It can run multiple DDDs at once to take advantage of parallelism, and it can make use of the caching mechanism built into IDDs, all of which simplifies the job of the DDD writer.

The DDD is responsible for establishing a connection to the data source (often nothing is required in this step), telling the PipeDriver it has made the connection, and waiting for a query to come in. When the DDD receives a query, it extracts the appropriate result from the data source and writes it out in a simple data exchange format. It then returns to waiting for the next query. This loop continues until the K2 server is brought down, at which point the PipeDriver tells the DDD to terminate.

DDDs have been written to connect with SRS, KEGG, and BLAST, as well as a number of Web-based sources. The time it takes to create a new DDD depends greatly on the capabilities of the data source for which it is being written and on how much intelligence is to be built into the DDD. For example, it only takes an hour or two to write a Perl script to connect to a simple document storage system; the script must be written to receive an ID, retrieve the document, and print it out in K2's exchange format, taking into account any error conditions that might occur.

However, the DDD writer has the flexibility to create special-purpose DDDs of any complexity. One DDD has been written that performs queries over a collection of documents that come from a remote Web site. This DDD maintains a local disk cache of the documents to speed access. It is responsible for downloading new versions of out-of-date documents, taking concurrency issues into account, and parsing and indexing documents on the fly. It also supports a complex language for querying the documents and for retrieving structured subsets of their components. This DDD was written over the course of two weeks and has been expanded periodically since.

8.6   QUERY OPTIMIZATION

K2 has a flexible, extensible query optimizer that uses both rewrite rules and a cost model. The rewrite rules are used to transform queries into structurally minimal forms whose execution is always faster than the original query; this is independent of the nature of the data. On the other hand, a cost model uses information about the nature of the data, such as the size of the data sets, selectivity of joins, available bandwidth, and latency of the data sources. The cost information is used to choose between minimal queries that are incomparable with respect to the rewrite rules.

After translating the query into an abstract syntax tree, which K2 uses to represent queries internally, it is manipulated by applying a series of rewrite rules. This is where the bulk of K2's optimization work is done.

First, a collection of rules is applied that simplifies the query by taking pieces expressed using certain kinds of tree nodes and replacing them with others. This reduces the number of types of nodes needed to deal with, thus reducing the number and complexity of the rewrite rules that follow.

Next, the query is normalized. Normalization rules include steps such as taking a function applied to a conditional structure and rewriting it so the function is applied to each of the expressions in the condition. Another normalization rule removes loops that range over collections known to be empty. Repeated application of the normalization rules, which currently number more than 20, reduces the query to a minimal, or *normal*, form.

A final set of rewrite rules are then applied to the normalized query. These rules include parallelizing the scanning of external data sources and pushing selections, projections, and joins down to the drivers where possible.

Even after all the rewrite rules have been applied, there may still be room for further optimization. In particular, a query may have a family of minimal forms rather than a single one. To choose between the minimal forms, the expected execution time of each version of the query is estimated using a cost model, and the fastest query form is chosen. The current cost model is still in the development stage. While it is functional and works well most of the time, it does not always choose the optimal form of the query.

8.7   USER INTERFACES

K2 has been developed using a client–server model. The K2 server listens for connections either through a socket or through Java RMI. It is easy to develop a client that can connect to K2 through one of these paths, issue queries, and receive results. Three basic clients come with K2: a text-based client, an RMI client, and one that runs as a servlet.

The interactive, text-based client connects to a K2 server through a socket connection. It accepts a query in OQL through a command-line-style interface, sends it to the server, gets the result back as formatted text, and displays it; then it waits for the next query to be entered. This simple client generally is used to test the socket connection to K2 and to issue simple queries during the development process. It is not intended to be an interface for end users.

The other type of user connection is through RMI. These connections are capable of executing multiple queries at once and can halt execution of queries in progress. This is the connection method that K2 servers use to connect to other K2 servers to distribute the execution of a query.

There is a client that makes an RMI connection to a K2 server with administrator privileges. The server restricts these connections to certain usernames connecting from certain IP addresses and requires a password. Currently, an administrator can examine the state of the server, add and remove connections to individual drivers, stop currently running queries, disconnect clients, and bring the server to a state where it can be stopped safely. More functionality is planned for administrators in the future.

A client that runs as a servlet is also included with K2. Using code developed at the Computational Biology and Informatics Laboratory (at the University of Pennsylvania), this servlet allows entry of *ad hoc* K2 queries and maintains the results for each username individually.

A major component of any user interface is the representation of the data to the user. As exemplified previously, a user (perhaps one serving a larger group) can define in K2MDL a transformed/integrated schema for a class of users and applications and can specify how the objects of this schema map to the underlying data sources. Users of this schema (called an *ontology* by some) can issue vastly simplified queries against it, without knowledge of the data sources themselves.

8.8   SCALABILITY

In theory, the K2 system can be used to interconnect an arbitrarily large number of data sources. In practice, the system has been configured with up to 30 data sources and software packages, including GUS, PubMed, MacOStat, GenBank, Swiss-Prot, BLAST, KEGG, and several relational databases maintained in Oracle, Sybase, and MySQL. Even when querying using this configuration, however, it has been rare to access more than five data sources and software packages in the same query.

The primary obstacles to scaling K2 to a larger system are (1) writing data drivers to connect to new data sources and (2) peak memory consumption. The difficulty of writing data drivers to external data sources has been mitigated to some extent by the fact that they are type specific rather than instance specific. For example, once an Oracle driver is written, it can be used for any Oracle data source. On the other hand, an AceDB driver must take into account the schema of the AceDB source and is therefore not generic. Drivers are also relatively simple in K2 because they merely perform data translation. Any complex, semantic transformations are performed by K2MDL code, which is high-level and therefore more maintainable.

While peak memory consumption has not as yet been an issue for the queries handled by K2 in the past, it could become a problem as applications become larger. K2 provides a means of limiting the number of queries it will run simultaneously and the number of data source connections it will maintain. This allows an administrator to tune the system to the capabilities of the machine on which it is running.

Because K2 is a view integration environment, it does not store any data locally; however, it may need to store intermediate results for operations that cannot be streamed (i.e., processed on the fly). At present, intermediate results are stored in main memory.

Examples of operations that cannot be streamed include sorting, set difference, nesting, and join. For example, when the difference of two data sets is taken, no output can be issued until both data sets are read; one of the data sets may have to be cached while the other is streamed and the difference is calculated. Similarly, although a join output can be produced as soon as a match is found between elements of the two data sets, data cannot be discarded until it is known not to match any future input from the other data set (see two *International Conference on the Management of Data (SIGMOD)* articles [21, 22] for discussions of implementations of operators in streaming environments). Thus, when data sets are large, these operations may need to cache temporary results in persistent memory. Such techniques are not currently part of the K2 system and require futher development.

Another difficulty of scaling to an arbitrarily large environment is the sheer complexity of understanding what information is available. To mitigate this, smaller mediated components can be composed to form larger mediated components. Thus, users need not be aware of the numerous underlying data sources and software systems, but they can interact with the system through a high-level interface representing the ontology of data.

8.9   IMPACT

As with Kleisli, a tremendous enhancement in productivity is gained by expressing complicated integrations in a few lines of K2MDL code as opposed to much larger programs written in Perl or C++. What this means for the system integrator is the ability to build central client–server or mix-and-match components that interoperate with other technologies. Among other things, K2 provides:

✦ Enhanced productivity (50 lines of K2MDL correspond to thousands of lines of C++)

✦ Maintainability and easy transitions (e.g., warehousing)

✦ Re-usability (structural changes easy to make at the mediation language level)

✦ Compliance with ODMG standards

K2 has been used extensively in applications within the pharmaceutical company GlaxoSmithKline. Some of the major benefits of the system exploited in these applications have been the ease with which ontologies can be represented in K2MDL and the ability to conveniently compose small mediators into larger mediators.

K2 was also used to implement a distributed genomic-neuroanatomical database. The system combines databases and software developed at the Center for Bioinformatics at the University of Pennsylvania—including databases of genetic and physical maps, genomic sequences, transcribed sequences, and gene expression data, all linked to external biology databases and internal project data (GUS [14])—with mouse brain atlas data and visualization packages developed at the Computer Vision Laboratory and Brain Mapping Center at Drexel University. The biological and medical value of the activity lay in the ability to correlate specific brain structures with molecular and physiological processes. The technological value of the activity was that K2 was ported to a Macintosh operating system environment (MacOSX), visualization packages were tightly integrated into the environment, and both an ethernet and a gigabit network were used in the application. The work was significantly facilitated by the fact that K2 is implemented in Java.

K2 now runs on Linux, Sun Solaris, Microsoft Windows, and Apple MacOS platforms.

8.10   SUMMARY

This chapter presented the K2 system for integrating heterogeneous data sources. K2 is general purpose, written in Java, and includes JDBC interfaces to relational sources as well as interfaces for a variety of special-format bioinformatics sources. The K2 system has a universal internal data model that allows for the direct representation of relational, nested complex value, object-oriented, information retrieval (dictionaries), and a variety of electronic data interchange (EDI) formats, including XML-based sources. The internal language features a set of equivalence laws on which an extensible optimizer is based. The system has user interfaces based on the ODMG standard, including a novel ODL–OQL combined design for high-level specifications of mediators.

ACKNOWLEDGMENTS

The authors would like to acknowledge the contributions of David Benton, Howard Bilofsky, Peter Buneman, Jonathan Crabtree, Carl Gustaufson, Kazem Lellahi, Yoni Nissanov, G. Christian Overton, Lucian Popa, and Limsoon Wong.

REFERENCES

[1] R. J. Robbins, ed. *Report of the Invitational DOE Workshop on Genome Informatics*, April 26–27, 1993. Baltimore, MD: DOE, 1993.

[2] J. Melton and A. Simon. *Understanding the New SQL*. San Francisco: Morgan Kaufman, 1993.

[3] R. G. G. Cattell and D. Barry, eds. *The Object Database Standard: ODMG 3.0*. San Francisco: Morgan Kaufmann, 2000.

[4] P. Buneman, L. Libkin, D. Suciu, et al. "Comprehension Syntax." *Special Interest Group on the Management of Data (SIGMOD) Record* 23, no. 1 (March 1994): 87–96.

[5] S. Chawathe, H. Garcia-Molina, J. Hammer, et al. "The TSIMMIS Project: Integration of Heterogeneous Information Sources." In *Proceedings of the 10th Meeting of the Information Processing Society of Japan Conference*. Tokyo, Japan: 1994.

[6] A. Levy, D. Srivastava, and T. Kirk. "Data Model and Query Evaluation in Global Information Systems." *Journal of Intelligent Information Systems* 5, no. 2 (1995): 121–143.

[7] L. M. Haas, P. M. Schwarz, P. Kodali, et al. "DiscoveryLink: A System for Integrated Access to Life Sciences Data Sources." *IBM Systems Journal* 40, no. 2 (2001): 489–511.

[8] I-M. A. Chen and V. M. Markowitz. "An Overview of the Object-Protocol Model (OPM) and OPM Data Management Tools." *Information Systems* 20, no. 5 (1995): 393–418.

[9] L. Popa and V. Tannen. "An Equational Chase for Path Conjunctive Queries, Constraints, and Views." In *Proceedings of the International Conference on Database Theory (ICDT), Lecture Notes in Computer Science*, 39–57. Heidelberg, Germany: Springer Verlag, 1999.

[10] K. Lellahi and V. Tannen. "A Calculus for Collections and Aggregates." In *Proceedings of the 7th International Conference on Category Theory and Computer Science, CTCS'97, Lecture Notes in Computer Science,* vol. 1290, edited by E. Moggi and G. Rosolini, 261–280. Heidelberg, Germany: Springer-Verlag, 1997.

[11] T. Etzold and P. Argos. "SRS: An Indexing and Retrieval Tool for Flat File Data Libraries." *Computer Applications of Biosciences* 9, no. 1 (1993): 49–57.

[12] W. Fujibuchi, S. Goto, H. Migimatsu, et al. "DBGET/LinkDB: An Integrated Database Retrieval System." In *Pacific Symposium on Biocomputing,* 683–694. 1998.

[13] M. Rebhan, V. Chalifa-Caspi, J. Prilusky, et al. *GeneCards: Encyclopedia for Genes, Proteins and Diseases.* Rehovot, Israel: Weizmann Institute of Science, Bioinformatics Unit and Genome Center, 1997. http://bioinformatics.weizmann.ac.il/cards.

[14] S. Davidson, J. Crabtree, B. Brunk, et al. "K2/Kleisli and GUS: Experiments in Integrated Access to Genomic Data Sources." *IBM Systems Journal* 40, no. 2 (2001): 512–531.

[15] S. Davidson, C. Overton, V. Tannen, et al. "BioKleisli: A Digital Library for Biomedical Researchers." *Journal of Digital Libraries* 1, no. 1 (November 1996): 36–53.

[16] G. Wiederhold. "Mediators in the Architecture of Future Information Systems." *IEEE Computer* 25, no. 3 (March 1992): 38–49.

[17] H. Garcia-Molina, Y. Papakonstantinou, D. Quass, et al. "The TSIMMIS Approach to Mediation: Data Models and Languages." In *Proceedings of the Second International Workshop on Next Generation Information Technologies and Systems,* 185–193. 1995.

[18] S. Abiteboul and A. Bonner. "Objects and Views." In *Proceedings of the 1991 ACM SIGMOD International Conference on the Management of Data,* 238–247. San Francisco: ACM Press, 1991.

[19] S. MacLane. *Categories for the Working Mathematician.* Berlin, Germany: Springer-Verlag, 1971.

[20] P. Buneman, S. Naqvi, V. Tannen, et al. "Principles of Programming with Collection Types." *Theoretical Computer Science* 149, no. 1 (1995): 3–48.

[21] Z. Ives, D. Florescu, M. Friedman, et al. "An Adaptive Query Execution System for Data Integration." In *Proceedings of the 1999 ACM SIGMOD International Conference on Management of Data,* 299–310. San Francisco: ACM Press, 1999.

[22] J. Chen, D. DeWitt, F. Tian, et al. "NiagaraCQ: A Scalable Continuous Query System for Internet Databases." In *Proceedings of the 2000 ACM SIGMOD International Conference on Management of Data,* 379–390. San Francisco: ACM Press, 2000.

P/FDM Mediator for a Bioinformatics Database Federation

Graham J. L. Kemp and Peter M. D. Gray

The Internet is an increasingly important research tool for scientists working in biotechnology and the biological sciences. Many collections of biological data can be accessed via the World Wide Web, including data on protein and genome sequences and structure, expression data, biological pathways, and molecular interactions. Scientists' ability to use these data resources effectively to explore hypotheses *in silico* is enhanced if it is easy to ask precise and complex questions that span across several different kinds of data resources to find the answer.

Some online data resources provide search facilities to enable scientists to find items of interest in a particular database more easily. However, working interactively with an Internet browser is extremely limited when one want to ask complex questions involving related data held at different locations and in different formats as one must formulate a series of data access requests, run these against the various databanks and databases, and then combine the results retrieved from the different sources. This is both awkward and time-consuming for the user.

To streamline this process, a federated architecture and the *P/FDM Mediator* are developed to integrate access to heterogeneous, distributed biological databases. The spectrum of choices for data integration is summarized in Figure 9.1. As advocated by Robbins, the approach presented in this chapter does not require that a common hardware platform or vendor database management system (DBMS) [1] be adopted by the participating sites. The approach presented here needs a "shared data model across participating sites," but does not require that the participating sites all use the same data model internally. Rather, it is sufficient for the mediator to hold descriptions of the participating sites that are expressed in a common data model; in this system, the P/FDM Mediator, the functional data model [2] is used for this purpose. Tasks performed by the P/FDM Mediator include determining which external databases are relevant in answering users' queries, dividing queries into parts that will be sent to different external databases,

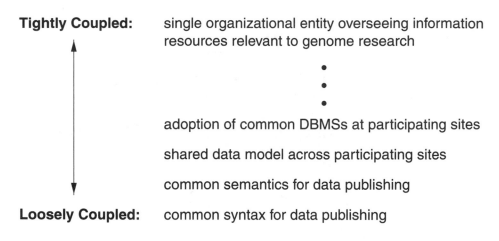

Tightly Coupled: single organizational entity overseeing information resources relevant to genome research

•
•
•

adoption of common DBMSs at participating sites

shared data model across participating sites

common semantics for data publishing

Loosely Coupled: common syntax for data publishing

9.1

FIGURE   Continuum from tightly coupled to loosely coupled distributed systems involving multiple databases [1].

translating these subqueries into the language(s) of the external databases, and combining the results for presentation.

9.1   APPROACH

9.1.1   Alternative Architectures for Integrating Databases

The aim is to develop a system that will provide uniform access to heterogeneous databases via a single high-level query language or graphical interface and will enable multi-database queries. This objective is illustrated in Figure 9.2. Data replication and multi-databases are two alternative approaches that could help to meet this objective.

Data Replication Approach

In a *data replication* architecture, all data from the various databases and databanks of interest would be copied to a single local data repository, under a single DBMS. This approach is taken by Rieche and Dittrich [3], who propose an architecture in which the contents of biological databanks including the EMBL nucleotide sequence databank and Swiss-Prot are imported into a central repository.

However, a data replication approach may not be appropriate for this application domain for several reasons. Significantly, by adopting a data repository

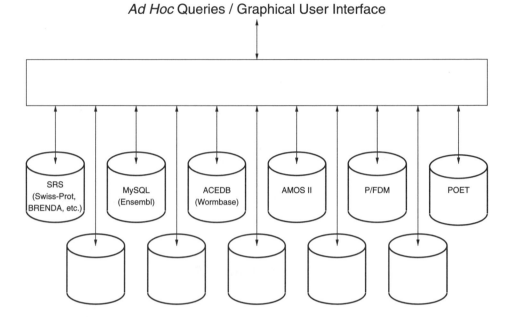

Ad Hoc Queries / Graphical User Interface

9.2

FIGURE

Users should be able to access heterogeneous, distributed bioinformatics resources via a single query language or graphical user interface.

approach, the advantages of the individual heterogeneous systems are lost. For example, many biological data resources have their own customized search capabilities tailored to the particular physical representation that best suits that data set. Rieche and Dittrich [3] acknowledge the need to use existing software and propose implementing *exporters* to export and convert data from the data repository into files that can be used as input to software tools.

Another disadvantage of a data replication approach is the time and effort required to maintain an up-to-date repository. Scientists want access to the most recent data as soon as they have been deposited in a databank. Therefore, whenever one of the contributing databases is updated, the same update should be made to the data repository.

Multi-Database Approach

A multi-database approach that makes use of existing remote data sources, with data described in terms of entities, attributes, and relationships in a high-level schema is favored. The schema is designed without regard to the physical storage format(s). Queries are expressed in terms of the conceptual schema, and it is the role of a complex software component called a *mediator* [4] to decide what component data sources need to be accessed to answer a particular query, organize the computation, and combine the results. Robbins [1] and Karp [5] have also advocated a federated, multi-database approach.

In contrast to a data replication approach, a multi-database approach takes advantage of the customized search capabilities of the component data sources in the federation by sending requests to these from the mediator. The component resources keep their autonomy, and users can continue to use them exactly as before. There is no local mirroring, and updates to the remote component databases are available immediately. A multi-database approach does not require that large data sets be imported from a variety of sources, and it is not necessary to convert all data for use with a single physical storage schema. However, extra effort is needed to achieve a mapping from the component databases onto the conceptual model.

9.1.2   The Functional Data Model

The system described in this chapter is based on the P/FDM object database system [6], which has been developed for storing and integrating protein structure data. P/FDM is itself based on the functional data model (FDM) [2], whose basic concepts are entities and functions. Entities are used to represent conceptual objects, while functions represent the properties of an object. Functions are used to model both scalar attributes, such as a protein structure's resolution and the number of amino acid residues in a protein chain, and relationships, such as the relationship between chains and the residues they contain. Functions may be single-valued or multi-valued, and their values can either be stored or computed on demand. Entity classes can be arranged in subtype hierarchies, with subclasses inheriting the properties of their superclass, as well as having their own specialized properties. Contrast this with the more widely used relational data model whose basic concept is the *relation*—a rectangular table of data. Unlike the FDM, the relational data model does not directly support class hierarchies or many-to-many relationships.

Daplex is the query language associated with the FDM and, to illustrate the syntax of the language, Figure 9.3 shows two Daplex queries expressed against an antibody database [7]. Query A prints "*the names of the amino acid residues found at the position identified by Kabat code number 88 in variable domains of antibody light chains (VL domains).*" This residue is located in the core of the VL domain and is spatially adjacent to the residue at Kabat position 23. Query B prints "*the names of the residues at these two positions together with the computed distance between the centers of their alpha-carbon (CA) atoms.*" Thus, one can explore a structural hypothesis about the spatial separation of these residues being

Query A:

```
for each d in ig_domain such that
    domain_type(d) = "variable" and chain_type(d) = "light"
  print(protein_code(domain_structure(d)),
        name(d),
        name(kabat_residue(d, "88")));
```

Query B:

```
for each s in structure
  for each vl in domain_structure_inv(s) such that
      domain_type(vl) = "variable" and chain_type(vl) = "light"
    print(protein_code(s),
          name(kabat_residue(vl, "23")),
          name(kabat_residue(vl, "88")),
          distance(atom(kabat_residue(vl, "23"), "CA"),
                   atom(kabat_residue(vl, "88"), "CA")));
```

9.3

FIGURE

Daplex queries against an antibody database. Query A: "*For each light chain variable domain, print the PDB entry code, the domain name, and the name of the residue occurring at Kabat position 88.*" Query B: "*For each VL domain, print the PDB entry code, the names of the residues at Kabat positions 23 and 88, and the distance between their alpha-carbon atoms.*"

related to the residue types occurring at these positions. In Query A, ig_domain is an entity class representing immunoglobulin domains, and the values that the variable d takes are the object identifiers of instances of that class. Domain_type and chain_type are string-valued functions defined on the class ig_domain. Domain_structure is a relationship function that returns the object identifier of the instance of the class *structure* that contains the ig_domain instance identified by the value of d. The expression protein_code(domain_structure(d)) illustrates an example of function composition. Query B shows nested loops in Daplex.

FDM had its origins in early work [2, 8], done before relational databases were a commercial product and before object-oriented programming (OOP) and windows, icons, menus, and pointers (WIMP) interfaces had caught on. Although it is an old model, it has adapted well to developments in computing because it was based on good principles. First, it was based on the use of values denoting persistent identifiers for instances of entity classes, as noted by Kulkarni and Atkinson [9].

This later became central to the object database manifesto [10]. Also, it had the notion of a subtype hierarchy, and it was not difficult to adapt this to include methods with overriding, as in OOP [11]. Second, the notions of an entity, a property, and a relationship (as represented by a function and its inverse) corresponded closely to the entity-relationship (ER) model and ER diagrams, which have stood the test of time. Third, it used a query language based on applicative expressions, which combined data extraction with computation. Thus, it was a mathematically well-formed language, based on the functional languages [12, 13], and it avoided the syntactic oddities of structured query language (SQL).

In developing the language since early work on the excluded function data model (EFDM) [9] Prolog was used as the implementation language because it is so good for pattern matching, program transformation, and code generation. Also, the data independence of the FDM, with its original roots in Multibase [14], allows to interface to a variety of kinds of data storage, instead of using a persistent programming language with its own data storage. Thus, unlike the relational or object-relational models, FDM does not have a particular notion of storage (row or tuple) built into it. Nor does it have fine details of arrays or record structures, as used in programming languages. Instead, it uses a mathematical notion of mapping from entity identifier to associated objects or values. Another change has been to strengthen the referential transparency of the original Daplex language by making it correspond more closely to Zermelo-Fraenkel set expressions (ZF-expressions), a name taken from the Miranda functional language [13, 15], and also called *list comprehensions*.

9.1.3   Schemas in the Federation

The design philosophy of P/FDM Mediator can be illustrated with reference to the *three-schema architecture* proposed by the ANSI Standards Planning And Requirements Committee (SPARC) [16]. This consists of the *internal level*, which describes the physical structure of the database; the *conceptual level*, which describes the database at a higher level and hides details of the physical storage; and the *external level*, which includes a number of *external schemas* or *user views*. This three-schema architecture promotes data independence by demanding that database systems be constructed so they provide both logical and physical data independence. Logical data independence provides that the conceptual data model must be able to *evolve* without changing external application programs. Only view definitions and mappings may need changing (e.g., to replace access to a stored field by access to a derived field calculated from others in the revised schema). Physical data independence allows to refine the internal schema for improved performance without needing to alter the way queries are formulated.

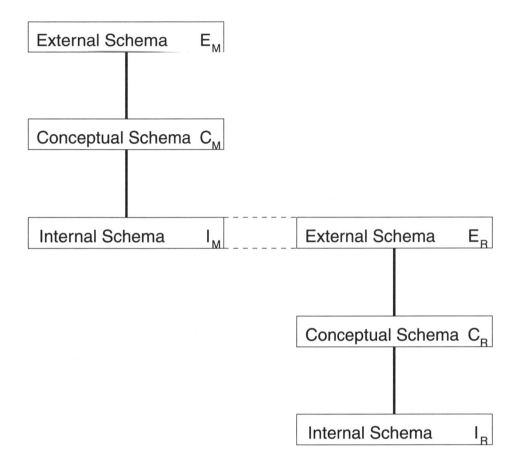

9.4

FIGURE
ANSI-SPARC schema architecture describing the mediator (left) and an external data resource (right).

The clear separation between schemas at different levels helps in building a database federation in a modular fashion. In Figure 9.4 the ANSI-SPARC three-schema architecture is shown in two situations: in each of the individual data resources and in the mediator itself.

First, consider an external data resource. The resource's *conceptual schema* (which is called C_R) describes the logical structure of the data contained in that resource. If the resource is a relational database, it will include information about table names, column names, and type information about stored values. With SRS [17], it is the databank names and field names. These systems also provide a mechanism for querying the data resource in terms of the table/class/databank names and column/tag/attribute/field names that are presented in the conceptual schema.

The *internal schema* (or *storage schema*, which is called I_R) contains details of allocation of data records to storage areas, placement strategy, use of indexes, set ordering, and internal data structures that impact efficiency and implementation details [6]. This chapter is not concerned with the internal schemas of individual data resources. The mapping from the conceptual schema to the internal schema has already been implemented by others within each of the individual resources and it is assumed this has been done to make best use of the resources' internal organization.

A resource's *external schema* (which is called E_R) describes a view onto the data resource's conceptual schema. At its simplest, the external schema could be identical to the conceptual schema. However, the ANSI-SPARC model allows for differences between the schemas at these layers so different users and application programmers can each be presented with a view that best suits their individual requirements and access privileges. Thus, there can be many external schemas, each providing users with a different view onto the resource's conceptual schema. A resource's external, conceptual, and internal schemas are represented on the right side of Figure 9.4.

The ANSI-SPARC three-layer model can also be used to describe the mediator central to the database federation, and this is shown on the left side of Figure 9.4. The mediator's conceptual schema (C_M), also referred to as the federation's *integration schema*, describes the content of the (virtual) data resources that are members of the federation, including the semantic relationships that hold between data items in these resources. This schema is expressed using the FDM because it makes computed data in a virtual resource. Both the derived results of arithmetic expressions and derived relationships look no different from stored data. Both are the result of functions—one calculates and the other extracts from storage. As far as possible, the C_M is designed based on the semantics of the domain, rather than consideration of the actual partitioning and organization of data in the external resources. Thus, through functional mappings, different attributes of the same conceptual entity can be spread across different external data resources, and subclass–superclass relationships between entities in the conceptual model of the domain might not be present explicitly in the external resources [18].

No one can expect scientists to agree on a single schema. Different scientists are interested in different aspects of the data and will want to see data structured in a way that matches the concepts, attributes, and relationships in their own personal model. This is made possible by following the ANSI-SPARC model; the principle of logical data independence means the system can provide different users with different views onto the integration schema. E_M is used to refer to an external schema presented to a user of the mediator. In this chapter, queries are expressed directly against an integration schema (C_M), but these could alternatively be expressed against an external schema (E_M). If so, an additional layer of mapping functions would be required to translate the query from E_M to C_M.

A vital task performed by the mediator is to map between C_M and the union of the different C_R. To facilitate this process, another schema layer that, in contrast to C_M, *is* based on the structure and content of the external data resources is introduced. This schema is internal to the mediator and is referred to as I_M. The mediator needs to have a view onto the data resource that matches this internal schema; thus, I_M and E_R should be the same. FDM is used to represent I_M/E_R. Having the same data model for C_M and I_M/E_R brings advantages in processing multi-database queries, as will be seen in Section 9.1.4.

By redrawing Figure 9.4, the situation where there are *u* different external schemas presented to users and *r* data resources, the relationship between schemas in the federation is as shown in Figure 9.5. The five-level schema shown there is similar to that described by Sheth and Larson [19].

From past experience in building the prototype system, designing the I_M/E_R schema in a way that most directly describes the structure of a particular external resource adds practical benefits. In adding a remote resource to C_M, one may focus mainly on those attributes in the remote resource that are important for making joins across to other resources. This clarifies the role of each schema level and the purpose of the query transformation task in transforming queries from one schema level to the next level down.

9.1.4 Mediator Architecture

The role of the mediator is to process queries expressed against the federation's integration schema (C_M). The mediator holds meta-data describing the integration schema and also the external schemas of each of the federation's data resources (E_R). In P/FDM, these meta-data are held, for convenience of pattern matching, as Prolog clauses compiled from high-level schema descriptions.

The architecture of the P/FDM Mediator is shown in Figure 9.6. The main components of the mediator are described in the following paragraphs.

The *parser* module reads a Daplex query (Daplex is the query language for the FDM), checks it for consistency against a schema (in this case the integration schema), and produces a *list comprehension* containing the essential elements of the query in a form that is easier to process than Daplex text (this internal form is called *ICode*).

The *simplifier*'s role is to produce shorter, more elegant, and more consistent ICode, mainly through removing redundant variables and expressions (e.g., if the ICode contains an expression equating two variables, that expression can be eliminated, provided that all references to one variable are replaced by references

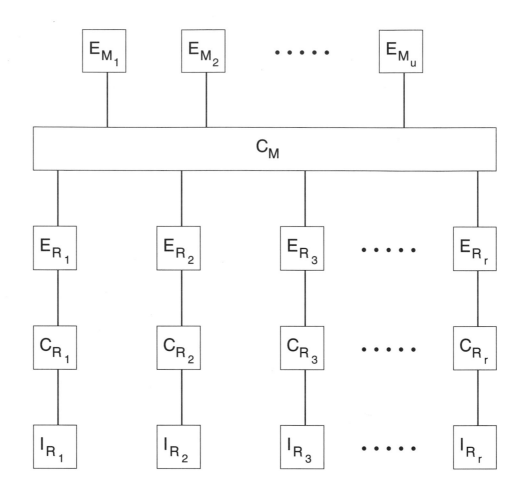

9.5

FIGURE

Schemas in a database federation.

to the other), and flattening out nested expressions where this does not change the meaning of the query. Essentially, simplifying the ICode form of a query makes the subsequent query processing steps more efficient by reducing the number of equivalent ICode combinations that need to be checked.

The *rule-based rewriter* matches expressions in the query with patterns present on the left-hand side of declarative rewrite rules and replaces these with the right-hand side of the rewrite rule after making appropriate variable substitutions. Rewrite rules can be used to perform *semantic query optimization*. This capability is important because graphical interfaces make it easy for users to express inefficient queries that cannot always be optimized using general purpose query

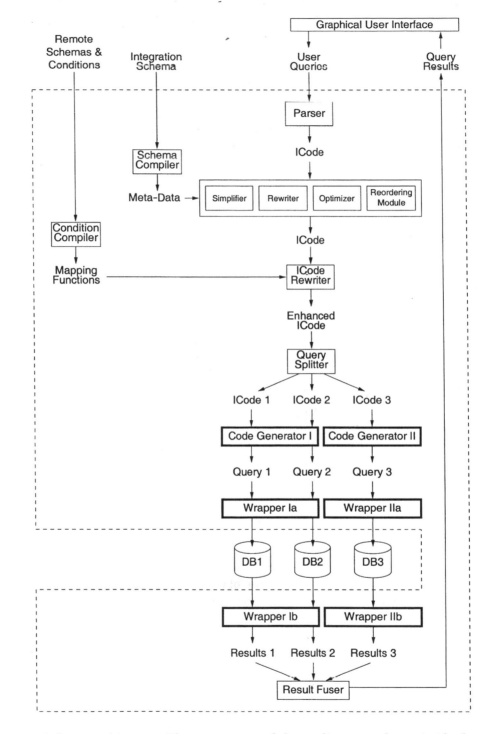

9.6

FIGURE

Mediator architecture. The components of the mediator are shown inside the dashed line.

optimization strategies. This is because transforming the original query to a more efficient one may require domain knowledge (e.g., two or more alternative navigation paths may exist between distantly related object classes but domain knowledge is needed to recognize that these are indeed equivalent).

A recent enhancement to the mediator is an extension to the Daplex compiler that allows generic rewrite rules to be expressed using a declarative high-level syntax [20]. This makes it easy to add new query optimization strategies to the mediator.

The *optimizer* module performs generic query optimization.

The *reordering module* reorders expressions in the ICode to ensure that all variable dependencies are observed.

The *condition compiler* reads declarative statements about conditions that must hold between data items in different external data resources so these values can be mapped onto the integration schema.

The *ICode rewriter* expands the original ICode by applying mapping functions that transform references to the integration schema into references to the federation's component databases. Essentially the same rewriter mentioned previously is used here, but with a different set of rewrite rules. These rewrite rules enhance the ICode by adding tags to indicate the actual data sources that contain particular entity classes and attribute values. Thus, the ICode rewriter transforms the query expressed against the C_M into a query expressed against the E_R of one or more external databases.

The crucial idea behind the *query splitter* is to move selective filter operations in the query down into the appropriate chunks so they can be applied early and efficiently using local search facilities as registered with the mediator [KIG94]. The mediator identifies which external databases hold data referred to by parts of an integrated query by inspecting the meta-data, and adjacent query elements referring to the same database are grouped together into chunks. Query chunks are shuffled and variable dependencies are checked to produce alternative execution plans. A generic description of costs is used to select a good schedule/sequence of instructions for accessing the remote databases.

Each ICode chunk is sent to one of several *code generators*. These translate ICode into queries that are executable by the remote databases, transforming query fragments from E_R to C_R. New code generators can be linked into the mediator at runtime.

Wrappers deal with communication with the external data resources. They consist of two parts: code responsible for sending queries to remote resources and code that receives and parses the results returned from the remote resources. Wrappers for new resources can be linked into the mediator at runtime. Note that a wrapper can only make use of whatever querying facilities are provided by the federation's component databases. Thus, the mediator's conceptual model (C_M) will only be able to map onto those data values that are identified in the remote resource's conceptual model (C_R). Thus, queries involving concepts like *gene* and *chromosome* in C_M can only be transformed into queries that run against a remote resource if that resource exports these concepts.

The *result fuser* provides a synchronization layer, which combines results retrieved from external databases so the rest of the query can proceed smoothly. The result fuser interacts tightly with the wrappers.

9.1.5 Example

A prototype mediator has been used to combine access databanks at the EBI via an SRS server [17] and (remote) P/FDM test servers. Remote access to a P/FDM database is provided through a CORBA server [21]. This example, using a small integration schema, illustrates the steps involved in processing multi-database queries.

In this example, three different databases are viewed through a unifying integration schema (C_M), which is shown in Figure 9.7(a). There are three classes in this schema: protein, enzyme, and swissprot_entry. A function representing the enzyme classification number (ec_number) is defined on the class enzyme, and enzymes inherit those functions that are declared on the superclass protein. Each instance of the class *protein* can be related to a set of swissprot_entry instances.

Figure 9.7(b) shows the actual distribution of data across the three databases; each of these three databases has its own external schema, E_R. Db I is a P/FDM database that contains the codes and name of proteins. Db II is also a P/FDM database and contains the protein code (here called *pdb_code*) and enzyme classification code of enzymes. To identify Swiss-Prot entries at the EBI that are related to a given protein instance, one must first identify the Protein Data Bank (PDB) entry whose ID matches the protein code and then follow further links to find related Swiss-Prot entries. Relationships between data in remote databases can be defined by conditions that must hold between the values of the related objects. Constraints on identifying values are represented by dashed arrows in Figure 9.7.

Figure 9.8 shows a Daplex query expressed against the integration schema, C_M. This query prints information about enzymes that satisfy certain selection criteria and their related Swiss-Prot entries. Figure 9.9 shows a pretty-printed version of the ICode produced when this query is compiled. This ICode is then processed by the query splitter, producing ICode (in terms of the resources' external schemas, E_R) that will be turned into queries that will be sent to the three external data resources (Figure 9.10). Note that the variable V1 is common to all three

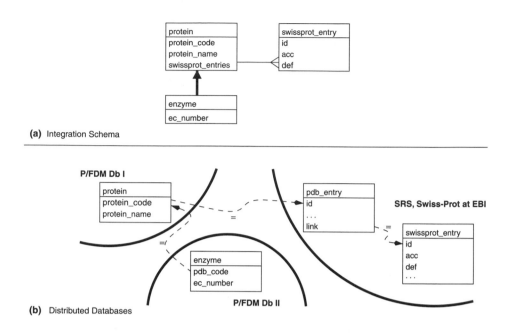

(a) Integration Schema

(b) Distributed Databases

9.7

FIGURE

Integration schema and distributed databases (P/FDM and SRS).

```
for each e in enzyme such that ec_number(e) = "1.1.1.1"
  for each s in swissprot_entries(e)
print(protein_name(e), def(s), acc(s));
```

9.8

FIGURE

Daplex query expressed against an integration schema.

```
[ V6, V4, V3 |   V1 ← enzyme ;
                 V2 ← swissprot_entries(V1) ;
                 V3 = acc(V2) ; V4 = def(V2) ;
                 V5 = ec_number(V1) ; V5 = "1.1.1.1" ;
                 V6 = protein_name(V1) ]
```

9.9

FIGURE

ICode corresponding to the query in Figure 9.8.

```
ICode for P/FDM Db II:
  [ V1 |    V2 ← enzyme ; V3 = ec_number(V2) ;
            V1 = pdb_code(V2) ; V3 = "1.1.1.1" ]

ICode for P/FDM Db I:
  [ V6 |    V4 ← protein ; V5 = protein_code(V4) ;
            V6 = protein_name(V4) ; V1 = V5 ]

ICode for SRS:
  [ V7, V8 |   V9 ← pdb_entry ; V10 = id(V9) ;
               V1 = V10 ; V11 ← link(V9) ;
               V12 ← swissprot_entry ;
               V13 = id(V12) ; V7 = def(V12) ;
               V8 = acc(V12) ; V13 in V11 ]
```

9.10

FIGURE

ICode sub-queries against the actual data resources that need to be accessed to answer the query in Figure 9.8.

```
foreign(swissprot_entries, [protein], srs_sprot, entity,
      KeyICode, ebi_db ) :-
    KeyICode = (V1,V2,[V3,V4,V5,V6],
                   [
                     restrict(ebi_db:id,[ebi_db:srs_sprot],[V2],V6),
                     restrict_subquery(some(V5),
                       [ generate(ebi_db:pdb_entry,V3),
                         restrict(ebi_db:id,[ebi_db:pdb_entry],[V3],V4),
                         restrict(protein_code,[protein],V1,V4),
                         restrict(ebi_db:link,[ebi_db:pdb_entry],[V3],V5) ],
                       [ expression([],[V6,V5],expr(=,V6,V5)) ]
                     )
                   ] ).
```

9.11

FIGURE

Mapping function used to expand the relationship swissprot_entries in the integration schema into ICode that refers to data held at the EBI.

query fragments. Values for this variable retrieved from P/FDM Db II are used in constructing queries to be sent to the other data resources.

In the example, the class protein is related to the class swissprot_entry in the integration schema by a multi-valued relationship function called swissprot_entries. The mapping function given in Figure 9.11 is used in transforming queries that contain this relationship into *enhanced* ICode that refers

```
for each d in ig_domain such that name(d) = "VH"
  for each r1 in kabat_residue(d,"34")
    for each r2 in kabat_residue(d,"78") such that
        distance(atom(r1,"CB"), atom(r2,"CB")) < 5.0
      for each s in swissprot_entries(domain_structure(d))
print(protein_code(d), name(r1), name(r2), def(s), acc(s));
```

9.12   Daplex query that combines computation and data retrieval.

FIGURE

to the external schemas, E_R, of the actual data resources. Mapping functions such as this can be compiled from high-level declarative rewrite rules and do not have to be written by hand.

9.1.6   Query Capabilities

Daplex is the query language of the system. The examples in Figure 9.3 and Figure 9.12 show how function calls can be composed in queries. The compositional form makes it easy to write complex queries and computations over the database that can be optimized by a query optimizer. This is a point often overlooked by the OOP community; Java and C++ have the necessary expressiveness, but because they lack referential transparency and a data model it is hard to make general optimizers for database applications in them. Daplex has greater expressive power that SQL (e.g., recursive functions can be defined directly in Daplex). This is particularly useful in many areas of bioinformatics, such as following transitive relationships through a sequence of reactions in a biochemical pathway or finding related biological terms in a hierarchical vocabulary.

As mentioned in Section 9.1.4, Daplex queries are converted into ICode for subsequent processing. The great advantage is that many important optimizations just involve reordering selection predicates and generators in the set expression. These, in turn, are conveniently implemented as rewrite rules in Prolog [6]. It was also shown how to expand definitions of derived functions in the course of optimizing set expressions [22]. This makes good use of the referential transparency of expressions in functional programming. By contrast, where the computation is embedded in C++ or Visual Basic with arbitrary assignments, it is very hard to do significant optimization. This has led to the widespread adoption of *comprehensions* (as ZF-expressions are now called). Buneman et al. [23] have shown the importance of distinguishing list, bag, and set comprehensions, so, strictly speaking, ZF-expressions compute bags but represent them by lists.

The Daplex query language enables arbitrary calculations to be combined with data retrieval operations [7]. For example, Figure 9.12 shows a query, expressed against an integration schema, that performs a geometric calculation on data in an antibody database and relates objects satisfying the given criteria with data in a remote Swiss-Prot database. The function *distance* computes a value rather than retrieving a stored value. As explained in Section 9.1.2, functions whose values are stored or derived look the same in the query, and the user cannot tell from looking at a result value whether it was retrieved from disc or computed.

Daplex queries frequently include calls to functions written in procedural languages. For example, when working with data on 3D protein structures, one often calls out to geometric code from within queries, including C routines to measure bond angles and torsion angles, and code to superpose one structural fragment on another to compare 3D similarity. Following the same approach, it would be possible to treat the results of a computation on a remote machine, such as a BLAST search, that are generated dynamically at run-time just like data values that are stored persistently on disc. The system does not yet have a wrapper for BLAST, but, in principle, such a wrapper would be implemented just like any other derived function in P/FDM.

The mediator does not currently cache query results, so subsequent queries cannot refer to a result set. However, both user interfaces described in Section 9.2.2 enable follow-on queries to be constructed incrementally based on the previous query.

For more complex P/FDM applications that cannot be expressed in Daplex, Prolog can be used [24]. However, unlike Daplex queries, these Prolog programs are not optimized automatically (see Section 9.2.1). The P/FDM system provides a set of Prolog routines that perform primitive data access operations, such as retrieving the object identifier of an instance of an entity class, retrieving the scalar value of an attribute of an object with a given object identifier, or retrieving the object identifier of a related object. Queries that require access to several data sources in the federation can be written directly in Prolog.

9.1.7 Data Sources

P/FDM was previously used with various data sources including hash files, relational databases, flat files (including some accessed via SRS), POET, AMOS II and AceDB.

There is no control over changes being made to remote resources, as remote sites retain their autonomy. Depending on the nature of the changes to a remote resource, this may require changes in one or both parts of the wrapper for that resource. Changes to a remote resource need not require changes to be made to the mediator's conceptual model (C_M), though they may require some changes to be made to the declarative mapping functions associated with data in the changed resource.

9.2   ANALYSIS

The use of mediators was originally proposed by Wiederhold [4] and became an important part of the knowledge sharing effort architecture [25]. Examples of such intelligent, information-seeking architectures are Infosleuth [26] and KRAFT [27]. In this architecture, the mediator can run on the client machine, or else be available as middleware on some shared machine, while the wrapper is on the remote machine containing the knowledge source. The idea behind this is that existing knowledge sources can evolve their schemas, yet present a consistent interface to the mediator via suitable changes to the wrapper. For this purpose the wrapper may be as simple as an SQL view, or it may be more complex, involving mapping of code. In any case, the site is able to preserve some local autonomy. Other mediators do not have to worry about how the site evolves internally. Also, new sites can join a growing network by registering themselves with a facilitator. All the mediator needs to know is how to contact the facilitator and that any knowledge sources the facilitator recommends will conform to the integration schema.

This chapter describes an alternative architecture, where the wrappers reside *with* the mediator. This has the advantage that there is no need to get the knowledge source to install and maintain custom-provided wrapper software.

In the architecture, shown in Figure 9.6, the code generators produce code in a different query language or constraint language. Thus, they are used in two directions. In one direction, they map queries or constraints into a language that can be used directly at the knowledge source. This can be crucial for efficiency because it allows one to move selection predicates closer to the knowledge source in a form that is capable of using local indexes. This can have a significant effect with database queries because it saves bringing many penny packets of data back through the interface, only to be filtered and rejected on the far side [28]. In the other direction, wrappers are used to map data values (e.g., by using scaling factors to change units or by using a lookup table to replace values by their new identifiers).

Note that building a so-called *global* integration schema is not advocated. These have often been criticized on the grounds that attempts to map every single concept in one all-embracing schema is both laborious and never-ending. Instead, an incrementally growing integration schema is visualized, driven by user needs. Ideally the schema would be built interactively using a GUI and rules that suggest various mappings, as proposed by Mitra et al. [29] in their ONION system for incremental development of ontology mappings. The crucial thing to realize is that the integration schema represents a *virtual* database, which allows it to evolve much more easily than a physical database.

Related work in the bioinformatics field includes the Kleisli system presented in Chapter 6 [30, 31]. The query language used in Kleisli is the Collection Programming Language (CPL), which is a comprehension-based language in which the *generators* are calls to library functions that request data from specific databases according to specific criteria. Thus, when writing queries, the user must be aware of how data are partitioned across external sites. This contrasts with the approach taken in the P/FDM Mediator, where references to particular resources do not feature in the integration schema or in user queries. Of course, an interface based on domain concepts and without references to particular resources could be built on top of Kleisli.

The TAMBIS system presented in Chapter 7 [32] writes query plans in CPL. Plans in TAMBIS are based on a classification hierarchy, whereas P/FDM plans are oriented toward *ad hoc* SQL3-like queries. However, the overall approach is similar to using a high-level intermediate code translated through wrappers.

Another related project is DiscoveryLink, presented in Chapter 11 [33]. The architecture of the DiscoveryLink system is similar to that presented in this chapter. DiscoveryLink uses the relational data model instead of FDM, and all the databases accessed via DiscoveryLink must present an SQL interface.

9.2.1 Optimization

Optimization takes a great advantage from using an easily transformable high-level representation based on functional composition.

Three kinds of optimization are done within the P/FDM Mediator. First, the rewriter can apply rules that perform *semantic query optimization*. Additionally, rewrite rules [20] can be used to implement the logical rules given by Jarke and Koch [34]. They can implement many forms of rewrites based on data semantics, as discussed in King [35]. They can spot opportunities to replace iteration by indexed search [33]. In experiments using the AMOS II system [36] as a remote resource, rewrite rules were able to implement flattening and un-nesting transformations that prevent wasting time compiling subqueries in AMOSQL. A similar approach could be adapted to features of other DBMSs. Most importantly, rewrites that change the relative workload between two processors in a distributed query can be performed. Finally, all these rewrites can be combined, as some of them will enable others to take place. Thus, one can deal with many combinations without having to foresee them and code them individually.

Second, the optimizer performs generic query optimizations. The philosophy of the optimizer is to use heuristics to improve queries. It examines alternative execution plans and, although it uses a simple cost model, it is successful in avoiding inefficient strategies, and it often selects the most effective approach [15]. The optimizer was subsequently rewritten using a simple heuristic to avoid the combinatorial problem of examining all possible execution plans for complex queries [22].

Third, the query splitter attempts to group together query elements into chunks that can be sent as single units to the external data resources, thus providing the remote system with as much information as possible to give it greater scope for optimizing the sub-query.

Outside the mediator, the approach is able to take advantage of the optimization capabilities of the external resources.

There is scope for introducing adaptive query processing techniques to improve the execution plans as execution proceeds and as results are returned to the mediator [37], but this has not yet been done in our prototype system.

9.2.2   User Interfaces

While queries against a P/FDM schema can be formulated directly in either Prolog or Daplex, this requires some programming ability and care must be taken to use the correct syntax. Therefore, two interfaces were developed to formulate queries without the user having to learn to program in either Prolog or Daplex. Both of these interfaces have a representation of the schema at their heart, and they enable the user to construct well-formed Daplex queries by clicking and typing values for attributes to restrict the result set.

A Java-based visual interface for P/FDM [38] was developed with a graphical representation of the database schema at its center. Figure 9.13 shows this interface in use with the schema for an antibody database [7]. Users construct queries by clicking on entity classes and relationships in the schema diagram and constraining the values of attributes selected from menus. As this is done, the Daplex text of the query under construction is built up in a sub-window (the query editor window). Queries are submitted to the database via a CORBA interface [21]. Results satisfying the selection criteria are displayed in a table in a separate result window. A particularly novel feature of the interface is *copy-and-drop*, which enables the user to select and copy data values in the result window and then drop these into the query editor window. When this is done, the selected values are merged into the original query automatically, in the appropriate place in the query text, to produce a more specialized query. The Java-based interface runs as a Java application, but it does not yet run within a Web browser.

In addition, a Web interface was developed with hypertext markup language (HTML) forms and accesses the mediator via a CGI program (Figure 9.14). Such interfaces can be generated automatically from a schema file. The interface's front

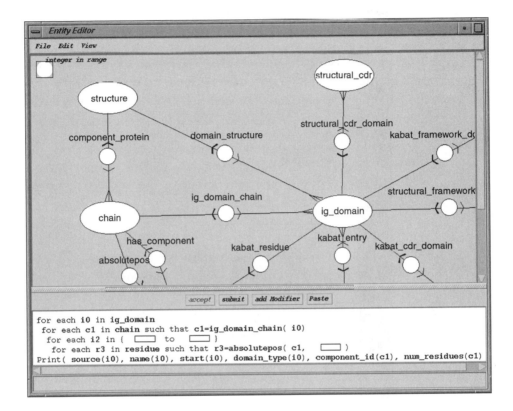

9.13
FIGURE
Visual Navigator query interface [38].

page lists the entity classes in the schema, and the user selects one of these as the starting point for the query. As the query is built up, checkboxes are used to indicate those attributes whose values are to be printed. The user can constrain the value of an attribute by typing into its entry box (e.g., 1.1.1.1 or <2.5) and can navigate to related objects using the selection box labeled *relationships* at the bottom of each object's representation within the Web page. Figure 9.14 shows the Web interface at the point where the user has formulated the query used in the example in Section 9.1.5. Pressing the *Submit* button will cause the equivalent Daplex query to be generated.

When using a graphical user interface that supports *ad hoc* querying, it is easy for naive queries that involve little or no data filtering to be expressed. This can result in queries that request huge result sets from remote resources. An alternative approach, as in TAMBIS, would be to provide only user interfaces that guide the user toward constructing queries with a particular structure and that have a suitable degree of filtering. However, such an interface would constrain the user to

Netscape: Mediator Demo

File   Edit   View   Go   Communicator   Help

Use the checkboxes on the left of the page to select the attributes that you want to print.
To constrain the value of an attribute, type into its entry box. To navigate to a related
object, use the selection box labelled **relationships**. Press **Submit** to proceed.

enzyme-1

☐ 1 . 1 . 1 . 1   **ec_number** (string)

☐   **protein_code** (string)

☑   **protein_name** (string)

Linked to **swissprot_entry-1** by relationship **swissprot_entries**.

| none   ⬜ | **relationships**

swissprot_entry-1

☐   **id** (string)

☑   **acc** (string)

☐   **dat** (string)

☑   **def** (string)

☐   **cc** (string)

| none   ⬜ | **relationships**

Submit   Reset

Mediator Home-Page | Bioinformatics Research Group | Computing Science | University of Aberdeen

{nicos,gjlk}@csd.abdn.ac.uk

9.14

FIGURE

Web-based query interface.

asking only parameterized variants of a set of canned queries anticipated by the interface designer. While such interfaces could be implemented easily, P/FDM design specification favors that users have the freedom to express arbitrary queries against a schema, and an area for future work is identifying and dealing with queries that could place unreasonable loads on the component databases in the federation.

Current interfaces do not provide personalization capabilities. It is, however, possible to provide users with their individual views of the federation (see E_M schemas in Figure 9.5), but this would be done by the database federation's administrator on behalf of users, rather than by users themselves.

9.2.3   Scalability

When a new external resource is added to the federation, the contents of that resource must be described in terms of entities, attributes, and relationships—the basic concepts in the FDM. For example, entity classes and attributes are used to describe the tables and columns in a relational database, the classes and tags in an AceDB database, and the databanks and fields accessed by SRS. The integration schema has to be extended to include concepts in the new resource, and mapping functions to be used by the ICode rewriter must be generated. Because the mediator has a modular architecture in which query transformation is done in stages, the only new software components that might have to be written are code generators and wrappers—the components shown with dark borders in Figure 9.6. All other components within the mediator are generic. However, the federation administrator might want to add declarative rewrite rules that can be used by the rewriter to improve the performance of queries involving the new resource. Code generators for new data sources can be written in one or two days when using existing code generators as a guide. In general, as the expressions being evaluated obey the principles of substitutability and referential integrity, expressions that match patterns in rewrite rules can be substituted with other expressions that have the same value. This means new mappings can be added without the risk of encountering special cases or some arbitrary limit on the complexity of expressions, as can happen with SQL.

A federated database system and a mediator system are similar architectures that differ in terms of how easily one can attach new database sources. In a federated architecture, the integration schema is relatively fixed and designed with particular database sources in mind. Extra databases can be added, with some effort, by the database administrator. A mediator tries to integrate new databases available at their given Web addresses on the basis of descriptions provided by the end user. The whole process is more dynamic. When dealing with a new source, a good mediator will try to spot heuristic optimization rules it can re-use from similar databases it knows about. In general, it is more intelligent and less reliant on human intervention. A long-term goal is that, as a suite of code generators and wrappers is added to the P/FDM Mediator, it will become easy to add new resources by presenting the mediator with new remote schemas and specifying which code generators and wrappers should be used.

9.3   CONCLUSIONS

The P/FDM Mediator is a computer program that supports transparent and integrated access to different data collections and resources. *Ad hoc* queries can be asked against an integration schema, which is a pre-defined collection of entity classes, attributes, and relationships. The integration schema can be extended at any time by adding declarative descriptions of new data resources to the mediator's set-up files.

Rather than building a data warehouse, the developed system brings data from remote sites on demand. The P/FDM Mediator arranges for this to happen without further human intervention. The presented approach preserves the autonomy of the external data resources and makes use of existing search capabilities implemented in those systems.

Bioinformatics faces a "crisis of data integration" [1], which is best addressed through federations that allow their constituent databases to develop autonomously and independently. The existence of schemas at different levels, as shown in Section 9.1.3, makes apparent the requirements for query transformation in a mediator in a database federation. The transformations in the system are all based on well-defined mathematical theory using function composition, as pioneered by Shipman [2] and Buneman [12]. This results in a modular design for the mediator that enables the federation to evolve incrementally.

ACKNOWLEDGMENT

The prototype P/FDM Mediator described in this chapter was implemented by Nicos Angelopoulos. This work was supported by a grant from the BBSRC/EPSRC Joint Programme in Bioinformatics (Grant Ref. 1/BIF06716).

REFERENCES

[1]   R. J. Robbins. "Bioinformatics: Essential Infrastructure for Global Biology." *Journal of Computational Biology* 3, no. 3 (1996): 465–478.

[2]   D. W. Shipman. "The Functional Data Model and the Data Language DAPLEX." *ACM Transactions on Database Systems* 6, no. 1 (1981): 140–173.

[3]   B. Rieche and K. R. Dittrich. "A Federated DBMS-Based Integrated Environment for Molecular Biology." *In Proceedings of the Seventh International Conference on Scientific and Statistical Database Management*, 118–127. Los Alamitos, CA: IEEE Computer Society, 1994.

[4]   G. Wiederhold. "Mediators in the Architecture of Future Information Systems." *IEEE Computer* 25, no. 3 (1992): 38–49.

[5]   P. D. Karp. "A Vision of DB Interoperation." In *Proceedings of the Second Meeting on the Interconnection of Molecular Biology Databases*, Cambridge, UK. July 20–22, 1995.

[6]   P. M. D. Gray, K. G. Kulkarni, and N. W. Paton. *Object-Oriented Databases: A Semantic Data Model Approach*. Hemel Hempstead, Hertfordshire: Prentice Hall International, 1992.

[7]   G. J. L. Kemp, Z. Jiao, P. M. D. Gray, et al. "Combining Computation with Database Accesss in Biomolecular Computing." In *Applications of Databases: Proceedings of the First International Conference*, edited by W. Litwin and T. Risch, 317–335. Heidelberg, Germany: Springer-Verlag, 1994.

[8]   L. Kerschberg and J. E. S. Pacheco. *A Functional Data Model*. Rio de Janeiro, Brazil: Department of Informatics, Universidade Catolica Rio de Janeiro, 1976.

[9]   K. G. Kulkarni and M. P. Atkinson. "EDFM: Extended Functional Data Model." *The Computer Journal* 29, no. 1 (1986): 38–46.

[10] M. Atkinson, F. Bancilhon, D. DeWitt, et al. "Deductive and Object-Oriented Databases." In *Proceedings of the 1st International Conference on Deductive and Object-Oriented Databases (DOOD '89)*, edited by W. Kim, J-M. Nicholas, and S. Nishio, 223–240. Amsterdam, The Netherlands: North-Holland, 1990.

[11] P. M. D. Gray, D. S. Moffat, and N. W. Paton. "A Prolog Interface to a Functional Data Model Database." In *Advances in Database Technology—EDBT '88*, edited by J. W. Smith, S. Ceri, and M. Missikoff, 34–48. Heidelberg, Germany: Springer-Verlag, 1988.

[12] P. Buneman and R. E. Frankel. "Fql: A Functional Query Language." In *Proceedings of the ACM SIGMOD International Conference on Management of Data*, edited by P. A. Bernstein, 52–58. Boston: ACM Press, 1979.

[13] D. A. Turner. "Miranda: A Non-Strict Functional Language with Polymorphic Types." In *Functional Programming Languages and Computer Architecture*, Lecture Notes in Computing Science, Vol. 201, edited by J-P. Jouannaud, 1–16. Heidelberg, Germany: Springer-Verlag, 1985.

[14] T. A. Landers and R. L. Rosenberg. "An Overview of MULTIBASE." In *Distributed Data Bases, Proceedings of the 2nd International Symposium on Distributed Data Bases*, edited by H-J. Schneider, 153–184. Amsterdam, The Netherlands: North-Holland, 1982.

[15] N. W. Paton and P. M. D. Gray. "Optimising and Executing Daplex Queries Using Prolog." *The Computer Journal* 33, no. 6 (1990): 547–555.

[16] ANSI Standards Planning and Requirements Committee. "Interim Report of the ANSI/X3/SPARC Study Group on Data Base Management Systems." *FDT—Bulletin of ACM SIGMOD* 7, no. 2 (1975): 1–140.

[17] T. Ezold and P. Argos. "SRS Indexing and Retrieval Tool for Flat File Data Libraries." *Computer Applications in the Biosciences* 9, no. 1 (1993): 49–57.

[18] S. Grufman, F. Samson, S. M. Embury, et al. "Distributing Semantic Constraints Between Heterogenous Databases." In *Proceedings of the 13th Annual Conference on Data Engineering*, edited by A. Gray and P-A. Larson, 33–42. New York: IEEE Computer Society Press, 1997.

[19] A. P. Sheth and J. A. Larson. "Federated Database Systems for Managing Distributed, Heterogenous and Autonomous Databases." *ACM Computing Surveys* 22, no. 3 (1990): 183–236.

[20] G. J. L. Kemp, P. M. D. Gray, and A. R. Sjöstedt. "Improving Federated Database Queries Using Declarative Rewrite Rules for Quantified Subqueries." *Journal of Intelligent Information Systems* 17, no. 2–3 (2001): 281–299.

[21] G. J. L. Kemp, C. J. Robertson, P. M. D. Gray, et al. "CORBA and XML: Design Choices for Database Federations." *In Proceedings of the Seventeenth British National Conference on Databases*, edited by B. Lings and K. Jeffery, 191–208. Heidelberg, Germany: Springer-Verlag, 2000.

[22] Z. Jiao and P. M. D. Gray. "Optimisation of Methods in a Navigational Query Language." In *Proceedings of the Second International Conference on Deductive and Object-Oriented Databases*, edited by C. Delobel, M, Kifer, and Y. Masunaga, 22–42. Heidelberg, Germany: Springer-Verlag, 1991.

[23] P. Buneman, L. Libkin, D. Suciu, et al. "Comprehension Syntax." *SIGMOD Record* 23, no. 1 (March 1994): 87–96.

[24] G. J. L. Kemp and P. M. D. Gray. "Finding Hydrophobic Microdomains Using an Object-Oriented Database." *Computer Applications in the Biosciences* 6, no. 4 (1990): 357–299.

[25] R. Neches, R. Fikes, T. Finin, et al. "Enabling Technology for Knowledge Sharing." *Artificial Intelligence* 12, no. 3 (1991): 36–56.

[26] R. J. Bayardo, B. Bohrer, R. S. Brice, et al. "InfoSleuth: Semantic Integration of Information in Open and Dynamic Environments." In *SIGMOD 1997, Proceedings of the ACM SIGMOD International Conference on Very Large Data Bases*, edited by J. Peckham, 195–206. New York: ACM Press, 1997.

[27] P. M. D. Gray, A. D. Preece, N. J. Fiddian, et al. "KRAFT: Knowledge Fusion from Distributed Databases and Knowledge Bases." In *Proceedings of the 8th International Workshop on Database and Expert Systems Applications*, edited by R. R. Wagner, 682–691. Los Alamitos, CA: IEEE Computer Society Press, 1997.

[28] G. J. L. Kemp, J. J. Iriarte, and P. M. D. Gray. "Efficient Access to FDM Objects Stored in a Relational Database." In *Directions in Databases: Proceedings of the Twelfth British National Conference on Databases*, edited by D. S. Bowers, 170–186. Heidelberg, Germany: Springer-Verlag, 1994.

[29] P. Mitra, G. Weiderhold, and M. Kersten. "A Graph-Oriented Model for Articulation of Ontology Interdependencies." In *Advanced Database Technology—EDBT 2000*, edited by C. Zaniolo, P. C. Lockerman, M. H. Scholl, et al., 86–100. Heidelberg, Germany: Springer-Verlag, 2000.

[30] P. Buneman, S. B. Davidson, K. Hart, et al. "A Data Transformation System for Biological Data Sources." In *VLDB '95, Proceedings of the 21st International Conference on Very Large Data Bases*, edited by U. Dayal, P. M. D. Gray, and S. Nishio, 158–169. San Francisco: Morgan Kaufmann, 1995.

[31] L. Wong. "Kleisli: Its Exchange Format, Supporting Tools and an Application in Protein Interaction Extraction." In *Proceedings of the IEEE International Symposium on Bio-Informatics and Biomedical Engineering*, 21–28. New York: IEEE Computer Society Press, 2000.

[32] N. W. Paton, R. Stevens, P. Baker, et al. "Query Processing in the TAMBIS Bioinformatics Source Integration System." In *Proceedings of the 11th International Conference on Scientific and Statistical Database Management*, 138–147. New York: IEEE Computer Society Press, 1999.

[33] L. M. Haas, P. Kodali, J. E. Rice, et al. "Integrating Life Sciences Data—With A Little Garlic." In *IEEE International Symposium on Bio-Informatics and Biomedical Engineering*, 5–12. New York: IEEE Computer Society Press, 2000.

[34] M. Jarke and J. Koch. "Range Nesting: A Fast Method to Evaluate Quantified Queries." In *SIGMOD '83, Proceedings of the Annual Meeting*, edited by D. J. DeWitt and G. Gandarin, 196–206. Boston: ACM Press, 1983.

[35] J. J. King. *Query Optimisation by Semantic Reasoning*. Ann Arbor, MI: University of Michigan Press, 1984.

[36] T. Risch, V. Josifovski, and T. Katchaounov. *AMOS II Concepts*, June 23, 2000, *http://www.dis.uu.se/~udbl/amos/doc/amos_concepts.html*.

[37] Z. G. Ives, D. Florescu, M. Friedman, et al. "An Adaptive Query Execution System for Data Integration." In *SIGMOD 1999, Proceedings of the ACM SIGMOD International Conference on Management of Data*, edited by A. Delis, C. Faloutsos, and S. Ghandeharizadeh, 299–310. Boston: ACM Press, 1999.

[38] I. Gil, P. M. D. Gray, and G. J. L. Kemp. "A Visual Interface and Navigator for the P/DFM Object Database." In *1999 User Interfaces to Data Intensive Systems*, edited by N. W. Paton and T. Griffiths, 54–63. Los Alamitos, CA: IEEE Computer Society Press, 1999.

10 Integration Challenges in Gene Expression Data Management

Victor M. Markowitz, John Campbell, I-Min A. Chen, Anthony Kosky, Krishna Palaniappan, and Thodoros Topaloglou

DNA microarrays have emerged as the leading technology for measuring gene expression, primarily because of their high throughput. A single microarray experiment provides measurements for the messenger RNA (mRNA) transcription level for tens of thousands of genes in parallel [1]. While this technology opens new opportunities for functional genomics and drug discovery applications, it also presents new bioinformatics and data management challenges arising from the need to capture, organize, interpret, and archive vast amounts of experimental data. Furthermore, to support meaningful biological reasoning, gene expression data need to be analyzed in the context of rich sample and gene annotations.

GeneExpress is a data management system that contains quantitative gene expression information for thousands of normal and diseased samples and for experimental animal model and cellular tissues generated under a variety of treatment conditions [2]. Initially the GeneExpress system was developed with the goal of supporting effective exploration, analysis, and management of gene expression data generated at Gene Logic using the Affymetrix GeneChip platform [3], integrated with comprehensive information on samples, clinical profiles, and rich gene annotations. Building such a system required resolving various data integration problems to associate gene expression data with sample data and gene annotations. A subsequent goal for the GeneExpress system was to provide support for incorporating gene expression data generated outside of Gene Logic. Addressing this additional goal required the resolution of various levels of syntactic and semantic heterogeneity of sample data, gene annotations, and gene expression data, which were often generated under different experimental conditions. These goals have been addressed using a data warehousing methodology adapted to the special requirements of the gene expression domain [4].

This chapter discusses the challenges associated with data integration in the context of a gene expression data management system and describes how the GeneExpress system addresses these challenges. Section 10.1 provides an overview of the area of gene expression data management. Section 10.2 provides a brief description of Gene Logic's GeneExpress system. Section 10.3 discusses the key semantic problems associated with integrating gene expression and related data and how they are addressed in the context of GeneExpress. Section 10.4 describes how third-party gene expression data can be integrated into GeneExpress. A summary and observations in Section 10.5 concludes the chapter.

10.1   GENE EXPRESSION DATA MANAGEMENT: BACKGROUND

The gene expression data application is reviewed briefly in this section. First discussed are the data spaces that need to be modeled by a gene expression data management system, then initiatives to establish standards for gene expression and related data.

10.1.1   Gene Expression Data Spaces

Gene expression systems measure mRNA transcription level of protein-coding genes in a cell. The mRNA *mix* used in gene expression experiments is derived from biomaterials (samples) such as tissues and cell lines. A microarray typically is designed to detect thousands of specific target sequences associated with these genes through hybridization. The reported measurements are meaningful only when something is known about the samples and the target sequences and their associated genes. The first goal of gene expression data management is to integrate expression data with sample and gene annotations and to allow users to use these annotations to explore, analyze, and interpret expression data [4, 5]. Typically, a gene expression data management system integrates data from three different data spaces: sample annotations, gene annotations, and gene expression measurements, each of which is described in the following sections.

Biological Sample Data Space

The main object in the sample data space is the *sample* representing the biological material that is the focus of an experiment. Samples originate from a variety of sources with different data standards and handling protocols. Annotations associated with each sample should address its physical features and quality, as well as the accuracy and extent of the information recorded. Ultimately, sample data are recorded in the sample data space of a gene expression system. A sample can be of tissue, cell, or processed RNA type, and it originates from a donor organism of a given species (e.g., human, mouse, rat). Attributes associated with samples describe their nature and condition (e.g., organ site, diagnosis, disease, stage of disease), as well as donor information (e.g., demographic and clinical record for human donors or strain, genetic modification, and treatment information for animal donors). Samples are commonly organized in groups that can be further grouped into studies or projects, such as time/dose studies. Information on how samples in such groups are related to one another is therefore a necessary annotation for the sample data space.

Gene Annotations Data Space

Gene annotations help to associate the expression data reported for sequence fragments on a microarray to biological entities such as genes and proteins. The main problem here is that sequence annotations, and annotations of the function of known genes, can change over time as the availability of more sequence, better computational tools, and new research lead to better gene prediction results. Furthermore, the sources for gene annotations are usually primary or consolidated databases that are heterogeneous and may contain inconsistent data. Consequently, the effort of keeping up-to-date gene annotation data for sequence fragments on microarrays combines the complexities of database integration with the ongoing research in the field of gene identification.

The main object in the gene annotation data space is the *gene fragment*, representing an entity for which the expression level is being determined. For microarray technologies, gene fragments are associated with a specific microarray type, such as a GeneChip human probe array (e.g., HG_U95A). The annotations associated with a gene fragment describe its biological context, including its associated primary expressed sequence tag (EST) sequence entry in GenBank; membership in a gene-oriented sequence cluster; association with a known gene (i.e., a gene that is recorded in an official nomenclature catalog, such as the Human Gene Nomenclature Database [HUGO] [6]); functional characterization, such as Gene Ontology (GO) annotations; and association to known metabolic and signaling pathways.

Gene Expression Measurement Data Space

Gene expression microarray systems are broadly classified into *single channel* and *two channel* systems. A single channel system takes a single sample of biological material and provides absolute measures of gene expression for that sample, while a two channel system takes a pair of samples and provides measurements of the difference in relative gene expression between them. Single channel systems are best represented by the Affymetrix GeneChip platform [7]. This chapter focuses on the management of gene expression data generated using the GeneChip platform. Note, however, that most data management and integration issues discussed in this chapter apply to gene expression data in general, regardless of the underlying technology platform.

Typically, data generated by a microarray system can be classified into three data types, each representing a different level of abstraction. This hierarchy of data types is common, with slight differences, to all microarray platforms and consists of:

1. *Raw data* consisting of binary image files generated by scanners

2. *Grid* or *probe intensity* data consisting of values associated with each probe or oligonucleotide sequence examined on a microarray

3. *Gene expression estimates* generated by combining data on related probes on a microarray

Each data type may have multiple data formats or representations associated with it, such as text or binary file-based formats or database representations.

The transformation between data types is carried out by platform-specific algorithms. It is not uncommon to use more than one algorithm to transform data from one data type to the next [8, 9]. The following paragraphs briefly describe the hierarchy of data types in the context of the GeneChip platform.

Affymetrix's GeneChip microarrays (also called *probe arrays*) are tiled with oligonucleotide sequences, each 25 base-pairs in length, known as *probes*. Each probe is designed to hybridize to a known mRNA fragment representing a target gene or EST. Probes are grouped into *probe pairs*, each of which consists of a *perfect-match* (PM) and a *mismatch* (MM) probe, with the MM probe being created from the PM probe by changing the middle (13th) base to measure non-specific binding. Each target gene or EST is represented by a *probe set* consisting of up to 20 probe pairs.

A GeneChip probe array experiment involves preparing the RNA sample, carrying out the probe array experiment (hybridization, washing, staining), and scanning the probe array [7]. The scanning process generates a file containing an image of the probe array, which constitutes the *raw data*.

The scanned images are interpreted using methods such as the GeneChip microarray suite (MAS) analysis algorithms. The MAS *cell averaging* algorithm averages pixel intensities and computes cell-level intensities in which each cell represents one probe on the probe array. The output from this process is a file containing the estimated intensities for each probe on the probe array, which constitutes the *probe data*. These intensities indicate the amount of hybridization that occurred for each oligonucleotide sequence on the array.

Probe intensity files can be further analyzed with methods such as the MAS *chip analysis* algorithms, which generate *gene expression estimates* by summarizing the intensities of each probe set that corresponds to a gene or EST fragment targeted by the probe array. Alternative gene expression estimates may be based on single or multiple (e.g., replicate) experiments.

The GeneChip Laboratory Information Management System (LIMS) provides support for transforming data between the different data types and for loading the gene expression estimates into a relational database based on the Affymetrix Analysis Data Model (AADM) [10].

The different data types and their associated formats result in files or data structures of different sizes. For example, for an experiment using an HG_U133 GeneChip probe array, the raw image file is around 45 megabytes in size, the probe intensity data file is around 12 megabytes, and the summarized gene expression data consists of roughly 22,000 values.

10.1.2   Standards: Benefits and Limitations

Effective exploration of microarray data has been hindered by the variety and heterogeneity of the data formats used. This problem has been recognized by several organizations, such as the European Bioinformatics Institute (EBI), the U.S. National Center for Biotechnology Information (NCBI), and the National Center for Genome Resources (NCGR), in their efforts to establish public data repositories for gene expression information. Microarray manufacturers have also proposed formats, such as the AADM used for the GeneChip LIMS relational database [10], to facilitate data exchange between different sources of gene expression data and the development of gene expression analysis packages.

Different standardization efforts have been consolidated by the Microarray Gene Expression Database Group (MGED), a consortium of academic and commercial organizations with the shared goal of defining standard formats that will allow gene expression data repositories to share and exchange data. MGED has recently published Minimum Information About a Microarray Experiment (MIAME), a recommendation for the minimum information required for a microarray experiment [5], and has developed a data exchange format (Microarray Gene Expression Markup Language [MAGE-ML]) and object model (Microarray Gene Expression Object Model [MAGE-OM]) for microarray experiment data.

Existing definitions and proposed standards for gene expression data provide useful guidelines for organizing expression data in systems such as GeneExpress. Adequate standards for the representation of sample and gene annotations, however, have not yet been established. MIAME's recommended standards for gene annotation for the fragments on a microarray are minimal to simplify compliance. For example, the suggested annotations for probes on a microarray consist of their identity, sequence, and the associated composite target sequence, along with gene symbol or reference to a model organism database. However, in-depth gene expression data analysis requires access to functional characteristics of these target gene fragments to interpret data analysis results.

Similarly, MIAME's minimum required sample annotations are not sufficient to establish the context needed for comprehensive gene expression data analysis. Clinical history, morphology, and pathology for samples are needed to interpret gene expression data. For example, it is necessary to know the precise stage of a tumor or medications taken during acquisition of a cancerous sample to interpret expression measurements for the sample.

For sample data, standardization involves establishing controlled vocabularies of terms for specific data domains, such as the Systematized Nomenclature for Medicine (SNOMED) [11] for anatomy or diseases. These efforts are usually sponsored by professional organizations within a specific field (e.g., SNOMED is supported by the College of American Pathologists) and are not easily accessible to academic organizations because of their associated costs.

For gene annotations, the most notable standardization effort is the development of the Gene Ontology (GO) by the GO Consortium [12]. The goal of GO is to provide a dynamic controlled vocabulary to describe the role of genes and gene products in terms of molecular function, biological process, and cellular components.

Data exchange formats or standards emphasize the syntactic aspects of expression data and, to a lesser degree, the meaning of the data in cases where the representation is well documented. However, these formats do not address the semantic issues regarding the comparability (or compatibility) of gene expression data. Data comparability is a prerequisite for analyzing expression data from multiple experiments or multiple sites together and is discussed in Section 10.3.

10.2   THE GENEEXPRESS SYSTEM

Gene Logic's GeneExpress system provides support for managing expression data generated using the Affymetrix GeneChip platform in a high throughput production environment. Sample, gene annotation, and gene expression data are collected from separate data sources: Sample data are collected and managed using a sample data management system; gene annotations are acquired from a variety of public and private genome databases and integrated into a gene annotation database; and the main source for gene expression data is an Affymetrix GeneChip LIMS database. GeneExpress was built using data warehousing and online analytical processing (OLAP) concepts adapted to the gene expression data domain [4].

10.2.1   GeneExpress System Components

The GeneExpress data store consists of the GeneExpress Data Warehouse (GXDW). GXDW is made up of component databases containing sample, gene annotation, and gene expression data and process information specific to the generation and analysis of the expression data [13].

The gene expression data in GXDW is represented by a three-dimensional array with expression values indexed by gene fragments (identified by their target sequence and the microarray type), samples, and algorithm or measurement type. This data structure is implemented by the Gene Expression Array (GXA) as a collection of matrices, each associated with a particular GeneChip probe array type (e.g., HG_U95A) and measurement type (e.g., a version of the MAS algorithm). Each matrix has axes representing samples and gene fragments. The GXA provides a basis for the GeneExpress Analysis Engine, which implements various analysis methods in a highly efficient manner.

The GXDW, GXA, and Analysis Engine applications reside on a GeneExpress server. The server also hosts the Workspace File System, which allows users to store analysis results and share them throughout an organization.

Data in GXDW can be accessed using the GeneExpress Explorer application, which provides support for specifying gene and sample sets of interest and for analyzing gene expression data in the context of such gene and sample sets using a variety of analysis tools. GeneExpress Explorer is implemented as a client-side Java application, which runs on desktops and accesses GXDW through Java DataBase Connectivity (JDBC) and the analysis server through a CORBA layer. The main components and architecture of the GeneExpress system are illustrated in Figure 10.1. The results of gene expression analysis can be examined in the context of gene annotations, such as pathways, and can be exported to third-party tools, such as Spotfire, GeneSpring, or Partek, for visualization or further analysis. The gene expression and associated data can also be accessed directly through Application Programming Interfaces (APIs), which are available for a number of popular programming languages and platforms.

10.2.2   GeneExpress Deployment and Update Issues

In most cases, a GeneExpress system for a particular customer resides on a dedicated server. These machines are either deployed at the customer site and connected to the customer's internal network, or they are located at Gene Logic and accessed via a Virtual Private Network (VPN) mechanism. The data content of each GeneExpress system, involving both GXDW and the GXA matrices, is updated on a regular schedule (e.g., bi-monthly or quarterly).

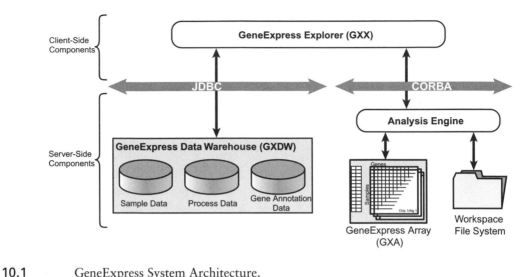

GeneExpress System Architecture.

FIGURE

The sample, process, and gene expression data components of GXDW are built by extracting the data for the relevant samples from a master *production* version of GXDW, which is maintained at Gene Logic. The subset of samples provided to each GeneExpress customer is determined by the specific GeneExpress product license for the customer and will usually contain new samples that have been processed by Gene Logic since the last content update. The sample, process, and gene expression portions of the production GXDW are maintained in an incremental fashion, with new samples and experiments being added as they become available. Similarly, the set of GXA matrices for a particular customer is built by extracting the portions of the internal production GXA matrices that pertain to the samples being supplied to the customer.

The update mechanism for the gene annotation data component of GXDW is somewhat different. To keep abreast of current genomic data available in the public domain, it is necessary to refresh the gene annotation database periodically. The static portion of the data, such as gene fragments and array design, will not change unless new arrays are introduced. However, links to genes and all the public genomic objects may change to reflect new versions of their data sources. Because of the complex interdependencies of the various genomic data sources and the fact that many such data sources do not provide incremental updates, it is not feasible to update the gene annotation database in an incremental fashion. Instead, it must be completely reloaded each time it is refreshed. This process is usually performed on a quarterly basis because of the high overhead involved.

10.3   MANAGING GENE EXPRESSION DATA: INTEGRATION CHALLENGES

This section presents some of the key challenges that arise from the management of gene expression and related data and briefly describes how each of these challenges is addressed in the GeneExpress system. Many of these challenges involve resolving semantic conflicts in gene expression, sample, and gene annotation data to integrate these data in a gene expression data management system. This section discusses the data management problems caused by differences in microarray versions, differences in algorithms and normalizations, and non-biological variability in expression data are discussed first, followed by challenges regarding sample data and gene annotation data.

10.3.1   Gene Expression Data: Array Versions

Microarray platforms keep evolving with new probe array versions benefiting from technological improvements (e.g., higher density arrays and better probe selection) and advances in deciphering the genome. For example, Affymetrix recently released the HG_U133 series of the human probe arrays, which replaced the previous HG_U95 series of arrays. Running the same or similar samples on two series of probe arrays doubles the amount of data generated. However, in many cases, this is necessary because the newer arrays may produce expression data for target transcript sequences that are not available on the previous versions. In addition, there may be multiple versions of a probe array within a particular array series if problems are discovered with a particular array (e.g., HG_U95A versions 1 and 2 within the HG_U95 series). Comparing data generated using different series of probe arrays entails addressing a complex semantic data integration problem, with gene annotation data providing only partial support for resolving it.

In general, data generated using different probe array series or versions are not comparable, nor can they be transformed to make them comparable. This is in part due to the selection of target genes and ESTs for new probe arrays, which are often based on newly published biological information. Furthermore, representative probes for the target genes on the new probe arrays may be different due to availability of better representative sequences or improved techniques for choosing oligos within a representative sequence. New probe arrays may also be associated with improved analysis algorithms for determining summary intensity values, which will not be directly comparable with older algorithms. Consequently, in order to allow comparison of gene expression data generated for new samples using new probe arrays with data for existing samples, it is necessary to re-run the existing samples using the new probe array versions and algorithms.

On the other hand, it is often still valuable to maintain data generated using older probe arrays because they may provide the basis for existing analyses or prediction models, which users do not wish to re-create, because sample material may no longer be available for re-running the experiments using new arrays, or because samples may no longer be considered important enough to warrant re-running them using new arrays. Further, older probe arrays may include gene fragments of interest that have been omitted or do not have a good representation on the newer arrays.

GeneExpress supports multiple probe array sets for each species and allows users to choose a probe array set in addition to a species when performing analyses. Annotations associating homologous or related gene fragments on different versions of a probe array are provided in the gene annotation database of GXDW and can be used to map fragments on a given probe array to fragments on another version of the probe array. Direct comparisons of gene expression data based on different probe arrays are not supported.

The amount of data generated with multiple probe array versions is kept manageable, in part, because GXDW and GXA contain only the estimates for expression measures and gene-level summary data. Images and probe intensity files are archived on an enterprise network-accessed storage system and are not incorporated into standard GeneExpress systems. When a new algorithm or a new probe array version needs to be supported within GeneExpress, the information describing the probe array design must be entered in the gene expression data space and a new matrix included in the GXA.

10.3.2   Gene Expression Data: Algorithms and Normalization

Different algorithms can be applied to generate gene expression data at different levels including image, probe level, and gene expression estimate data. For example, recently several alternative methods have been developed to estimate expression measures from probe data ([8, 9]) in addition to Affymetrix' GeneChip MAS algorithms. For GeneChip, the MAS 5.0 algorithm has recently replaced the MAS 4.0 algorithm and is required for analyzing the data generated with the newer versions of probe arrays. To take advantage of a new or alternative algorithm, it is necessary to re-analyze raw or probe data and generate new estimates of gene expression. It is important to note that expression estimates generated by different algorithms are not directly comparable. Furthermore, some algorithms depend on certain parameters that may also affect the generated expression estimates.

In GeneExpress, a number of factors are recorded that may determine the comparability of expression data, including the following.

1. The algorithms employed to generate expression estimates, namely MAS 4.0 (employed for all probe arrays through the end of 2001) or MAS 5.0 (required for the new HG-U133 probe arrays and optional for other probe arrays) are recorded. Data generated using different algorithms are not comparable.

2. Scaling factors used to reduce discrepancies caused by sample preparation or probe array lot variability are also recorded. Data generated using different scaling factors are transformed to a common factor using straightforward multiplication.

3. Normalizations, that may be applied to the values generated by the MAS or other algorithms are recorded: GeneExpress provides support for several normalization methods including Standard Curve Normalization, based on using spike-ins of known concentrations for certain (bacterial) genes when preparing samples for experiments [14]. Data must be generated using the same normalization to be comparable.

The Gene Expression analysis software, GeneExpress Explorer, ensures that data analyzed together have been generated using the same algorithms and normalization methods.

10.3.3   Gene Expression Data: Variability

Determining if gene expression data from two or more sources, such as different organizations or different sites within an organization, are comparable involves assessing non-biological differences that may affect analysis results. While gene-to-gene differences and sample-to-sample differences will be present in any set of experimental data, it is important to determine if there are other significant sources of variability. Many factors may contribute to such variability, including differences in the processes for obtaining and storing samples; differences in experimental practices and techniques; differences in adjustment of equipment, such as scanners; and so on.

Statistical methods are used to identify the magnitude and qualitative nature of non-biological variability. Initial exploration ideally involves samples collected from the same type of tissue (i.e., from the same type of organ and a similar location in the organ) and with the same pathology. In this case, data comparability can be assessed using the entire set of genes involved in the experiments. If samples are from the same type of tissue but with different pathologies, data comparability can be assessed using only genes that are not likely to be involved in the biological difference between the two groups of samples.

Exploratory statistical techniques employed for assessing the comparability of such samples include univariate (single experiment) and bivariate (pairs of experiments) analyses. One simple way to compare numerous univariate distributions is by displaying boxplots of the distributions side by side [15]. Such boxplots would indicate whether there are significant effects due to, for example, scaling or saturation, which would result in a shift in the distribution of expression values. Further exploration would involve assessing the reproducibility of expression values between experiments and the variability of expression values within each group of experiments and between groups of experiments.

Gene Logic limits non-biological sources of variability in the gene expression data it generates by following strictly controlled procedures and monitoring the quality control measures, both for running experiments and for the collection and preparation of samples. Once data are generated from experiments, quality control procedures based on statistical methods are used to ensure that data included in GeneExpress are not unduly affected by non-biological factors.

10.3.4 Sample Data

Accurate and consistent characterization of samples is essential in dealing with gene expression data because errors can have a substantial effect on expression analysis. It is not sufficient to base sample classification solely on annotations provided by the supplier because (1) samples may be mis-labeled (e.g., a diseased tissue being labeled as normal) and (2) there may be inconsistencies of classification due to the perspective of the pathologist or scientist who did the initial labeling. In the GeneExpress system sample classification validation involves a careful review of the micro-section images by a pathologist and a thorough review of the clinical information accompanying each sample. Using SNOMED [11], the sample can be further characterized by topography, morphology, disease, and disease stage. The use of SNOMED and other controlled vocabularies in the GeneExpress system leads to a more robust classification of samples and provides a consistent representation of the data to users. However, even with an established controlled vocabulary such as SNOMED, the choice of terms to characterize a tissue type or disease may be ambiguous, so Gene Logic's pathologists use a consistent system of rules to determine which SNOMED terms to use.

10.3.5   Gene Annotations

Associating gene fragments with annotations from various public and private data sources provides the genomic context for interpreting gene expression data. Integrating such annotations into a data warehouse, as opposed to accessing the remote data sources through a federated database approach (see, for example, Eckman et al.'s article in *Bioinformatics* [16]), allows better representation of the semantics, powerful query expression, improved query performance, and also allows the quality of the data to be checked during the integration process (a similar conclusion is reached in an *IBM Systems Journal* article by Davidson et al. [17]). Acquiring gene annotations from various data sources involves identifying important and reliable data sources, regularly querying these sources, parsing and interpreting the results, and establishing associations between related entities, such as the correlation of gene fragments and known genes.

Gene annotation or gene index databases are generally based on data collected from well-established and reliable public data sources. For example, gene fragments can be organized in non-redundant classes based on UniGene, and associated with known genes recorded in LocusLink. However, such data sources may not contain genomic information for all species: Some may provide good human and mouse gene annotations but not cover other species such as yeast or rat. In such cases, it is necessary either to find alternative data sources or to derive gene annotations for these species by finding homologous genes on better annotated species, such as human or mouse. The choice of which approach to use may change from time to time depending on the availability of annotations.

Gene fragments are further associated with gene products (e.g., protein data from Swiss-Prot), GO ontology terms, enzymes, metabolic and signaling pathways, chromosome maps, genomic contigs, and cross-species gene homologies. For genomic information such as pathways, there is no unique data source that satisfies all needs. For example, the Kyoto Encyclopedia of Genes and Genomes (KEGG) provides good metabolic pathways, but it is not complete, while other public or private pathway data sources provide valuable additional data. Integration of similar or potentially overlapping data from two or more data sources requires the potential problems of redundant and inconsistent data to be addressed.

Genomic data sources are usually updated on different schedules, and the size of such data sources usually prohibits all versions of a data source from being loaded into a data warehouse. The gene annotation component of the GeneExpress data warehouse contains more than 5 gigabytes of data with only the most current version of data collected from various data sources. However, storing data from only one version of a data source may lead to inconsistencies; one source may reference entities in a different version of another data source, which may have been updated or may no longer exist. Further, data sources may change their data structure or schema between versions (e.g., adding, removing, or modifying attributes or fields). In addition, keywords can be changed, and data files can be reorganized. Such changes necessitate revisions of data collection tools and reconciliation of data mappings.

The gene annotation component of GXDW provides an integrated view of the genomic data space, based on a unified schema that spans the various object spaces relevant to each of the public or private data sources used. One key feature of the schema is that it models the primary objects from the genomic data space in a generic way, though such objects originate from a wide variety of data sources. This minimizes the frequency of schema changes needed, even as the structures of the primary data sources evolve.

To keep up-to-date with the evolving gene annotation data sources, the gene annotation component of GXDW is refreshed periodically. Each refresh involves extracting data from the latest versions of more than a dozen relevant public and private data sources, including UniGene, LocusLink, Swiss-Prot, Online Mendelian Inheritance in Man (OMIM), Enzyme, GO, KEGG, proprietary pathway databases, and model organism genome databases for organisms such as *E. coli* and yeast. During the integration and the assembly process, various data transformations and data cleansing operations are performed to resolve conflicts and correct data errors. Due to the rapidly evolving nature of these data sources, their content may change, both syntactically and semantically, between refreshes. Consequently, establishing cross-database links often requires manual curation to deal with orphans and links to retired entries. For example, LocusLink may refer to an Enzyme Commission (EC) number that is obsolete in the Enzyme catalog database, in which case it will be necessary to identify the correct, current EC number and update the data sources.

The data-warehousing strategy employed for constructing and maintaining GXDW supports various derived annotations such as cross-species homology relations between genes of different organisms and other objects. This is particularly valuable for comparative expression analysis between model organisms. The integration of genomic data sources helps uncover non-obvious relationships between genes, such as co-clustered gene fragments, and covers large parts of the genome -> transcriptome -> proteome -> metabolome information needed for gene expression analysis.

Due to the rapidly changing nature of the gene annotation data and data sources, it is important to search continually for new sources of gene annotation data and to re-evaluate existing data sources. When a new data source is considered for GeneExpress, decisions must be made regarding whether the new data source can or will replace any existing data source, whether existing curation methods must be modified, whether the data model or schema needs to be revised, and how existing data should be associated with data from the new source.

10.4 INTEGRATING THIRD-PARTY GENE EXPRESSION DATA IN GENEEXPRESS

The GeneExpress system was originally developed for the purpose of managing, exploring, and analyzing gene expression data generated at Gene Logic, primarily using the Affymetrix GeneChip platform. However, as the system has been adopted by various customers, some of which have their own internal efforts to generate gene expression data, the need to integrate customer data into the GXDW, so as to enable analysis of Gene Logic and customer gene expression data together, has become apparent.

To support the integration of customer sample and gene expression data into GeneExpress, the GX Connect tool has been developed at Gene Logic. GX Connect supports integration of gene expression data residing in an AADM-based GeneChip LIMS database and sample data conforming to the Gene Express Sample Data Exchange Format into GXDW. When there is a need to integrate gene annotation data,[1] gene expression data represented using alternative formats, or data that, for other reasons, cannot be integrated using GX Connect, custom data integration tools are developed.

The following section discusses some of the challenges involved in integrating customer gene expression data with Gene Logic data and how these challenges have been addressed in the context of GeneExpress. First, data exchange formats that simplify the tasks of developing and maintaining mappings of customer data to GXDW are described. Next described are some of the structural and semantic data transformation issues involved in developing such mappings. Finally, some of the data management issues associated with data loading and updating the Gene Logic content of a system containing both Gene Logic and customer data conclude the discussion.

10.4.1 Data Exchange Formats

To avoid developing and maintaining multiple data migration and loading tools for each external data source considered for integration, *data exchange formats* serve as intermediate representations for data being transferred from various data sources to the GeneExpress data warehouse. The process of integrating external data is then divided into two phases: (1) structural transformations and semantic mappings need to be applied to the external data to convert them into the data

---

1. Data exchange formats and integration tools for gene annotation data are planned for future versions of GX Connect.

exchange formats; (2) the data in the data exchange formats needs to be loaded into the warehouse. Note that developing and maintaining tools that convert data from sources into a well-defined data format, such as one based on extensible markup language (XML) or a similar notation, is generally easier than developing tools to transform data and populate a target data warehouse.

A number of formats have been proposed for gene expression data, as mentioned in Section 10.1.2. Because the focus was so far on integrating Affymetrix GeneChip expression data into GeneExpress, Affymetrix model AADM [10] was used as the data exchange format for gene expression data. In this format, expression data are associated with samples, gene fragments, analysis methods, and various experimental parameters.

For sample and clinical data, standard formats such as AADM have not yet been established. Consequently, data exchange formats that satisfy GX requirements were defined.

The central object class of the sample data exchange format is *sample*, representing the biological materials (e.g., tissue or cell-line) investigated using probe arrays (see Figure 10.2). Attributes associated with samples may describe their structural and morphological characteristics (e.g., organ site, diagnosis, disease, stage of disease). A sample is associated with a *donor* (e.g., a human or an animal model), which may in turn be qualified by various *treatments* and has additional attributes (e.g., clinical records and demographics for human donors or strain and genetic modification for animal donors). Each sample may be associated with several experiments (e.g., using different chip types). Samples may be grouped into *studies*, which may be further subdivided into *study groups* based on time or treatment parameters.

Various classes in the sample data exchange format include *catch-all* attributes that can accommodate any data, represented as tagged-value pairs, that do not otherwise fit the format.

For data represented in the data exchange formats described previously, the GX Connect tool can be used to control and automate the process of data transfer into the GeneExpress warehouse [2]. This tool can be deployed at customer sites

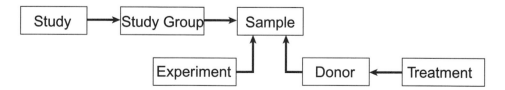

10.2   Sample data exchange format.

FIGURE

and be used to perform incremental (e.g., nightly) updates to GXDW. Consequently, the main task associated with integrating customer data becomes defining and implementing the semantic and structural transformations necessary to convert customer data into the data exchange formats, to prepare them for loading into GXDW.

10.4.2   Structural Data Transformation Issues

Data from individual data sources may be supplied in a flattened or un-normalized form, such as Microsoft Excel spreadsheets, so determining their structure and how to map them to the various data exchange formats is often a complex and involved task. First, it is necessary to determine the dependencies and correlations between individual data objects, which may be provided during the data export process or may need to be determined by searching for patterns in the data. In either case, it is necessary to confirm that the correlations found are consistent with the intended semantics of the data.

Data dependencies and correlations can be used to form an object model for the source data and to define a mapping from this model to the data exchange formats. Defining such a mapping requires structural conflicts between the models to be resolved, and in some cases, it may be necessary to choose between several possible solutions.

For example, the GeneExpress sample data exchange format classifies samples in a two-level hierarchy, with the levels represented by the classes *Study* and *Study-Group*. Sample data exported from an external data source might employ a three-level hierarchy, such as *Project, Study*, and *Treatment*. There are two possible ways to resolve such a difference in structure: Either combine the exported *Study* and *Treatment* classes into the sample data exchange format *Study-Group* class and map the exported *Project* class to sample data exchange format *Study* class or map the exported *Project* and *Study* classes to sample data exchange format *Study* class and the *Treatment* class to the *Study-Group* class.

In addition, it is necessary to deal with the evolution of databases and formats over time. Both the external data sources and the GeneExpress data warehouse may change either their structure or their controlled vocabularies or data formats to reflect changes in requirements. These changes require updates to the mappings.

10.4.3   Semantic Data Mapping Issues

For gene expression data, the semantic challenges of integrating data from multiple sources are similar to those described in section 10.3. Experimental data from different platforms are generally not comparable. Even if experiments are from the same platform, expression values may have to be adjusted (e.g., to compensate for different scanner settings) before they can be compared. Moreover, expression data will not be comparable unless they are analyzed using the same version of a probe array and the same algorithm.

The mappings for sample data are usually the most difficult because there is no widely accepted standard for representing clinical data [18]. In the following sections some of the problems of mapping sample and gene annotation data are discussed.

Sample Data Mapping: Studies

Expression data are often organized into studies. For Gene Logic data, *studies* are used to group data that address specific questions about the effects of certain variables (such as treatment conditions, disease stage, time, and so on) on gene expression levels. Studies may be further divided into *study groups*, which represent samples grouped according to certain attributes, such as specific treatment conditions, time points, or disease stages.

The structure and nature of a study performed outside of Gene Logic may be conceptually different from studies defined in the context of GeneExpress data. To group customer samples into studies or study groups, it is necessary to identify an equivalent structure in the source sample data model, which may use different terminology or organize data along different principles. If there is no appropriate concept in the source data model, rules can be incorporated into the mapping from the source data model into the sample data exchange format, allowing studies and study groups to be created based on other source data attributes, such as tissue type or treatment. Alternatively, customer data can be organized into studies and study groups manually by editing the data once they have been converted to the sample data exchange format.

Sample Data Mapping: Nomenclature

To map individual sample data values to the sample data exchange format, differences of nomenclature, units, and formatting must be resolved. Differences in nomenclature are the most difficult to deal with, and often there is no single, optimal resolution for such differences. Various attributes in the data exchange formats are represented using controlled vocabularies. In particular, in the sample data exchange format, sample organ types, pathologies, and disease diagnoses are represented using subsets of the SNOMED vocabulary [11].

External sample data repositories often use their own vocabularies for such concepts, and even within a given standard such as SNOMED, different pathologists or other experts may not agree on which term should be used for a certain disease or organ type. For example, in a recent integration project, a customer included samples with the diagnosis labeled DIABETES. The SNOMED vocabulary includes several varieties of diabetes and related complications, so it was necessary to consult with the customer to determine the best choice of mapping. After some discussion it was determined that, given the differences of interpretations, the best choice was to map this to the term OBESITY in GeneExpress. Similarly, the customer data might include abbreviations, such as DRG, which was mapped to DORSAL ROOT GANGLION, or common terms, such as FAT, which was mapped to ADIPOSE TISSUE. Moreover, a SNOMED term code is usually associated with one primary term and one or more synonyms. Some customers may prefer a different synonym than the one chosen by Gene Logic.

Sample data may also differ in the choice of units: For example, drug treatments can use units such as μMol or ng/ml, while age can be provided in days, weeks, or years. A conversion table is required to map any units to comparable units in the sample data exchange format.

Formatting of individual items also needs to be resolved. For example, the sample data exchange format uses the terms *Male* and *Female* to represent the sex of a donor, while a customer database may use *male* and *female* or just *M* and *F*. Further, data may contain typographic errors, such as misspelling the name of a supplier. When vocabularies are small, or for controlled vocabularies, it may be possible to spot and correct such errors manually, but in general, these errors can go undetected.

All these conflicts need to be resolved as part of the mapping from the source data format to the sample data exchange format. In some cases, it is not possible to implement rules to resolve such conflicts automatically, so manual inspection and curation of the data must be performed before mapping it. In general, if the source data are consistent in their use of controlled vocabularies, formatting, and units, it is possible to hardwire the correct mappings into the mapping implementation. However, whenever a new conflict arises, it is necessary to find a resolution and adapt the mapping implementation.

Sample mapping provides consistency between Gene Logic and customer sample classifications rather than finding an optimal classification. Sample classification in GeneExpress is based on sound clinical and pathology principles in the strict framework of the SNOMED nomenclature. However, not all medical concepts map straightforwardly to SNOMED terms, and therefore, there may not be a best classification for a concept but rather several reasonable ones.

Gene Annotation Data Mapping

In general, gene annotations are not involved in the integration of expression data from multiple sources. In certain cases, however, it is necessary to integrate gene annotations associated with non-Gene Logic expression data (e.g., to extend the system to include custom probe arrays with proprietary gene fragments or to support a customer's proprietary gene annotation data).

Gene annotations generally have well understood semantics; although, there are ambiguities with regard to the classification of some of these annotations (see Pearson's article in *Nature* [19] for a discussion of problems associated with gene nomenclature and identification).

Because gene annotations are often stored in proprietary databases, a possible approach is to provide links to these annotations, instead of importing them into GXDW. This approach supports neither the ability to query the contents of these databases directly nor superimpose expression data on these annotations (e.g., superimpose expression levels associated with genes displayed on a pathway or chromosome map), but it can make the information readily accessible from Gene Express. In such cases, individual genes within GeneExpress are linked to network-accessible reports or interactive services. When query access is required, custom gene annotations can be integrated into GeneExpress using a mechanism similar to that used for sample data. Defining a mapping remains non-trivial, but as gene annotation data are often more rigorously structured than clinical information, the problem is usually less severe.

Another problem specific to gene annotations is the fact that related but different annotations are likely to reside in multiple sources. This introduces a key challenge: reconciling differences between different gene annotation sources. When different versions of a single source (e.g., UniGene) conflict, it is usually acceptable to defer to the newer version. When different sources conflict, there may not be an ideal way of resolving the differences.

In addition, a customer may prefer alternative sources for gene annotation data (e.g., protein data sources other than Swiss-Prot or sequence clusters other than those provided in UniGene) rather than those used in GeneExpress. Even when the same data sources are used, different refresh policies may lead to the use of different versions or different builds of the same data source. Furthermore, there may be multiple ways to associate two related biological objects (e.g., links from gene fragments to known gene clusters may be based on data supplied by the probe array manufacturer or on homology searches using the fragment's target sequence). Consequently, integrating customer gene annotations with Gene Logic gene annotations requires resolving potentially complex data discrepancies.

10.4.4   Data Loading Issues

Once data from external data sources have been mapped to the data exchange formats, additional processing and curation may be required before integrating and loading them into the warehouse. First, it is necessary to detect invalid data, such as missing clinical data associated with samples or inconsistent associations of sample and gene expression data. In general, data migration tools, such as GX Connect, handle such cases by skipping the data affected by errors and issuing warning messages in a log file. Data editing can be used to correct problems not resolved during the mapping process.

Next, differences between identifiers of external objects and objects already in the warehouse must be resolved to maintain database consistency. Transformations of this type are carried out using *staging databases* before loading data into the warehouse itself. In addition, it is necessary to keep track of any identifiers created for customer data so that if customer data objects are dropped and reloaded (e.g., to allow the data to be edited), they do not reappear with different identifiers.

Finally *derived data*, such as quality control data (e.g., measures of saturation for the scanners), are also computed during the final loading stage.

10.4.5   Update Issues

Section 10.2.2 describes the process of updating a GeneExpress system containing only Gene Logic data. The content update becomes more complex if the system contains both Gene Logic data and customer data, either loaded with the GX Connect tool or with custom tools. Both Gene Logic data and customer data change over time, therefore content update procedures must ensure that new data from both sources are maintained correctly in the GeneExpress data warehouse.

Data in the GXDW can be classified into: (1) data *shared* by Gene Logic and customer data stores, such as controlled vocabularies; and (2) data that are not shared, that is, data generated by either Gene Logic or the customer only. Examples of shared data include SNOMED terms and species information in the sample database and probe array types and algorithm types in the expression database. When performing a content update, shared data occurring in both the Gene Logic and the customer data contents must be consolidated. Examples of data that are not shared include data pertaining to an individual sample in the sample database and experiment expression values in the expression database. It is not necessary to merge these data because customer sample and experiment objects are always distinct from Gene Logic sample and experiment objects. Instead, separate spaces of object identities are maintained for customer and Gene Logic data.

Depending on the nature of the data, a variety of techniques can be used for handling updates. For example, because it is not necessary to merge data for individual experiments or samples, such as expression values, from different data sources, these data can reside in different database partitions. In this case, content update is as simple as replacing a database partition. On the other hand, controlled vocabularies and other shared data must be consolidated; therefore, special tools are required to reconcile terms in customer and Gene Logic data and to make sure they are consistent in the integrated warehouse (e.g., having the same ID values). The consolidation process involves resolving the identification of objects and terms, as well as object references.

10.5   SUMMARY

This chapter provided discussion of the data integration challenges involved in building a system for managing gene expression data and how these challenges have been addressed in the GeneExpress system and in the context of several GeneExpress integration projects.

A data warehouse approach and tools were used in developing GeneExpress and were found to provide an effective environment for developing a system to support the integration and management of data from diverse sources, in which data may be imprecise and may evolve over time. Other non-warehouse (i.e., non-materialized view) approaches were also briefly considered, based on previous experience with developing genomic data management systems using the Object Protocol Model (OPM) tools [20], but they were not adopted for reasons similar to those described by Davidson et al. [17]. The data warehouse approach has proven well suited for systems such as GeneExpress that need to integrate data from multiple data sources, with data requiring validation and cleansing, and in cases where system performance and robustness are critical. However, the general data warehouse approach cannot be applied *as is* to the gene expression domain and needs to be adapted [14]. Also, coping with issues of data semantics in the area of genomic applications remains complex and difficult and often requires manual solutions.

Because good performance is a critical requirement for GeneExpress, a comprehensive set of benchmarks has been devised to assess system performance continuously as its data content grows. The benchmarks involve running typical queries and expression analysis operations on a series of data sets, using various configurations of Sun SparcUltra II- and III-based servers and Pentium-based clients. These benchmarks first measure the single-user performance of query and analysis operations, then measure multi-user performance with up to 300 simulated concurrent users, each running analysis steps across all available array types. It was found that, given sufficient server system memory, performance for multiple users scaled linearly with the number of processors and number of concurrent users.

Though this chapter has focused on the GeneExpress system and the Affymetrix GeneChip platform, the challenges addressed by the GeneExpress system are shared by other systems for managing and analyzing gene expression data. In particular, for all gene expression platforms, the problems associated with relating the data to gene and sample annotations and issues such as compatibility of array versions and analysis algorithms are similar.

The first version of GeneExpress was released in early 2000. Through the end of 2002, the GeneExpress system has evolved through several versions and has been deployed at more than 25 biotech and pharmaceutical companies worldwide, and at several academic institutions. Based on the experience gained in developing tools for incorporating customer data into GeneExpress, the GX Connect tool has been developed to provide support for interactive extraction, transformation, and loading of gene expression data generated using the Affymetrix GeneChip platform and related clinical data into GeneExpress. GeneExpress and GX Connect are deployed together as part of the Genesis Enterprise System [2].

Five data integration systems that provide support for integrating gene expression data from both Gene Logic and customer sources have been deployed through the end of 2002. All these systems provide support for integrating sample (clinical) data based on proprietary data formats and allow regular incremental updates of customer data; two of these systems provide support for custom Affymetrix GeneChip probe arrays; and one system also provides support for proprietary gene annotations.

ACKNOWLEDGMENTS

We want to thank our past and present colleagues at Gene Logic who have been involved in the development of the GeneExpress and Genesis systems for their outstanding work. Doug Dolginow initiated the development of GeneExpress at Gene Logic and had the key role in defining user requirements for GeneExpress Explorer. Kevin McLoughlin has led the development of GeneExpress Explorer, probably the best known part of GeneExpress. Special thanks to Mike Cariaso, François Collin, William Craven, Michael Elashoff, Aaron Hechmer, and Dmitry Krylov for their feedback and contributions to this paper.

TRADEMARKS

GeneExpress®, GX^{TM}, and Genesis Enterprise System^{TM} are trademarks owned by Gene Logic Inc. Affymetrix® and GeneChip® are trademarks owned by Affymetrix, Inc.

REFERENCES

[1]   D. J. Lockhart and A. E. Winzeler. "Genomics, Gene Expression, and DNA Arrays." *Nature* 405 (2000): 827–836.

[2]   Gene Logic Products. http://www.genelogic.com/products.htm. See GeneExpress product line and Genesis Enterprise software.

[3]   D. J. Lockhart, H. Dong, M. C. Byrne, et al. "Expression Monitoring by Hybridization to High-Density Oligonucleotide Arrays." *Nature Biotechnology* 14 (1996): 1675–1680.

[4]   V. M. Markowitz and T. Topaloglou. "Applying Data Warehousing Concepts to Gene Expression Data Management." In *Proceedings of the 2nd IEEE International Symposium on Bioinformatics and Bioengineering*, 65–72. Bethesda, MD: IEEE Computer Society, 2001.

[5]   A. Brazma, P. Hingamp, and L Quackenbush. "Minimum Information About a Microarray Experiment (MIAME): Towards Standards for Microarray Data." *Nature Genetics* 29, no. 4 (2001): 365–371.

[6]   Human Gene Nomenclature Database. http://www.gene.ucl.ac.uk/nomenclature/.

[7]   *Affymetrix GeneChip Analysis Suite User Guide*. Affymetrix, 2000.

[8]   R. A. Irizarry, B. Bolstad, F. Collin, et al. "Summaries of Affymetrix GeneChip Probe Level Data." *Nucleic Acids Research* 31, no. 4 (2003).

[9]   C. Li and W. Wong. "Model-Based Analysis of Oligonucleotide Arrays: Expression Index Computation and Outlier Detection." *Proceedings of the National Academy of Science* 98 (1998): 31–36.

[10]   Affymetrix. Affymetrix Analysis Data Model. http://www.affymetrix.com/support/.

[11]   SNOMED. Systematized Nomenclature for Medicine. http://www.snomed.org/.

[12]   The Gene Ontology Consortium. "Gene Ontology: Tool for the Unification of Biology." *Nature Genetics* 25 (2000): 25–29. http://www.geneontology.org.

[13]   V. M. Markowitz, I. A. Chen, and A. Kosky. "Gene Expression Data Management: A Case Study." In *Proceedings of the 8th International Conference on Extending Database Technology (EDBT)*, from the series Lecture Notes in Computer Science, edited by C. S. Jensen, K. G. Jeffery, L Pokorny, et al., 722–731. Heidelberg, Germany: Springer-Verlag, 2002.

[14]   A. A. Hill, E. L. Brown, M. Z. Whitley, et al. "Evaluation of Normalization Procedures for Oligonucleotide Array Data Based on Spiked cRNA Controls." *Genome Biology* 2, no. 12 (2001): 0055.1–0055.13. http://www.genomebiology.com/2001/2/12/research/0055/.

[15]   D. C. Hoaglin, F. Mosteller, and J. W. Tukey. *Understanding Robust and Exploratory Data Analysis*. New York: John Wiley, 1983.

[16] B. A. Eckman, A. S. Kosky, and A. L. Laroco. "Extending Traditional Query-Based Integration Approaches for Functional Characterization of Post-Genomic Data." *Bioinformatics* 17, no. 7 (2001): 587 601.

[17] S. B. Davidson, J. Crabtree, B. Brunk, et al. "K2/Kleisli and GUS: Experiments in Integrated Access to Genomic Data Sources." *IBM Systems Journal* 40, no. 2 (2001): 512–531.

[18] *Presentations. Third International Meeting on Microarray Data Standards, Annotations, Ontologies, and Databases*, March 25–31, 2001. Palo Alto, CA: Stanford University, 2001. http://www.mgedsourceforye.net/ontologies/index.php.

[19] H. Pearson. "Biology's Name Game." *Nature* 417 (2001): 631–632.

[20] V. M. Markowitz, I. A. Chen, A. Kosky, et al. "OPM: Object-Protocol Model Data Management Tools." In *Bioinformatics: Databases and Systems*, edited by S. I. Letovsky, 187–199. Boston: Kluwer Academic, 1999.

DiscoveryLink

**Laura M. Haas, Barbara A. Eckman, Prasad Kodali,
Eileen T. Lin, Julia E. Rice, and Peter M. Schwarz**

DiscoveryLink enables the integration of diverse data from diverse sources into a single, virtual database, with the goal of making it easier for scientists to find the information they need to prevent and cure diseases. To progress in this quest, scientists need to answer questions that relate data about genomics, proteomics, chemical compounds, and assay results, which are found in relational databases, flat files, extensible markup language (XML), Web sites, document management systems, applications, and special-purpose systems. They need to search through large volumes of data and correlate information in complex ways.

In bioinformatics research in the post-genomic era, the sheer volume of data and number of techniques available for use in the identification and characterization of regions of functional interest in the genomic sequence is increasing too quickly to be managed by traditional methods. Investigators must deal with the enormous influx of genomic sequence data from human and other organisms. The results of analysis applications such as the Basic Local Alignment Search Tool (BLAST) [1], PROSITE [2], and GeneWise [3] must be integrated with a large variety of sequence annotations found in data sources such as GenBank [4], Swiss-Prot [5], and PubMed [6]. Public and private repositories of experimental results, such as the Jackson Laboratory's Gene Expression Database (GXD) [7], must also be integrated. Deriving the greatest advantage from this data requires full, query-based access to the most up-to-date information available, irrespective of where it is stored or its format, with the flexibility to customize queries easily to meet the needs of a variety of individual investigators and protein families.

In an industrial setting, mergers and acquisitions increase the need for data integration in the life science industry in general and the pharmaceutical industry in particular. Even without mergers, in a typical pharmaceutical company, the research groups are geographically dispersed and divided into groups based on therapeutic areas. Scientists in each of these therapeutic areas might be involved in various stages of the drug discovery process such as target identification, target validation, lead identification, lead validation, and lead optimization. During each of these stages, they need to access diverse data sources, some specific to the therapeutic area of interest and the particular stage of the process and others that are of value to many therapeutic areas and at many stages of the process. Providing the data integration infrastructure to support this research environment (geographically dispersed research groups accessing different sets of diverse data sources depending on their area of research and the stage of the drug discovery process) is a daunting task for any information technology (IT) group.

As pharmaceutical companies try to shorten the drug-discovery cycle, they must identify new drug candidates more quickly by increasing the efficiency of the research processes and eliminating the false positives earlier in the discovery process. Providing scientists with easy access to the relevant information is essential. Researchers working in the gene expression domain may gain valuable insights if they have access to data from comparative genomics, biological pathways, or cheminformatics. This is also true for a scientist working in the lead identification or optimization areas.

This case can be illustrated by an example. A research group in a pharmaceutical company working in a particular therapeutic area might be interested in looking at all the compounds active in biological assays that have been generated and tested for a given receptor. In addition, the researchers might be interested in looking at similar compounds and their activities against similar receptors. This will help them understand the specificity and selectivity of the compounds identified. The knowledge that a particular set of compounds was considered for a different therapeutic area by another team could help them develop new leads or eliminate compounds that are not specific in their activities. To answer these queries, the research group must correlate information from multiple databases, some relational (e.g., the assay data may be stored relationally), some not (the chemical structure data might be stored by a special-purpose system), and use specialized functions of the data sources (e.g., similarity searches involving compound structures or DNA sequences).

There are many different approaches to integrating diverse data sources. Often, integration is provided by applications that can talk to one of several data sources, depending on the user's request. In these systems, access to the data sources is typically "hard-wired." Replacing one data source with another means rewriting a portion of the application. In addition, data from different sources cannot be compared in response to a single request unless the comparison is likewise wired into the application. Moving all relevant data to a warehouse allows greater flexibility in retrieving and comparing data, but at the cost of re-implementing or losing the specialized functions of the original source, as well as the cost of maintenance. A third approach is to create a homogeneous object layer to encapsulate diverse sources. This encapsulation makes applications easier to write and more extensible, but it does not solve the problem of comparing data from multiple sources.

To return to the example, today this problem would be addressed by writing an application that accesses chemical structure databases (with specific functionality such as similarity or substructure searches), assay databases (maybe in relational format), and sequence databases (flat file, relational, or XML format). Answering the previous question requires multiple queries against these data sources:

1. *"Show me all the active compounds for each of the assays for a particular receptor."*

2. *"Show me all the compounds that are similar to the top five compounds from the previous query"* (may require multiple requests, one per compound, depending on the sophistication of the data store and application).

3. *"Do a BLAST* [1] *run to find similar receptors."*

4. *"Show me the results of the compounds from Query 2 from all the assays against the receptors of Query 3"* (may require multiple requests, one per compound or one per compound-receptor pair, depending on the sophistication of the data store and application).

5. *"Sort the result set by order of the specificity or selectivity information"* (if multiple queries were needed in Step 4).

Depending on the activities of the set of compounds, various scenarios emerge that tell the researchers how best to continue their research. However, the application is hard to write and may need to be extended if additional sources are needed (e.g., if a new source of compound or assay information is acquired).

A virtual database, on the other hand, offers users the ability to combine data from multiple sources in a single query without creating a physical warehouse. DiscoveryLink [8] uses federated database technology to provide integrated access to data sources used in the life sciences industry. The federated middleware *wraps* the actual data sources, providing an extensible framework and encapsulating the details of the sources and how they are accessed. In this way, DiscoveryLink provides users with a virtual database to which they can pose arbitrarily complex queries in the high-level, non-procedural query language SQL. DiscoveryLink focuses on efficiently answering these queries, even though the necessary data may be scattered across several different sources, and those sources may not themselves possess all the functionality needed to answer such a query. In other words, DiscoveryLink is able to optimize queries and compensate for SQL functions that may be lacking in a data source. Additionally, queries can exploit the specialized functions of a data source, so no functionality is lost in accessing the source through DiscoveryLink.

Using DiscoveryLink in the example, a single query could retrieve the structures of compounds that are active in multiple assays against different receptors.

Views could be defined to create a canonical representation of the data. Furthermore, the query would be optimized for efficient execution. DiscoveryLink's goal is to give the end user the perspective of a single data source, saving effort and frustration. In a real scenario, before researchers propose the synthesis and testing of an interesting compound they have found, they would like to know the toxicity profile of the compound and related compounds and also the pathways in which the compound or related compounds might be involved. This would require gathering information from a (proprietary) toxicity database, as well as one with information on metabolic pathways, such as the Kyoto Encyclopedia of Genes and Genomes (KEGG), and using the structures and names of the compounds to look up the data—another series of potentially tricky queries without an engine such as DiscoveryLink.

This chapter presents an overview of DiscoveryLink and shows how it can be used to integrate life sciences data from heterogeneous data sources. The next section provides an overview of the DiscoveryLink approach, discussing the data representation, query capability, architecture, and the integration of data sources, as well as providing a brief comparison to other systems for data integration. Section 11.2 focuses on query processing and optimization. Section 11.3 addresses performance, scalability, and ease of use. The final section concludes with some thoughts on the current status and success of the system, as well as some directions for future enhancements.

11.1   APPROACH

DiscoveryLink is based on federated database technology, which offers powerful facilities for combining information from multiple data sources. Built on technology from an earlier product, DB2 DataJoiner [9], and enhanced with additional features for extensibility and performance from the Garlic research project [10, 11], DiscoveryLink's federated database capabilities provide a single, virtual database to users. DB2 DataJoiner first introduced the concept of a virtual database, which is created by federating together multiple heterogeneous, relational data sources. Users of DB2 DataJoiner could pose arbitrary queries over data stored anywhere in the federated system without worrying about the data's location, the SQL dialect of the actual data store(s), or the capabilities of those stores. Instead, users had the full capabilities of DB2 against any data in the federation. The Garlic project demonstrated the feasibility of extending this idea to build a federated database system that effectively exploits the query capabilities of diverse, often non-relational data sources. In both of these systems, as in DiscoveryLink, a middleware query processor develops optimized execution plans and compensates for any functionality the data sources may lack.

There are many advantages of a federated database approach to integrating life science data. In particular, this approach is characterized by transparency (the degree to which it hides all details of data location and management), heterogeneity (the extent to which it tolerates data source diversity), a high degree of function providing the benefits of both SQL and the underlying data source capabilities, autonomy for the underlying federated sources, easy extensibility, openness, and optimized performance. All other approaches fall short in one or another of these categories. These other approaches are numerous, including domain-specific solutions, language-based frameworks, dictionary-based solutions, frameworks based on an object model, and data warehousing approaches.

For example, companies like Informax provide data retrieval and data integration for biological databases. Their system, and many like it, benefits from being created specifically for bioinformatics data, but as a result, it cannot readily exploit advances in query processing (e.g., in the relational database industry). Kleisli's [12] Collection Programming Language (CPL) presented in Chapter 6 allows the expression of complicated transformations across heterogeneous data sources, but it provides no global schema, making query formulation and optimization difficult. The Sequence Retrieval System (SRS) [13, 14] presented in Chapter 5 provides fast access to a vast number of text files, and LION provides a rich biology workbench of integrated tools built on SRS. SRS has its own proprietary query language, which offers excellent support for navigational access but less power for cross-source querying than SQL. In fact, LION's DiscoveryCenter uses DiscoveryLink to extend its database integration capabilities. Biomax provides similar functionality in its Biological Databanks Retrieval System (BioRS) tool, with cleanly structured interfaces based on the Common Object Request Broker Architecture (CORBA) for scalability on both multi-processors and workstations alike. BioRS also offers a curated and annotated database of the human genome and a number of powerful analysis tools. Again, while the domain-specific tooling makes this a great package for biologists, the language used for queries is more limited than SQL. Accelrys provides a relational data management and analysis package, SeqStore, and a rich set of bioinformatics tools for sequence analysis—the Genetics Computer Group (GCG) Wisconsin Package. SeqStore includes a relational data warehouse for sequence data, coupled with tools to receive automated updates, to analyze sequences with the wide range of analyses available in the GCG Wisconsin Package, and to create automated sequence analysis pipelines. The warehousing approach requires that data be moved (or copied), interfering with source autonomy and limiting the extensibility of the system—or at least making it harder to extend. The object frameworks, such as that provided by Tripos, provide only limited transparency. Similar arguments apply to most other bioinformatics integration engines.

The two biology-focused integration engines that come closest to DiscoveryLink's vision are Gene Logic's Object Protocol Model (OPM) [15] and the Transparent Access to Multiple Biological Information Systems (TAMBIS) [16], presented in Chapter 7. OPM provides a virtual, *object-oriented* database, with queries in the proprietary OPM-MQL query language over diverse query sources. OPM's query optimization is rule-based and somewhat limited, because of the difficulties of optimizing over its more complex data model. While an object-oriented model is a natural choice for modeling life sciences data, and OPM's class methods have been demonstrated to add significant scientific value [17], DiscoveryLink follows an industry standard (relational), believing that the virtues of openness and the benefits of riding on technology that is constantly evolving and growing in power (due to the large number of users and uses) outweighed the annoyances of modeling data as relations. In fact, the database industry is now rapidly adding support for XML and XQuery to its once purely relational products; DiscoveryLink will exploit these capabilities as they become available, alleviating any modeling issues substantially. For example, the DiscoveryLink engine already supports SQL/XML functions that allow it to return XML documents instead of tuples.

TAMBIS is unique in its use of an ontology to guide query formulation, query processing, and data integration. It also offers users a virtual database and deals with a great deal of heterogeneity. Originally based on CPL *wrappers* for accessing data sources, TAMBIS now uses a more general Java wrapper mechanism. TAMBIS focuses on supporting direct user interactions, unlike DiscoveryLink, which is meant to be a general infrastructure against which many different query tools can be used. Again, DiscoveryLink benefits from its open, industry-standard interfaces for both queries and wrappers. However, the use of an ontology to generate and refine queries is a powerful mechanism, and the marriage of such techniques to DiscoveryLink middleware could be explored to provide a more biology-centric experience for users.

Because DiscoveryLink is a general platform for data integration, it also can be compared to other database integration offerings. Most of the major database vendors offer some sort of cross-database query product, often called a *gateway*. For example, Oracle offers both *dblinks* (for cross-Oracle queries) and Oracle Transparent Gateway (for more heterogeneous data sources). DiscoveryLink differs from these and other products in three fundamental ways: (1) It offers an open application programming interface (API) for wrapper construction; (2) it allows the use of data source functions in queries that span multiple data sources; and (3) it has the most powerful optimization capabilities available (it is the only system

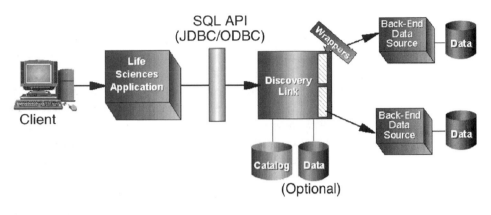

11.1

FIGURE

DiscoveryLink architecture.

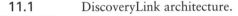

that takes query-specific input from wrappers during query planning). Few systems offer the same degree of transparency and the same query processing power against heterogeneous sources.

11.1.1   Architecture

The overall architecture of DiscoveryLink, shown in Figure 11.1, is common to many heterogeneous database systems, including the Stanford-IBM Manager of Multiple Information Sources (TSIMMIS) [18], Distributed Information Search Component (DISCO) [19], Pegasus [20], Distributed Interoperable Object Model (DIOM) [21], Heterogeneous Reasoning and Mediator System (HERMES) [22], and Garlic [10, 11]. Applications connect to the DiscoveryLink server using any of a variety of standard database client interfaces, such as Call Level Interface (CLI) [23], Object Database Connectivity (ODBC), or Java Database Connectivity (JDBC), and submit queries to DiscoveryLink in standard SQL (specifically SQL3 [24]). The information required to answer the query comes from the local database and/or from one or more data sources, which have been identified to DiscoveryLink through a process called *registration*. The data from the sources is modeled as relational tables in DiscoveryLink. The user sees a single, virtual relational database, with the original locations and formats of the sources hidden. The full power of SQL is supported against all the data in this virtual database, regardless of where the data is actually stored and whether the data source actually supports the SQL operations.

When an application submits a query to the DiscoveryLink server, the server identifies the relevant data sources and develops a query execution plan for obtaining the requested data. The plan typically breaks the original query into fragments that represent work to be delegated to individual data sources and additional processing to be performed by the DiscoveryLink server to filter, aggregate, or merge the data. The ability of the DiscoveryLink server to further process data received from sources allows applications to take advantage of the full power of the SQL language, even if some of the information they request comes from data sources with little or no native query processing capability, such as simple text files. The local data store allows query results to be stored for further processing and refinement, if desired, and also provides temporary storage for partial results during query processing.

The DiscoveryLink server communicates with a data source by means of a *wrapper* [11], a software module tailored to a particular family of data sources. The wrapper for a data source is responsible for four tasks:

+ Mapping the information stored by the data source into DiscoveryLink's relational data model

+ Informing DiscoveryLink about the data source's query processing capabilities by analyzing plan fragments during query optimization

+ Mapping the query fragments submitted to the wrapper into requests that can be processed using the native query language or programming interface of the data source

+ Executing such requests and returning results

The interface between the DiscoveryLink server and the wrapper supports the International Standards Organization/Structured Query Language/Management of External Data (ISO SQL/MED) standard [25].

Wrappers are the key to extensibility in DiscoveryLink, so one of the primary goals for the wrapper architecture was to allow wrappers for the widest possible variety of data sources to be produced with a minimum of effort. Past experience has shown that this is feasible. To make the range of data sources that can be integrated using DiscoveryLink as broad as possible, a data (or application) source only needs to have some form of programmatic interface that can respond to queries and, at a minimum, return unfiltered data that can be modeled (by the wrapper) as rows of (one or more) tables. The author of a wrapper need not implement a standard query interface that may be too high-level or low-level for the underlying data source. Instead, a wrapper provides information about a data source's query processing capabilities and specialized search facilities to the DiscoveryLink server, which dynamically determines how much of a given query the data source is capable of handling. This approach allows wrappers for simple data sources to be built quickly, while retaining the ability to exploit the unique query processing capabilities of non-traditional data sources such as search engines for chemical structures or images. For DiscoveryLink, this design was validated by wrapping a diverse set of data sources including flat files, relational databases, Web sites, a specialized search engine for text, and the BLAST search engine.

To make wrapper authoring as simple as possible, only a small set of key services from a wrapper is required, and the approach ensures that a wrapper can be written with very little knowledge of DiscoveryLink's internal structure. As a result, the cost of writing a basic wrapper is small. In past experience, a wrapper that just makes the data at a new source available to DiscoveryLink, without attempting to exploit much of the data source's native query processing capability, can be prototyped in a matter of days by someone familiar with the data source interfaces and the wrapper concepts. Because the DiscoveryLink server can compensate for missing functionality at the data sources, even such a simple wrapper allows applications to apply the full power of SQL to retrieve the new data and integrate it with information from other sources, albeit with perhaps less-than-optimal performance. Once a basic wrapper is written, it can be incrementally improved to exploit more of the data source's query processing capability, leading to better performance and increased functionality as specialized search algorithms or other novel query processing facilities of the data source are exposed.

A DiscoveryLink wrapper is a C++ program, packaged as a shared library that can be loaded dynamically by the DiscoveryLink server when needed. Often a single wrapper is capable of accessing several data sources, as long as they share a common or similar API. For one thing, wrappers do not need to encode information on the schema of data in the source. For example, the Oracle wrapper provided with DiscoveryLink can be used to access any number of Oracle databases, each having a different schema. In fact, the same wrapper supports several Oracle release levels as well. This has a side benefit, namely that schemas can evolve without requiring any change in the wrapper as long as the source's API remains unchanged. In addition, wrappers can get connection information for individual servers from SQL Data Definition Language (DDL) statements, even if the schemas are identical. On the other hand, there is a tradeoff between flexibility and ease of configuration (the more flexible the wrapper, the more it needs to be told during registration). For that reason, it is sometimes more practical to encode (parts of) the schema in the wrapper. For example, the BLAST wrapper defines many fixed columns, but allows the user to specify others that are appropriate for their instantiation of BLAST.

This architecture has many benefits, as described previously. However, there are some controversial aspects. First and foremost, much biology data is semi-structured, and the current implementation forces data to be modeled relationally. While this does complicate wrapper writing somewhat, there are several examples of wrappers today that deal with nested and semi-structured data, including an XML wrapper. These wrappers expose their data as multiple relations, which can be joined to get back the full structure (note that the data is still stored in its nested form, and the joins are often translated into simple retrievals as a result). Future direction is to support XML and XQuery natively in DiscoveryLink's engine and to allow wrapper writers their choice of a relational or an XML model. That will make the modeling issues less painful.

A second issue is the use of C++, a general purpose and somewhat arcane programming language, for writing wrappers, as opposed to a simpler scripting language or a specialized wrapper construction mechanism. A general-purpose language was chosen for several reasons. First, DiscoveryLink is meant to handle large-scale queries over many data sources and large volumes of data. C++ is an efficient language, suitable for such applications. Second, DiscoveryLink wrappers are required to do more than ordinary connectors or adaptors, and the general-purpose programming language allows the wrapper writer complete flexibility in accomplishing the wrapper tasks. A toolkit and tools for wrapper development can ease the pain of programming by providing template functions, automatic generation of parts of the code, error checking, and so on. Last but not least, the DiscoveryLink engine happens to be written in C++, so this was by far the easiest to interface with the engine initially. A Java version of the toolkit is currently produced, as well as a set of generic wrappers for particular styles of data source access (e.g., a Web services wrapper, ODBC and JDBC wrappers, maybe even a Perl script wrapper). These facilities should increase the ease of adding new wrappers.

Related to the ease of wrapper writing is the ease of changing wrappers when (if) the interface to the data source changes. For most data sources, such changes are uncommon (note that this is not in reference to schema changes but to changes in the API or language used by the data source). When changes do occur, they are often additions to the existing interface, and the wrapper can continue to function as-is, only needing modification if exploiting the new feature(s) is desired. Most commercial data sources, for example, try to maintain upward compatibility in interface between one release and the next. But for some classes of sources (especially Web data sources), change is much more common. To deal with these sources, it is particularly desirable to have some non-programmatic or scripted way of creating wrappers. Our explorations into generic wrappers that can be easily tailored will address this concern.

11.1.2 Registration

The process of using a wrapper to access a data source begins with *registration*, the means by which a wrapper is defined to DiscoveryLink and configured to provide access to selected collections of data managed by a particular data source. Registration consists of several steps, each taking the form of a DDL statement. Each registration statement stores configuration meta-data in system catalogs maintained by the DiscoveryLink server.

The first step in registration is to define the wrapper itself and identify the shared library that must be loaded before the wrapper can be used. The CREATE WRAPPER statement serves this purpose. BLAST [1] is a search engine for finding nucleotide or peptide sequences similar to a given pattern sequence. A wrapper for BLAST might be created as follows:

```
CREATE WRAPPER BlastWrapper LIBRARY 'libblast.a'
```

Note that a particular data source has not yet been identified, only the software required to access any data source of this kind. The next step of the registration process is to define specific data sources using the CREATE SERVER statement. If several sources of the same type are to be used, only one CREATE WRAPPER statement is needed, but a separate CREATE SERVER would be needed for each source. For a particular BLAST service, the statement might be as follows:

```
CREATE SERVER TBlastNServ TYPE 'tBLASTn' VERSION '2.1.2'
WRAPPER BlastWrapper
OPTIONS(NODE 'myblast.bigpharma.com', PORT '2003')
```

This statement registers a data source that will be known to DiscoveryLink as TBlastNServ and indicates that it is to be accessed using the previously registered wrapper, BlastWrapper. It further identifies that this BLAST server is doing a tBLASTn search (i.e., comparing an amino acid sequence, the input, to a database of nucleotide sequences) and that it is using version 2.1.2 of the BLAST software. The additional information specified in the OPTIONS clause is a set of pairs (option name, option value) that are stored in the DiscoveryLink catalogs but are meaningful only to the relevant wrapper. In this case, they indicate to the wrapper that the TBlastNServ data source can be contacted via a particular Internet Protocol (IP) address and port number. In general, the set of valid option names and option values will vary from wrapper to wrapper because different data sources require different configuration information. Options can be specified on each of the registration DDL statements and provide a simple but powerful form of extensible meta-data. Because a wrapper understands the options it defines, only that wrapper can validate that the option names and values specified on a registration statement are meaningful and mutually compatible. As a result, wrappers participate in each step of the registration process and may reject, alter, or augment the option information provided in the registration DDL statement.

The third registration step is to identify, for each data source, particular collections of data that will be exposed to DiscoveryLink applications as tables. This is done using the CREATE NICKNAME statement. Collectively, these statements define the schema of each data source and form the basis of the integrated schema seen by applications.

For example, suppose there are three data sources. One is a relational database system providing data on protein targets. The second is a Web site storing information about technical publications. The third is a BLAST server that has the ability to compare an input sequence to a file of stored sequences as described previously. For this example, three sets of CREATE NICKNAME statements are needed, one set for each of the three data sources. Figure 11.2 shows representative CREATE NICKNAME statements that define partial schemas for each source.

The protein sequence source exports two relations. The first is Proteins, with columns representing the unique identifier for a protein, the common (print) name, the amino acid sequence associated with the protein, the function of the protein, and a list of diseases with which the protein has been associated. In real

Protein Sequence Source Schema (Relational Database)	Publications Source Schema (Web Site)	BLAST Source Schema (Search Engine)
| ```
CREATE NICKNAME Proteins
(protein_id varchar(30)not null,
  name varchar(60),
  sequence varchar(32000),
  function  varchar(100),
  diseases varchar(256))
FOR proteindb.bio.swpdata

CREATE NICKNAME Prot-Pubs
  (prot_id varchar(30) not null,
  pub_ref varchar(10) not null)
FOR proteindb.bio.swppubs
``` | ```
CREATE NICKNAME Pubs
  (pub_id varchar(10) not null,
  pub_title varchar(30)not null,
  pub_date date,
  keywords varchar(256))
FOR SERVER pubdb
OPTIONS (
  URL 'http://www.pubsite.org')

CREATE FUNCTION
MAPPING FOR
  contains(varchar(10),
           varchar(30),
           varchar(256))
RETURNS char(1)
FOR SERVER pubdb
``` | ```
CREATE NICKNAME
Protein_blast
  (query_seq varchar(32000),
  accession varchar(10)
   options(index '1', delimit ' '),
  definition varchar(100)
   options(index '2'),
  hsp_info varchar(100)
  )
FOR SERVER TBlastNServ
OPTIONS (datasource 'gbest')
``` |

11.2

FIGURE

Representative configuration statements (syntax simplified for illustration).
```

life, a Database Administrator (DBA) would likely declare a fuller set of columns, representing more of the information contained in the source; the schema is simplified in the interest of space only. Also, because the data source—a relational Database Management System (DBMS)—has a self-describing schema, the DBA would not actually need to put the column information in the CREATE NICKNAME statement. That information could be read from the data source catalogs automatically. The second relation exported from this source is a mapping table that maps proteins to publications that reference them. The FOR clause identifies, via a three-part name, the server, schema, and remote table referenced by the nickname. This syntax may be used with relational data sources.

Similarly, the DBA makes visible a single table, Pubs, from the publication source, for which only four columns are shown: the publication identifier, the title of the article, the date the article was published, and a list of keywords for the publication. Note that the nickname definitions give the types of attributes in terms of standard SQL data types. This represents a commitment on the part of the wrapper to translate types used by the data source to these types as necessary.

Finally, the BLAST search engine is modeled as a virtual table, indexed on the input sequence and with columns representing both input parameters and the results of the BLAST search. Again, only a subset of the schema is shown. Here are shown the input column, query_seq, and output columns for the accession number, definition, and hsp_info (the information string computed for a given *high-scoring segment pair* containing information about the number of nucleotides or amino acids that matched between the query and the hit sequences). Note the use of options clauses on both the CREATE NICKNAME statement and on the definition of individual columns. These give the DBA the ability to specify information needed by the wrapper. For the BLAST wrapper, the options on the individual columns tell the wrapper how to parse the BLAST *defline* into these columns. In this case, the defline is assumed to contain the accession number, followed by the definition, delimited by white space. (Columns whose values do not come from the defline have no parsing options specified.) The option on the overall CREATE NICKNAME tells the wrapper which data source to blast against (in this case GenBank's *gbest*). Actually, the BLAST wrapper supports so many different input and output columns that part of the schema is hard-wired so a DBA does not have to re-type all the columns in the CREATE NICKNAME statement. Further details on this wrapper can be found in the *IBM DB2 Life Sciences Data Connect Planning, Installation and Configuration Guide* [26].

Specialized search or data manipulation capabilities of a data source also can be modeled as user-defined functions, and identifying these functions by means of CREATE FUNCTION MAPPING statements is the fourth step in registration. Thus, the definition of the publications data source in Figure 11.2 also includes a CREATE FUNCTION MAPPING statement, registering that source's function contains(A, B, C). This function returns 'Y' if the publication identified by A contains the string C in column B, for example, contains ('ML546', 'keywords', 'ovarian cyst'). The mapping identifies this function to the query processor and declares its signature and return type in terms of standard SQL data types. As with nicknames, the wrapper must convert values of these types to and from the corresponding types used by the data source. This function models the underlying data source's Boolean search capability.

Finally, user mappings are defined. A user mapping tells DiscoveryLink how to connect a particular local user to a data source. For example, if a DiscoveryLink user identified by LAURA connects to the protein database as ITNerd, using the password DLRocks, the following DDL statement might be issued:

```
CREATE USER MAPPING FOR LAURA SERVER proteindb
OPTIONS (REMOTE_AUTHID'ITNerd', REMOTE_PASSWORD'DLRocks')
```

With these five steps, registration is complete. The new data source is ready to use. Queries can combine data from all the registered sources and use the specialized capabilities of these sources; in the example, two techniques for modeling these special capabilities were shown: using a virtual table, as done for the BLAST source, and using a function mapping, as done for the contains function of the publications source. Note that additional sources can be added at any time without affecting the ongoing operations of the federated system. The system need not be quiesced, and existing applications and queries need not be altered. However, new queries that combine information from the preexisting sources and the new source can now be asked.

If data source schemas or functions change, they must be re-registered. DiscoveryLink currently has no mechanism to detect changes in the sources, though an application that periodically compares the DiscoveryLink and source schemas could be written.

11.2 QUERY PROCESSING OVERVIEW

Once registration is completed, the newly defined nicknames and functions can be used in queries. When an application issues a query, the DiscoveryLink server uses the meta-data in the catalogs to determine which data sources hold the requested information. Then it optimizes the query, looking for an efficient execution plan.

It explores the space of possible query plans, using dynamic programming to enumerate plans for joins. The optimizer first generates plans for single table accesses, then for two-way joins, and so on. With each round of planning, the optimizer considers various join orders and join methods, and if all the tables are located at a common data source, it tries to generate plans for performing the join either at the data source or at the federated server.

Once the optimizer has chosen a plan for a query, query fragments are distributed to the data sources for execution. Each wrapper maps the query fragment it receives into a sequence of operations that make use of its data source's native programming interface and/or query language. Once the plan has been translated, it can be executed immediately or saved for later execution. The DiscoveryLink server's execution engine is pipelined and employs a fixed set of functions (open, fetch, close) that each wrapper must implement to control the execution of a query fragment. When accepting parameters from the server or returning results, the wrapper is responsible for converting values from the data source type system to DiscoveryLink's SQL-based type system.

DiscoveryLink includes a full database engine that can execute arbitrary (DB2) SQL queries. Features useful for life sciences applications include support for long data types (e.g., Binary Large Object [BLOB], Character Large Object [CLOB]) and user-defined functions. Applications also benefit from the ability to update information at relational data sources via SQL statements submitted to DiscoveryLink (and in the future, full transaction management for data sources that comply with the X/Open XA-interface specification), the ability to invoke stored procedures that reference nicknames, and the ability to use DiscoveryLink DDL statements to create new data collections at relational data sources. Another feature allows certain queries to be answered using pre-materialized automatic summary tables stored by DiscoveryLink, with little or no access to the data sources themselves. Joins, subqueries, table expressions, aggregation, statistical functions, and many other SQL constructs are supported against data, whether the data is locally stored or retrieved from remote data sources.

11.2.1 Query Optimization

During the planning process, the DiscoveryLink server takes into account the query processing power of each data source. As it identifies query fragments to be performed at a data source, it must ensure that the fragments are executable by that source. If a fragment cannot be performed by the source, the optimizer builds a plan to *compensate* for the missing function by doing that piece of work in the DiscoveryLink server. For example, if the data source does not do joins, but it is necessary to join together data from two nicknames at that source, the data will be retrieved from both nicknames (typically after restricting it with any predicates the source can apply), and then joined by DiscoveryLink.

The DiscoveryLink server has two ways of obtaining information about query processing power. Wrappers provided by IBM for relational data sources (and for other sources that are similar to a relational source in function) provide a *server attributes table* (SAT). The SAT contains a long list of parameters that are set to appropriate values by the wrapper. For example, if the parameter PUSHDOWN is set to "N", DiscoveryLink will not request that the data source perform query fragments more complex than:

```
SELECT <column_list> FROM <nickname>
```

Note: In this chapter, SQL is used as a concise way of expressing the work to be done by a remote data source. This work is actually represented internally by various data structures for efficient data processing.

If PUSHDOWN is set to 'Y', more complex requests may be generated, depending on the nature of the query and the values of other SAT parameters. For example, if the value of the BASIC_PRED parameter in the SAT is 'Y', requests may include predicates such as:

```
...WHERE pub_date > '12/31/1995'
```

The parameter MAX_TABS is used to indicate a data source's ability to perform joins. If it is set to 1, no joins are supported. Otherwise MAX_TABS indicates the maximum number of nicknames that can appear in a single FROM clause of the query fragment to be sent to the data source.

Information about the cost of query processing by a data source is supplied to the DiscoveryLink optimizer in a similar way, using a fixed set of parameters such as CPU_RATIO, which is the relative speed of the data source's processor relative to the one hosting the DiscoveryLink server. Additional parameters, such as average number of instructions per invocation and average number of Input/Output (I/O) operations per invocation, can be provided for data source functions defined to DiscoveryLink with function mappings, as can statistics about tables defined as nicknames. Once defined, these parameters and statistics can be easily updated whenever necessary.

This approach has proven satisfactory for describing the query processing capabilities and costs of the relational database engines supported by DiscoveryLink; although even for these superficially similar sources, a large set (hundreds) of parameters is needed. However, it is difficult to extend this approach to more idiosyncratic data sources. Web servers, for example, may be able to supply many pieces of information about some entity, but frequently they will only allow certain attributes to be used as search criteria. This sort of restriction is difficult to express using a fixed set of parameters. Similarly, the cost of executing a query fragment at a data source may not be easily expressed in terms of fixed parameters, if, for example, the cost depends on the value of an argument to a function. For instance, a BLAST function asked to do a *BLASTp* comparison against a moderate amount of data will return in seconds, whereas if it is asked to do a *tBLASTn* comparison against a large dataset, it may need hours.

The solution, validated in the Garlic prototype, is to involve the wrappers directly in planning of individual queries. Instead of attempting to model the behavior of a data source using a fixed set of parameters with statically determined values, the DiscoveryLink server will generate requests for the wrapper to process specific query fragments. In return, the wrapper will produce one or more wrapper plans, each describing a specific portion of the fragment that can be processed, along with an estimate for the cost of computing the result and its estimated size.

11.2.2 An Example

Voltage-sensitive calcium channel proteins mediate the entry of calcium ions into cells and are involved in such processes as neurotransmitter release. They respond to electric changes, which are a prominent feature of the neural system. The discovery of a novel gene that codes for a calcium channel protein would potentially be of great interest to pharmaceutical researchers seeking new drug targets for a neuropsychological disease. A popular method of novel gene discovery is to search Expressed Sequence Tag (EST) databases for (expressed) sequences similar to known genes or proteins. For example, a scientist with access to the data sources just described might like to see the results of the following query:

"Return accession numbers and definitions of EST sequences that are similar (60% identical over 50 amino acids) to calcium channel sequences in the protein data source that reference papers published since 1995 mentioning 'brain'."

The hsp_info column holds a condensed form of the equivalent data in the XML specification for BLAST provided by the National Center for Biotechnology Information. But to answer the above query, one needs direct access to the percentage of identities within the hsp alignment and the length of that alignment. Assume that two user-defined functions are defined to extract this information from the hsp_info string:

```
CREATE FUNCTION percent_identity(varchar(100))
RETURNS float EXTERNAL NAME 'hsp_info.a'
```

and

```
CREATE FUNCTION align_length(varchar(100))
RETURNS integer EXTERNAL NAME 'hsp_info.a'
```

Then, this request can be expressed as a single SQL statement that combines data from all three data sources:

```
SELECT b.name, c.accession, c.definition, a.pub_id
FROM Pubs a, Proteins b, Protein_blast c, Prot-Pubs d
WHERE a.pub_id = d.pub_ref
    AND d.prot_id = b.protein_id
    AND b.sequence = c.Query_seq
    AND b.function = 'calcium channel'
    AND a.pub_date > '12/31/1995'
    AND contains(a.pub_id, 'Keyword', 'brain') = 'Y'
    AND percent_identity(c.hsp_info) >0.6
    AND align_length(c.hsp_info) > 50
```

Many possible evaluation plans exist for this query. One plan is shown in Figure 11.3. In this figure, each box represents an operator. The leaves represent actions at a data source. Because DiscoveryLink does not model the details of those actions,

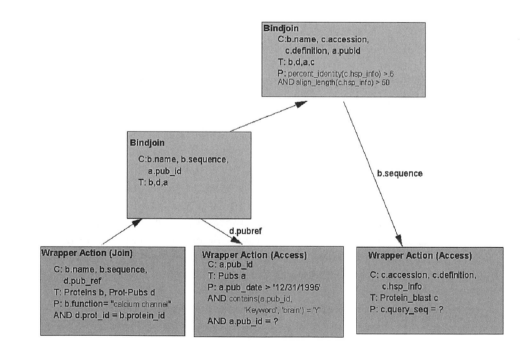

11.3

FIGURE

One evaluation plan for the query.

each action is modeled as a single operator, even if it might involve a series of operators at the source. (For relational sources, DiscoveryLink does in fact model the individual operators, but to simplify the figures, details are omitted. Thus the *join* of Proteins and Prot-Pubs is modeled in the figures as a single operator.) For non-relational sources, DiscoveryLink would not know whether a logical join action was an actual join or whether, in fact, the data was stored in a nested, pre-joined format. Nor does DiscoveryLink know whether the data are scanned and then predicates applied, or there is an indexed access, and so on. Instead, DiscoveryLink keeps track of the work that has been done by recording the *properties* of each operator. The properties include the set of columns available (C), the set of tables accessed (T), and the set of predicates applied (P), as shown in Figure 11.3. Non-leaf nodes represent individual operations at the DiscoveryLink server. The optimizer models these local operations separately.

This plan first accesses the protein data source, retrieving protein names and sequences and corresponding publication identifiers, for all proteins that serve as calcium channels. This information is returned to DiscoveryLink, where the *bindjoin* operator sends the publication references to the publications source one at a time. At the publications source, these publication identifiers are used to find relevant publications, and those publications are further checked for compliance with the query restrictions on keyword and pub_date. For those publications that pass all the tests, the identifier is returned to DiscoveryLink. There, the second bindjoin operator sends the sequence for any surviving proteins to BLAST, where they are compared against gbest, and the results are returned to DiscoveryLink where each hsp_info is analyzed to see if the sequence is sufficiently similar.

A second, superficially similar plan is shown in Figure 11.4. In this plan, the publications with appropriate dates and keywords are sent to DiscoveryLink, where a *hash table* is built. The data from the protein data source are also sent to DiscoveryLink and used to probe the hash table. Matches are passed to the bindjoin operator, which BLASTs the sequences against gbest, then returns them to DiscoveryLink to check the quality of the match.

It is not obvious which plan is best. The first plan results in one query of the protein database, but many queries (one for each qualified protein) of the publications database. The second plan only queries each of these sources once, but potentially returns many publication entries for proteins that will not qualify.

Either of these plans is likely to be better than the one shown in Figure 11.5. In this plan, the protein data is extracted first and all calcium channel proteins are BLASTed against gbest, regardless of what publications they reference. DiscoveryLink then filters the sequences using the similarity criterion, and the remaining proteins are passed to the nested loop join operator. This join compares each protein's referenced publications with a temporary table created by storing in

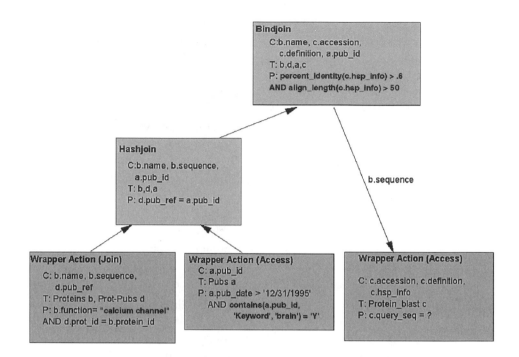

11.4   A second query evaluation plan.

FIGURE

DiscoveryLink those recent publications that discuss the brain. This plan could only win if there were very few recent publications with *brain* as a keyword (so the cost of the query to make the temporary table is small), and yet virtually every calcium channel protein in the protein database referenced at least one of them (so there is no benefit to doing the join of proteins with publications early). While that is unlikely for this example, if there were a more restrictive set of predicates (e.g., a recently discovered protein of interest and papers within the last two months), this plan could, in fact, be a sensible one.

11.2.3   Determining Costs

Accurately determining the cost of the various possible plans for this or any query is difficult for several reasons. One challenge is estimating the cost of evaluating the wrapper actions. For example, the DiscoveryLink engine has no notion of what must actually be done to find similar sequences or how the costs will vary depending on the input parameters (the bound columns). For BLAST, the actual algorithm used can change the costs dramatically, as can the data set

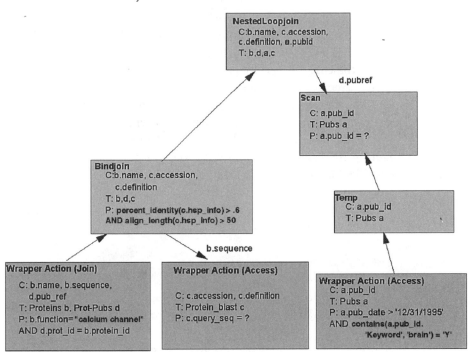

A third plan for the query.

being searched. As a second challenge, the query processor has no way to estimate the number of results that may be returned by the data sources. While the wrapper could, perhaps, provide some statistics to DiscoveryLink, purely relational statistics may not be sufficient. For example, cardinalities, as well as costs, for search engines like BLAST may vary depending on the inputs. A third challenge is to estimate the cost of the functions in the query. The costing parameters maintained by relational wrappers in DiscoveryLink for a function implemented by a data source include a cost for the initial invocation and a per-row cost for each additional invocation. However, the only way to take the value of a function argument into account is through a cost adjustment based on the size of the argument value in bytes. While this may be acceptable for simple functions like `percent_identity` and `align_length`, it is unlikely, in general, to give accurate results. For example, if *contains* actually has to search in different ways depending on the type of the column passed as an argument (e.g., a simple scan for keyword but an index lookup for the paper itself), the cost parameters must be set to reflect some amalgamation of all the search techniques. A simple case statement, easily written by the wrapper provider, could model the differences and allow more sensible choices of plans. While the costs of powerful functions in some cases can be hard to predict, many vendors do, in fact, know quite a bit about the costs of their functions. They often model costs themselves to improve their systems' performance.

The challenges of accurately estimating costs are met by letting the wrapper examine possible plan fragments to provide information about what the data source can do and how much it will cost. Consider our example query once again. During the first phase of optimization, when single-table access plans are being considered, the publications database will receive the following fragment for consideration (again, query fragments are represented in SQL; the actual wrapper interface uses an equivalent data structure that does not require parsing by the wrapper).

```
SELECT a.pub_id, a.pub_date
FROM Pubs a
WHERE a.pub_date > '12/31/1995'
 AND contains(a.pub_id, 'keyword', 'brain') = 'Y'
```

Assume that, in a single operation, the publications database can apply either the predicate on publication date or the contains predicate, but not both. Many Web sites can handle only a single predicate at a time or only restricted combinations. (Note: In the previous illustrative plans, it was assumed that the publications database *could* do both. Either assumption might be true. This one is adopted here to illustrate the point.) Further assume that it is possible to invoke the contains function separately later (this is like asking a new, very restrictive query of the Web site). Many Web sites do allow such follow-on queries to retrieve additional information about an object or do some more complex computation. The wrapper might return two wrapper plans for this fragment. The first would indicate that the data source could perform the following portion of the fragment:

```
SELECT a.pub_id, a.pub_date
FROM pubs a
WHERE a.pub_date > '12/31/1995'
```

with an estimated execution cost of 3.2 seconds and an estimated result size of 500 publications (in reality, of course, the result size would be much bigger). To estimate the total cost of the query fragment using this wrapper plan, the DiscoveryLink optimizer would add to the cost for the wrapper plan the cost of invoking the *contains* function on each of the 500 publications returned. If each invocation costs a second (because of the high overhead of going out to the World Wide Web), the total cost of this portion of the query, using this plan, would be 503.2 seconds.

The second wrapper plan would indicate that the data source could perform the following portion of the fragment:

```
SELECT a.pub_id, a.pub_date
FROM Pubs a
WHERE contains(a.pub_id, 'keyword', 'brain') = 'Y'
```

with an estimated execution cost of 18 seconds and an estimated result size of 1000 publications (i.e., entries for all the publications in the database with the keyword *brain*). To compute the total cost in this case, the optimizer would augment the cost for the wrapper plan with the cost of using the DiscoveryLink engine to apply the predicate on publication date to each of the 1000 publications. If filtering one publication takes a $1/_{100}$ of a second, the total cost for this portion of the query, using this plan, would be 28 seconds—a clear winner.

Wrappers participate in query planning in the same way during the join enumeration portion of optimization. In the example, the wrapper might be asked to consider the following query fragment:

```
SELECT a.pub_id, a.pub_date
FROM Pubs a
WHERE a.pub_date > '12/31/1995'
  AND contains(a.pub_id, 'keyword', 'brain') = 'Y'
  AND a.pub_id = :H0
```

This is essentially a single-table access, but the third predicate would not be considered during single-table access planning because the value being compared to pub_id comes from a different table. For each pub_id produced by the rest of the query (represented above by the host variable :H0), the publications database is asked to find the important properties of the corresponding publication, if it matches the other criteria. As before, the wrapper would return one or more plans and indicate in each one which of the predicates would be evaluated.

Only a few of the plans that DiscoveryLink would consider in optimizing this query were shown. The goal was not to give an exhaustive list of alternatives, but rather to illustrate the process. As well, the chapter has demonstrated the critical role an optimizer plays for complex queries. It is neither obvious nor intuitive which plan will ultimately be the best; the answer depends on many factors including data volumes, data distributions, the speeds of different processors and network connections, and so on. Simple heuristics generally cannot arrive at the right answer. Only a cost-based process with input on specific data source characteristics can hope to choose the right plans for the vast array of possible queries.

As a wrapper may be asked to consider many query fragments during the planning of a single query, it is important that communication with the wrapper be efficient. This is achieved easily in DiscoveryLink because the shared library that contains a wrapper's query planning code is loaded on demand into the address space of the DiscoveryLink server process handling the query. The overhead for communicating with a wrapper is, therefore, merely the cost of a local procedure call.

This approach to query planning has many benefits. It is both simple and extremely flexible. Instead of using an ever-expanding set of parameters to invest the DiscoveryLink server with detailed knowledge of each data source's capabilities, this knowledge resides where it falls more naturally, in the wrapper for the source in question. This allows to exploit the special functionality of the underlying source, as was done for the BLAST server (by modeling the search algorithm as a virtual table) and the publications source (using a template function). The wrapper only responds to specific requests in the context of a specific query. As the previous examples have shown, sources that only support searches on the values of certain fields or on combinations of fields are easily accommodated. In a similar way, one can accommodate sources that can only sort results under certain circumstances or can only perform certain computations in combination with others. Because a wrapper needs only to respond to a request with a single plan, or in some cases no plan at all, it is possible to start with a simple wrapper that evolves to reflect more of the underlying data source's query processing power.

This approach to query planning need not place too much of a burden on the wrapper writer, either. In a paper presented at the annual conference on very large databases [27], Roth et al. showed that it is possible to provide a simple default cost model and costing functions along with a utility to gather and update all necessary cost parameters. The default model did an excellent job of modeling simple data sources and did a good job predicting costs, even for sources that could apply quite complex predicates. This same paper further showed that even an approximate cost model dramatically improved the choice of plans over no information or fixed default values [27]. Therefore, it is believed that this method of query planning is not only viable, but necessary. With this advanced system for optimization, DiscoveryLink has the extensibility, flexibility, and performance required to meet the needs of life sciences applications.

11.3   EASE OF USE, SCALABILITY, AND PERFORMANCE

DiscoveryLink provides a flexible platform for building life sciences applications. It is not intended for the scientist, but rather for the application programmer, an IT worker, or a vendor who creates the tools that the actual scientists will use. While it provides only a simple user interface, it supports multiple programming interfaces, including such *de facto* industry standards as ODBC and JDBC. It, therefore, can be used with any commercially available tool that supports these interfaces, including popular query builders, such as those by Brio or Cognos, application-building frameworks such as VisualAge from IBM, or industry-specific applications including LabBook, Spotfire, and so on. Alternatively, in-house applications can be developed that meet the needs of specific organizations. IBM has a number of business partners who are including DiscoveryLink in their offerings to create a more complete scientific workbench for their customers.

Database Administrators will also be users of DiscoveryLink. For these, DiscoveryLink has a Graphical User Interface (GUI) to help with the registration process. Yet life sciences applications require more support than this. Complete integration also requires the development of tools to bridge between different models of data. The life sciences research community is not a homogeneous one. Different groups use different terms for the same concept or describe different concepts similarly. Semantic mappings must be created, and applications for particular communities must be developed. DiscoveryLink provides features that help with these tasks, but it does not solve either. For example, *views* can help with the problems of semantic integration by hiding mappings from one data representation to another, but the views still must be created manually by the DBAs.

Another characteristic of life sciences data and research environments is frequent change, both in the amounts of data and in the schemas in which data is stored (causing more work for DBAs). Further, new sources of information are always appearing as new technologies and informatics companies evolve. In such an environment, flexibility is essential. DiscoveryLink's powerful query processor and non-procedural SQL interface protect applications (to the extent possible) from changes in the underlying data source via the principle of logical data independence. New sources of information require a new server definition, however, and perhaps a new wrapper, and may also require adjusting view definitions to reference their data. Changes in a data source's interfaces often can be hidden from the application by modifying the translation portion of the wrapper or installing a new wrapper with the new version of the source. The query processing technology is built to handle complex queries and to scale to terabytes of data. Thus, the database middleware concept itself allows DiscoveryLink to deal with the changes in this environment, but it puts a burden on the DBA to administer these changes.

Wrapper writers are a third group of users. The wrapper architecture has been designed for extensibility. Only a small number of functions need to be written to create a working wrapper. Simple sources can be wrapped quickly, in a week or two; more complex sources may require from a few weeks to a few months to completely model, but even for these a working wrapper, perhaps with limited functionality, can be completed quickly. Template code for each part of the wrapper and default cost modeling code are provided for wrapper writers. Wrappers are built to enable as much sharing of code as possible, so that one wrapper can be written to handle multiple versions of a data source, and so that wrappers for similar sources can build on existing wrappers. The ability to separate schema information from wrapper code means that changes in the schema of a data source require no code changes in the wrappers. The addition of a new data source requires no change to any existing wrappers. Thus, the wrappers also help the system adapt to the many changes possible in the environment, and the wrapper architecture eases the wrapper writer's task.

Scalability is a fundamental goal of DiscoveryLink. There is no *a priori* limit to the number of different sources it can handle, because sources are independent and consume little in the way of system resources when not in use. (Wrapper code is loaded dynamically; when not in use, the only trace of the source is a set of catalog entries.) There may be limitations in practice if many sources of different types are used at the same time, depending on how much memory is available. This is akin to the limits on query complexity in relational database management systems today, which are not typically hit until several hundred tables are used in the same query. Because DiscoveryLink is built on robust and scalable relational database technology, there should also be no *a priori* limit on the amount of data the system can handle. Because the data are left in the native stores until needed, they can still be updated by directly modifying those stores. (That is, updates do not need to go through DiscoveryLink, though it may be convenient to do so for relational data sources.) In this case, the update rate is only limited by the update rate of the data sources; DiscoveryLink is not a bottleneck.

As with all database management systems, DiscoveryLink needs to be able to handle complex queries over large volumes of data swiftly and efficiently. For DiscoveryLink, this task is further complicated by the fact that much, if not all, of the data resides in other data sources, which may be distributed over a wide geographic area. Query optimization, which is described in this chapter, is the main tool DiscoveryLink uses to ensure good performance. There are other aspects of the system that also help. For example, before optimization, the query is passed to a rewrite engine. This engine applies a variety of transformations that can greatly improve the ultimate performance. Transforms can, for example, eliminate unnecessary operations such as sorts or even joins. Others can derive new predicates that restrict operations or allow the use of a different access path, again enhancing performance. In addition to query rewrite, wrappers are carefully tuned to use the most efficient programming interfaces provided by the source (e.g., taking advantage of bulk reads and writes to efficiently transport data between sources). Additional constructs such as automatic summary tables (materialized views over local and/or remote data that can be automatically substituted into a query to save remote data access and re-computation) provide a simple form of caching.

How well does the system perform? There are no benchmarks yet for this style of federated data access, and IBM experience to date is limited to a few customers and some experiments in IBM's lab. But, some statements can be made. For example, it is known that DiscoveryLink adds little if any overhead. A simple experiment compares queries that can be run against a single source with the same query submitted against a DiscoveryLink nickname for that source. In most cases, the native performance and performance via DiscoveryLink are indistinguishable [8]. In a few cases, due either to the sophisticated rewrite engine or just the addition of more hardware power, performance using this three-tiered approach (client→Discovery Link→source) is *better* than performance using the source directly (client→source). This has been borne out by repeated experiments on both customer and standard TPC-H workloads. For queries that involve data from multiple sources, it is harder to make broad claims, as there is no clear standard for comparison. The overall experience so far shows that performance depends heavily on the complexity of the query and the amounts of data that must be transported to complete the query. Overall, performance seems to be meeting customers' needs; that is, it is normally good enough that they do not shy away from distributed queries and often are not even aware of the distribution. There are areas for improvement, however, including better exploitation of parallelism when available and some form of automated caching.

11.4   CONCLUSIONS

This chapter described IBM's DiscoveryLink offering. DiscoveryLink allows users to query data that may be physically stored in many disparate, specialized data stores as if that data were all co-located in a single virtual database. Queries against this data may exploit all of the power of SQL, regardless of how much or how little SQL function the data sources provide. In addition, queries may employ any additional functionality provided by individual data stores, allowing users the best of both the SQL and the specialized data source worlds. A sophisticated query optimization facility ensures that the query is executed as efficiently as possible. The interfaces, performance, and scalability of DiscoveryLink were also discussed.

DiscoveryLink is a new offering, but it is based on a fusion of well-tested technologies from DB2 Universal Database (UDB), DB2 DataJoiner, and the Garlic research project. Both DB2 UDB (originally DB2 Client/Server [C/S]) and DB2 DataJoiner have been available as products since the early 1990s, and they have been used by thousands of customers in the past decade. The Garlic project began in 1994, and much of its technology was developed as the result of joint studies with customers, including an early study with Merck Research Laboratories. DiscoveryLink's extensible wrapper architecture and the interactions between wrapper and optimizer during query planning derive from Garlic. As part of Garlic, wrappers were successfully built and queried for a diverse set of data sources, including two relational database systems (DB2 and Oracle), a patent server stored in Lotus Notes, searchable sites on the World Wide Web (including a database of business listings and a hotel guide), and specialized search engines for collections of images, chemical structures, and text.

Currently, IBM is working on building a portfolio of wrappers specific to the life sciences industry. In addition to key relational data sources such as Oracle and Microsoft's SQL Server, wrappers are available for application sources such as BLAST and general sources of interest to the industry such as Microsoft Excel, flat files, Documentum for text management, and XML. IBM is also working with key industry vendors to wrap the data sources they supply. This will provide access to many key biological and chemical sources. While wrappers will be created as quickly as possible, it is anticipated that most installations will require one or more new wrappers to be created because of the sheer number of data sources that exist and the fact that many potential users have their own proprietary sources as well. Hence, a set of tools is being developed for writing wrappers and training a staff of wrapper writers who will be able to build new wrappers as part of the DiscoveryLink software and services offering model. As DiscoveryLink supports the SQL/MED standard [25] for accessing external data sources, those who would rather create their own wrappers (customers, universities, and business partners) may do so, too. Hopefully, in this way a rich set of wrappers will quickly become available for use with DiscoveryLink.

From the preceding pages, hopefully it is clear that DiscoveryLink plays an essential role in integrating life science data. DiscoveryLink provides the plumbing, or infrastructure, that enables data to be brought together, synthesized, and transformed. This plumbing provides a high-level interface, a virtual database against which sophisticated queries can be posed and from which results are returned with excellent performance. It allows querying of heterogeneous collections of data from diverse data sources without regard to where they are stored or how they are accessed.

While not a complete solution to all heterogeneous data source woes, DiscoveryLink is well suited to the life sciences environment. It serves as a platform for data integration, allowing complex cross-source queries and optimizing them for high performance. In addition, several of its features can help in the resolution of semantic discrepancies by providing mechanisms DBAs can use to bridge the gaps between data representations. Finally, the high-level SQL interface and the flexibility and careful design of the wrapper architecture make it easy to accommodate the many types of change prevalent in this environment.

Of course, there are plenty of areas in which further research is needed. For the query engine, key topics are the exploitation of parallelism to enhance performance and richer support for modeling of object features in foreign data sources. There is also a need for additional tools and facilities that enhance the basic DiscoveryLink offering. Some preliminary work was done on a system for data annotation that provides a rich model of annotations, while exploiting the DiscoveryLink engine to allow querying of annotations and data separately and in conjunction. A tool is also being built to help users create mappings between source data and a target, integrated schema [28, 29] to ease the burden of view definition and reconciliation of schemas and data that plagues today's system administrators. Hopefully, as DiscoveryLink matures it will serve as a basis for more advanced solutions that will distill information from the oceans of data in which life sciences researchers are currently drowning, for the advancement of human health and for basic scientific understanding.

REFERENCES

[1] S. Altschul, W. Gish, W. Miller, et al. "Basic Local Alignment Search Tool." *Journal of Molecular Biology* 215, no. 3 (1990): 403–410.

[2] L. Falquet, M. Pagni, P. Bucher, et al. "The PROSITE Database, Its Status in 2002." *Nucleic Acids Research* 30, no. 1 (2002): 235–238.

[3] E. Birney and R. Durbin. "Using GeneWise in the Drosophilia Annotation Experiment" [see comments]. *Genome Research* 10, no. 4 (2000): 547–548.

[4] D. A. Benson, I. Karsch-Mizrachi, D. J. Lipman, et al. "GenBank." *Nucleic Acids Research* 30, no. 1 (2002): 17–20.

[5] A. Bairoch and R. Apweiler. "The SWISS-PROT Protein Sequence Database and Its Supplement TrEMBL in 2000." *Nucleic Acids Research* 28, no. 1 (2000): 45–48.

[6] D. L. Wheeler, D. M. Church, A. E. Lash, et al. "Database Resources of the National Center for Biotechnology Information: 2002 Update." *Nucleic Acids Research* 20, no. 1 (2002): 13–16.

[7] M. Ringwald, J. T. Epping, D. A. Begley, et al. "The Mouse Gene Expression Database (GXD)." *Nucleic Acids Research* 29, no. 1 (2001): 98–101.

[8] L. M. Haas, P. M. Schwarz, P. Kodali, et al. "DiscoveryLink: A System for Integrated Access to Life Sciences Data Sources." *IBM Systems Journal* 40, no. 2 (February 2001): 489–511.

[9] P. Gupta and E. T. Lin. "Datajoiner: A Practical Approach to Multi-Database Access." In *Proceedings of the International IEEE Conference on Parallel and Distributed Information Systems*, 264. Los Alamitos, CA: IEEE Computer Society, 1994.

[10] L. M. Haas, D. Kossmann, E. L. Wimmers, et al. "Optimizing Queries Across Diverse Data Sources." In *Proceedings of the Conference on Very Large Databases (VLDB)*, 276–285. San Francisco: Morgan Kaufmann, 1997.

[11] M. T. Roth and P. M. Schwarz. "Don't Scrap It, Wrap It! A Wrapper Architecture for Legacy Data Sources."In *Proceedings of the Conference on Very Large Data Bases (VLDB)*, 266–275. San Francisco: Morgan Kaufmann, 1997.

[12] S. Davidson, C. Overton, V. Tannen, et al. "BioKleisli: A Digital Library for Biomedical Researchers." *International Journal of Digital Libraries* 1, no. 1 (January 1997): 36–53.

[13] T. Etzold and P. Argos. "SRS: An Indexing and Retrieval Tool for Flat File Data Libraries." *Computer Applications in the Biosciences* 9, no. 1 (1993): 49–57.

[14] P. Carter, T. Coupaye, D. Kreil, et al. "SRS: Analyzing and Using Data from Heterogeneous Textual Databanks." In *Bioinformatics: Databases and Systems*, edited by S. Letovsky. Boston: Kluwer Academic, 1998.

[15] I-M. A. Chen, A. S. Kosky, V. M. Markowitz, et al. "Constructing and Maintaining Scientific Database Views in the Framework of the Object-Protocol Model." In *Proceedings of the Ninth International Conference on Scientific and Statistical Database Management*, 237–248. Los Alamitos, CA: IEEE Computer Society, 1997.

[16] R. Stevens, C. Goble, N. W. Paton, et al. "Complex Query Formulation Over Diverse Information Sources in TAMBIS." In Z. Lacroix and T. Critchlow (eds). *Bioinformatics: Managing Scientific Data*, 189–223. San Francisco: Morgan Kaufmann, 2004.

[17] B. A. Eckman, A. S. Kosky, and L. A. Laroco Jr. "Extending Traditional Query-Based Integration Approaches for Functional Characterization of Post-Genomic Data." *Bioinformatics* 17, no. 7 (2001): 587–601.

[18] Y. Papakonstantinou, H. Garcia-Molina, and J. Widom. "Object Exchange Across Heterogeneous Information Sources." In *Proceedings of the IEEE Conference on Data Engineering*, 251–260. Los Alamitos, CA: IEEE Computer Society, 1995.

[19] A. Tomasic, L. Raschid, and P. Valduriez. "Scaling Heterogeneous Databases and the Design of DISCO." In *Proceedings of International Conference on Distributed Computing Systems (ICDCS)*, 449–457. Los Alamitos, CA: IEEE Computer Society, 1996.

[20] M-C. Shan, R. Ahmed, J. Davis, et al. "Pegasus: A Heterogeneous Information Management System." In *Modern Database Systems*, edited by W. Kim, 664–682. Reading, MA: Addison-Wesley, 1995.

[21] L. Liu and C. Pu. "The Distributed Interoperable Object Model and its Application to Large-Scale Interoperable Database Systems." In *Proceedings of the ACM International Conference on Information and Knowledge Management*, 105–112. New York: Association for Computing Machinery, 1995.

[22] S. Adali, K. Candan, Y. Papakonstantinou, et al. "Query Caching and Optimization in Distributed Mediator Systems." In *Proceedings of the ACM SIGMOD Conference on Management of Data*, 137–148. New York: Association for Computing Machinery, 1996.

[23] International Organization for Standardization. "Information Technology—Database Languages—SQL—Part 3: Call Level Interface (SQL/CLI)." *ISO/IEC 9075-3*. Geneva, Switzerland: International Organization for Standardization, 1999.

[24] International Organization for Standardization. "Information Technology—Database Languages—SQL—Part 2: Foundation (SQL/Foundation)." *ISO/IEC 9075-2*. Geneva, Switzerland: International Organization for Standardization, 1999.

[25] International Organization for Standardization. "Information Technology—Database Languages—SQL—Part 9: Management of External Data (SQL/MED)." *ISO/IEC 9075-9*. Geneva, Switzerland: International Organization for Standardization, 2000.

[26] IBM. *IBM DB2 Life Sciences Data Connect Planning, Installation and Configuration Guide*, Version 7.2 FP 5. White Plains, NY: IBM, 2001. http://www-3.ibm.com/software/data/db2/lifesciencesdataconnect/db2ls-pdf.html.

[27] M. T. Roth, F. Ozcan, and L. M. Haas. "Cost Models Do Matter: Providing Cost Information for Diverse Data Sources in a Federated System." In *Proceedings of the Conference on Very Large Data Bases (VLDB)*, 559–610. San Francisco: Morgan Kaufmann, 1999.

[28] L. M. Haas, R. J. Miller, B. Niswonger, et al. "Transforming Heterogeneous Data with Database Middleware: Beyond Integration." *IEEE Data Engineering Bulletin* 22, no. 1 (1999): 31–36.

[29] R. J. Miller, L. M. Haas, and M. Hernandez. "Schema Mapping as Query Discovery." In *Proceedings of the Conference on Very Large Data Bases (VLDB)*, 77–88. San Francisco: Morgan Kaufmann, 2000.

A Model-Based Mediator System for Scientific Data Management

Bertram Ludäscher, Amarnath Gupta, and Maryann E. Martone

A database mediator system combines information from multiple existing source databases and creates a new virtual, mediated database that comprises the integrated entities and their relationships. When mediating scientific data, the technically challenging problem of mediator query processing is further complicated by the complexity of the source data and the relationships between them. In particular, one is often confronted with complex multiple-world scenarios in which the semantics of individual sources, as well as the knowledge to link them, require a deeper modeling than is offered by current database mediator systems. Based on experiences with federation of brain data, this chapter presents an extension called *model-based mediation* (MBM). In MBM, data sources export not only raw data and schema information but also *conceptual models* (CMs), including domain semantics, to the mediator, effectively lifting data sources to *knowledge sources*. This allows a mediation engineer to define integrated views based on (1) the local CMs of registered sources and (2) auxiliary domain knowledge sources called *domain maps* (DMs) and *process maps* (PMs), respectively, which act as sources of *glue knowledge*. For complex scientific data sources, semantically rich CMs are *necessary* to represent and reason with scientific rationale for linking a wide variety of heterogeneous experimental assumptions, observations, and conclusions that together constitute an experimental study. This chapter illustrates the challenges using real-world examples from a complex neuroscience integration problem and presents the methodology and some tools, in particular the knowledge-based integration of neuroscience data (KIND) mediator prototype for model-based mediation of scientific data.

12.1   BACKGROUND

Seamless data access and sharing, handling of large amounts of data, federation and integration of heterogeneous data, distributed query processing and application integration, data mining, and visualization are among the common and recurring broad themes of scientific data management. A main stream of activity in the bioinformatics domain is concerned with sequence and structural databases such as GenBank, the Protein Data Bank (PDB), and Swiss-Prot, and much work is devoted to algorithmic challenges stemming from problems (e.g., efficient sequence alignment and structure prediction). However, in addition to the well-known challenges of bioinformatics applications such as algorithmic complexity and scalability (e.g., in genomics), there are other major challenges that are sometimes overlooked, particularly when considering scientific data beyond the level of sequence and protein data (e.g., brain imagery data). These challenges arise in the context of *information integration of scientific data* and have to do with the inherent semantic complexity of (1) the actual source data and (2) the *glue knowledge* necessary to link the source data in meaningful ways. Traditional federated database system architectures, and those of the more recent database mediators developed by the database community, need to be extended to handle adequately information integration of complex scientific data from multiple sources. This extension is a combination of knowledge representation and mediator technology. In a nutshell:

Model-Based Mediation = Database Mediation + Knowledge Representation

With respect to their *semantic heterogeneity* (ignoring syntactic and system aspects), information integration/mediation scenarios (scientific or otherwise) can be roughly classified along a spectrum as follows: On one end, there are *simple one-world scenarios*; somewhere in the middle are *simple multiple-world scenarios*; and at the other end of the spectrum are *complex multiple-world scenarios*. An example of a *simple one-world scenario* (i.e., in which the modeled real-world entities can be related easily to one another and come from a single domain) is *comparison shopping* for books. A typical query is to find the cheapest price for a given book from a number of sources such as amazon.com and bn.com. An example of a *simple multiple-world scenario* is the integration of realtor and census data to annotate and rank real estate by neighborhood quality. Here, the approach combines and relates quite different kinds of information, but the relations between the multiple worlds are simple enough to be understood without deep domain knowledge. Examples of *complex multiple-world scenarios* are often found in scientific data management and are the subject of this chapter. Thus, *simple* and *complex* here refer to the degree in which specific *domain semantics* is required to formalize or even state meaningful associations and linkages between data objects of interest; it does not mean that the database and mediation technology for realizing such mediators is simple.[1] For example, to state the problem of what the result of an integrated comparison shopping view should be, a basic understanding of a *books schema* (title, authors, publisher, price, etc.) is sufficient. In particular, the association operation that links objects of interest across sources can be executed (at least in principle) as a *syntactic join* on the ISBN. Similarly, in the realtor example, data can be joined based on the ZIP code, latitude and longitude, or street address (i.e., by *spatial joins* that can be modeled as atomic function calls to a *spatial oracle*). To understand the basic linkage of information objects, no insight into the details of the spatial join is required.

This is fundamentally different for complex multiple-world scenarios as found in many scientific domains. There, even if data is stored in state-of-the-art (often Web accessible) databases, significant domain knowledge is required to articulate meaningful queries across disciplines (or within different micro-worlds of a single discipline); further examples are offered in the next section.

Outline

In this chapter, these challenges are illustrated with examples from ongoing collaborations with users and providers of scientific data sets, in particular from the neuroscience domain (see Section 12.2). Then a methodology called *model-based mediation*, which extends current database mediator technology by incorporating knowledge representation (KR) techniques to create explicit representations of domain experts' knowledge that can be used in various ways by mediation engineers and by the MBM system itself, is presented in Section 12.3. The goal of MBM could be paraphrased as:

Turning scientists' *questions* into executable database *queries*.

Section 12.4 introduces some of the KR formalisms (e.g., for domain maps and process maps) and describes their use in MBM. In Section 12.5 the KIND mediator prototype and other tools being developed at the San Diego Supercomputer Center (SDSC) and the University of California at San Diego (UCSD) are presented primarily in the context of the neuroscience domain. Section 12.6 discusses related work and concludes the chapter.

---

1. Such *simple* mediation scenarios often pose very difficult technical challenges (e.g., query processing in the presence of limited source capabilities) [1, 2].

12.2   SCIENTIFIC DATA INTEGRATION ACROSS MULTIPLE WORLDS: EXAMPLES AND CHALLENGES FROM THE NEUROSCIENCES

Some of the challenges of scientific data integration in complex multiple-world scenarios are illustrated using examples that involve different neuroscience worlds. Such examples occur regularly when trying to federate brain data across multiple sites, scales, and even species [3] and have led to new research and development projects aimed at overcoming the current limitations of biomedical data sharing and mediation [4].

Example 12.2.1 (Two Neuroscience Worlds). Consider two neuro-science laboratories, SYNAPSE and NCMIR[2], that perform experiments on two different brain regions. The first laboratory, SYNAPSE, studies dendritic spines of pyramidal cells in the hippocampus. The primary schema elements are thus the anatomical entities reconstructed from 3D serial sections. For each entity (e.g., spines, dendrites), researchers make a number of measurements and study how these measurements change across age and species under several experimental conditions.

In contrast, the NCMIR laboratory studies a different cell type, the Purkinje cells of the cerebellum. They inspect the branching patterns from the dendrites of filled neurons and the localization of various proteins in neuron compartments. The schema used by this group consists of a number of measurements of the dendrite branches (e.g., segment diameter) and the amount of different proteins found in each of these subdivisions. Assume each of the two schemas has a class C with a location attribute that has the value `Pyramidal Cell dendrite` and `Purkinje Cell`, respectively.

How are the schemas of SYNAPSE and NCMIR related? Evidently, they carry distinctly different information and do not even enter the purview of the schema conflicts usually studied in databases [5]. To the scientist, however, they are related for the following reason: Like pyramidal neurons, Purkinje cells also possess dendritic spines. Release of calcium in spiny dendrites occurs as a result of neurotransmission and causes changes in spine morphology (sizes and shapes obtained from SYNAPSE). Propagation of calcium signals throughout a neuron depends on the morphology of the dendrites, the distribution of calcium stored in a neuron,

---

2. Information about the two laboratories SYNAPSE and NCMIR is respectively available at *http://synapses.bu.edu* and *http://www-ncmir.ucsd.edu.*

and the distribution of calcium binding proteins, whose subcellular distribution for Purkinje cells are measured by NCMIR.

Thus, a researcher who wanted to model the effects of neurotransmission in hippocampal spines would get structural information on hippocampal spines from SYNAPSE and information about the types of calcium binding proteins found in spines from NCMIR. Note that neither of the sources contains information that would allow a mediator system to bridge the *semantic gap* between them. Therefore, *additional domain knowledge*—independent of the observed experimental raw data of each source—is needed to connect the two sources. The domain expert, here a neuroscientist, it is easy to provide the necessary *glue knowledge*

> Purkinje cells and Pyramidal cells have dendrites that have higher-order branches that contain spines. Dendritic spines are ion (calcium) regulating components. Spines have ion binding proteins. Neurotransmission involves ionic activity (release). Ion-binding proteins control ion activity (propagation) in a cell. Ion-regulating components of cells affect ionic activity (release).

To capture such domain knowledge and make it available to the system, the proposed approach employs two kinds of *ontologies*, called *domain maps* and *process maps*, respectively. The former are aimed at capturing the basic domain terminology, and the latter are used to model different process contexts. Ontologies, such as the domain map in Figure 12.1, are often formalized in logic (in this case statements in *description logic* [6]; see Section 12.4.1). Together with additional inference rules (e.g., capturing transitivity of has), logic axioms like these formally capture the domain knowledge and allow mediator systems to work with this knowledge (e.g., a concept or class hierarchy can be used to determine whether the system should retrieve objects of class C' when the user is looking for instances of C).

Domain maps not only provide a concept-oriented browsing and data exploration tool for the end user, but—even more importantly—they can be used for defining and executing integrated view definitions (IVDs) at the mediator. The previous real-world example illustrates a fundamental difference in the nature of information integration as studied in most of the database literature and as is necessary for scientific data management. In the latter, seemingly unconnected schema can be semantically close *when situated in the scientific context*, which, in this case, is the neuroanatomy and neurophysiological setting described previously. Therefore, this is called *mediation across multiple worlds* and it is facilitated using domain maps such as the one shown (see Figure 12.1).

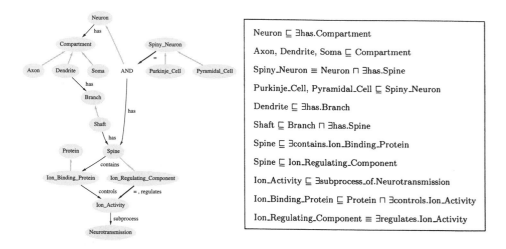

Neuron $\sqsubseteq$ $\exists$has.Compartment

Axon, Dendrite, Soma $\sqsubseteq$ Compartment

Spiny_Neuron $\equiv$ Neuron $\sqcap$ $\exists$has.Spine

Purkinje_Cell, Pyramidal_Cell $\sqsubseteq$ Spiny_Neuron

Dendrite $\sqsubseteq$ $\exists$has.Branch

Shaft $\sqsubseteq$ Branch $\sqcap$ $\exists$has.Spine

Spine $\sqsubseteq$ $\exists$contains.Ion_Binding_Protein

Spine $\sqsubseteq$ Ion_Regulating_Component

Ion_Activity $\sqsubseteq$ $\exists$subprocess_of.Neurotransmission

Ion_Binding_Protein $\sqsubseteq$ Protein $\sqcap$ $\exists$controls.Ion_Activity

Ion_Regulating_Component $\equiv$ $\exists$regulates.Ion_Activity

12.1 FIGURE   A *domain map* for SYNAPSE and NCMIR (left) and its formalization in description logic (right). Unlabeled, gray edges $\approx$ "isa" $\approx$ "$\sqsubseteq$".

12.2.1   From Terminology and Static Knowledge to Process Context

While domain maps are useful to put data into a terminological and thus somewhat static knowledge context, a different knowledge representation has to be devised when trying to put data into a dynamic or process context. Consider, for example, the groups of neuroscientists who study the science of mammalian memory and learning. Many of these groups study a phenomena called *long-term potentiation* (LTP) in nerve cells, in which repeated or sustained input to nerves in specific brain regions (such as the hippocampus) conditions them in such a manner that after some time, the neuron produces a large output even with a small amount of *known* input. Given this general commonality of purpose, however, individual scientists study and collect observational data for very different aspects of the phenomena.

Example 12.2.2 (Capturing Process Knowledge). Consider a group [7] that studies the role of a specific protein *N-Cadherin* in the context of *synapse formation* during *late-phase long-term potentiation* (L-LTP), a subprocess of LTP. The data collected by the group consists of measurements that illustrate how the amount of *N-Cadherin* and the number of synapses (nerve junctions) both simultaneously increase in cells during L-LTP. Now consider that a different group [8] studies a new enzyme called CAMK-IV and its impact on a chemical reaction called *phosphorylation* of a protein called CREB. Their data are collected to show how modulating the amounts of CAMK-IV and other related enzymes affect the amount of CREB production, and how this, in turn, affects other products in the nucleus of the neurons. Ideally, the goal of *mediating* between experimental information from these two sources would be to produce an integrated view that enables an end-user scientist to get a deeper understanding of the LTP phenomena. Specifically, the end user should be able to ask queries (and get answers) that exploit the scientific interrelationship between these experiments. In this way, the integrated access provided by a mediator system can lead to new observations and questions, thus eventually driving new experiments.

At the risk of oversimplification, the first group looks at synapse formation and is only interested in the fact that some proteins (including *N-Cadherin*) bring about the formation of synapses. They do not look at the processes leading to the production of these proteins. The second group looks at a specific chain of events leading up to the production of the proteins but does not identify which proteins are produced. The *semantic connection* between these two sources can be constructed in terms of the underlying *event structure* and the way the two groups elaborate on different parts of it. Figure 12.2 depicts a simplified view of the relationship explained previously and shows the cyclic progression of events leading to synapse formation during LTP: Red edges situate the first source with respect to the overall process, and blue edges situate the second source. In either case, the dashed lines show the subsequence of events the sources *glossed over*, or abstracted. Thus, the first source does not have any information pertaining to *phosphorylates (CAMK-IV, CREB)*, and the second source does not have any data related to *forms (protein, synapse)*. Neither source has any data about the (black) edge *synthesizes (gene, protein)*.

Domain maps allow data providers to put their source data into a static/terminological context, and process maps allow them to do the same for a dynamic/process context. Together, they capture valuable *glue knowledge* that resides at the mediator and facilitates integration of hard-to-correlate sources: in particular, concept-oriented data discovery (semantic browsing) [9], view definition, and semantic query optimization [10]. To make model-based mediation effective, it is also necessary to *hook* the elements of the source schema to the domain map and the process map. This process, called the *contextualization* mechanism, is central to the MBM framework.

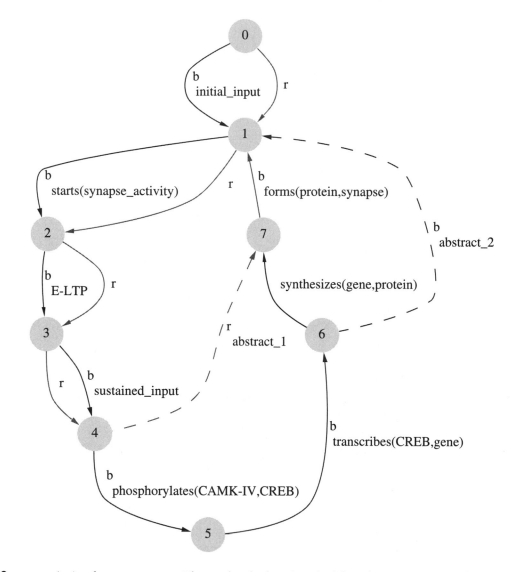

A simple process map. Blue and red edges (marked *b* and *r*, respectively) depict processes about which two data sources/research groups have observational data; dashed edges indicate abstractions (short cuts). No observational data is available for the edge 6–7; hence, this edge is shown in black (unmarked).

12.3   MODEL-BASED MEDIATION

In mediator systems, differences in syntax and data models of sources $S_1, S_2, \ldots$ are resolved by wrappers that translate the raw data into a common data format, typically extensible markup language (XML). In most current mediator systems, all other differences, in particular schema heterogeneities, are then handled by an appropriate integrated view definition (IVD), which is defined using an XML query language [11, 12]. This architecture is extended by lifting exported source data from the level of uninterpreted, semistructured data in XML syntax to the semantically rich level of *conceptual models (CMs) with domain knowledge*. Then, the mediator's views can be defined in terms of CMs (i.e., IVDs are defined in a *global-as-view* fashion) and thus make use of a semantically richer model involving class hierarchies, complex object structure, and properties of relationships (relational constraints, cardinalities).

12.3.1   Model-Based Mediation: The Protagonists

The underlying methodology and procedures of MBM involve users in different roles and at different levels:

+ *Data providers* are typically domain experts, such as bench scientists who would like to make their data from experimental studies available to the community. In MBM, data providers can not only export an XML-queriable version of their data, but they can also export *domain semantics* by lifting the exported data and schema information from a structural level (e.g., XML DTDs [Document Type Definitions]) to the level of CMs.[3] Allowing data providers to situate or *contextualize* (see Example 12.3.2) their primary data themselves has significant benefits. First, data providers know best where their data fit on the glue maps. Second, even without the IVDs defined by mediation engineers, data are automatically associated across different sources via their domain/process map contexts.

+ *View providers* specify integrated view definitions (IVDs), that is, they program complex views in an expressive, declarative rule language. The IVDs are defined over the registered complex sources CM(S$_1$), CM(S$_2$),... and the glue knowledge sources in the mediator's repository. Thus, view providers are

---

3. The w3c working group XML Schema (*http://www.w3.org/XML/Schema*) and similar efforts like RELAX NG (*http://www.oasis-open.org/committees/relax-ng/*) play an intermediate role between purely structure-based models (DTDs) and richer semantic models with constraint mechanisms.

the actual mediation engineers and they bring together (as a team or individually) expertise in the application domain and in databases and knowledge representation.

The new fused objects defined by an IVD can be contextualized, based on the contexts provided by the source conceptual models (see right side of Figure 12.6). In this way, an integrated, virtual view exported by the mediator becomes a first-class citizen of the federation; it is considered a conceptual level source CM(M) itself and can be used just like any original CM-wrapped source.

✦ *End users* can start with *semantic browsing* of CMs, by navigating the domain and process ontologies in the style of topic maps, in which a user navigates through a concept space by following certain relationships, going up and down concept hierarchies and so on. Users may also focus their view by issuing graph queries over domain or process maps, which return only the subgraphs of interest. Eventually, the user can access raw data from different sources, which is (due to contextualization) automatically organized by context [9], and access derived data resulting from user queries against the mediated views.

12.3.2   Conceptual Models and Registration of Sources at the Mediator

The following components of the conceptual model CM of a source S can be distinguished:

$$CM(S) = OM(S) \cup ONT(S) \cup CON(S)$$

The different logical components and their dependencies are depicted in Figure 12.3:

✦ $OM(S)$ is the *object model* of the source S and provides signatures for *classes*, *associations* between classes, and *functions*. $OM(S)$ structures can be defined extensionally by facts (EDB), or intensionally via rules (IDB).

✦ $ONT(S)$ is the *local ontology* of the source S. It defines concepts and their relationships from the source's perspective.

✦ $ONTG(S)$ is the *ontological grounding* of $OM(S)$ in $ONT(S)$, that is, a mapping between the object model $OM(S)$ (classes, attributes, associations) and the concepts and relationships of $ONT(S)$.

✦ $CON(S)$ is the *contextualization* of the local source ontology relative to a mediator ontology, $ONT(M)$.

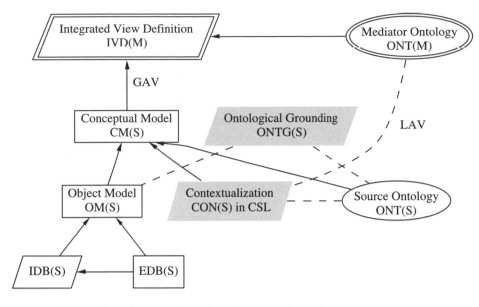

Model-based mediation: dependencies among logical components.

✦ IVD(*M*) is the mediator's *integrated view definition* and comprises logic view definitions in terms of the sources' object models OM(*S*) and the mediator's ontology ONT(*M*). By posing queries against the mediator's IVD(*M*), the user has the illusion of interacting with a single, semantically integrated source instead of interacting with independent, unrelated sources.

In the following, the local parts of CM(*S*) (OM(*S*), ONT(*S*), and ONTG(*S*)) are presented through a running example. For details on the contextualization CON(*S*) see Example 12.3.2 and the related work on registering scientific data sources [13].

Example 12.3.1   (Cell-Centered Database [CCDB]). Figure 12.4 shows pieces of a simplified version of the conceptual model CM(CCDB) of a real-world scientific information source called the *Cell-Centered Database* [14]. The database consists of a set of EXPERIMENTS objects. Each experiment collects a number of cell IMAGES from one or more instruments. For each image, the scientists mark out cellular STRUCTURES in the image and perform measurements on them [14]. They also identify a second set of regions, called DEPOSITs, in images that show the deposition of molecules of proteins or genetic markers. In general, a region marked as *deposit* does not necessarily coincide with a region marked as a *structure*.

Classes in OM(CCDB)

EXPERIMENT(<u>id</u>:id, date:date, cell_type:string, images:SET(image)).

IMAGE(<u>id</u>:id, instrument:ENUM{c_microscope, e_microscope}, resolution:float, size_x:int, size_y:int,

    depth:int, structures:SET(structure), regions:SET(deposit)).

STRUCTURE(<u>id</u>:id, name:string, length:float, surface_area:float, volume:float, bounding_box:Cube).

DEPOSIT(<u>id</u>:id, substance_name:string, deposit_type:string, relative_intesity:ENUM{dark,normal,bright},

    amount:float, bounding_box:Cube).

. . .

Associations in OM(CCDB)

co_localizes_with(DEPOSIT.substance_name, DEPOSIT.substance_name, STRUCTURE.name).

surrounds(s1:STRUCTURE, s2:STRUCTURE).

. . .

Functions in OM(CCDB)

deposit_in_structure(DEPOSIT.id) $\longrightarrow$ SET(STRUCTURE.name)

. . .

Source Ontology – ONT(CCDB)

brain $\xrightarrow{has(co)}$ cerebellum $\xrightarrow{has(co)}$ cerebellar cortex $\xrightarrow{has(co)}$ vermis   (ONT1)

dendrite $\xrightarrow{has(co)}$ spine process $\xrightarrow{has(pm)}$ spine   (ONT2)

cell $\xrightarrow{projects\_to}$ brain_region

globus_pallidus $\xrightarrow{isa}$ brain_region.   . . .   denaturation $\xrightarrow{isa}$ process.   (ONT3)

$tc\_has(co) :=$ **transitive_closure**$(has(co))$. $tc\_has(pm) :=$ **transitive_closure**$(has(pm))$. (ONT4)

$has\_co\_pm :=$ **chain**$(tc\_has(co), tc\_has(pm))$ (ONT5)

. . .

Ontological Grounding – ONTG(CCDB)

domain(STRUCTURE.volume) **in** [0,300]

domain(STRUCTURE.name) **in** $tc\_has(co)$(cerebellum)   (OG1)

domain(EXPERIMENT.cell_type) **in** $tc\_has(co)$(cerebellum)   (OG2)

EXPERIMENT.cell_type $\xrightarrow{projects\_to}$ globus_pallidus   (OG3)

DENATURED_PROTEIN $\xrightarrow{exhibits}$ denaturation.   (OG4)

. . .

12.4

FIGURE

Conceptual model for registering the Cell-Centered Database [14].

Note that OM(CCDB) in Figure 12.4 includes classes that are instantiated with observed data, that is, the extensional database EDB(CCDB). In addition to classes, OM(CCDB) stores *associations*, which are *n*-ary relationships between object classes. The association co_localizes_with specifies which pairs of substances occur together in a specific structure. The object model also contains *functions*, such as the domain specific methods that can be invoked by a user as part of a query. For example, when the mediator or another client calls the function CCDB.deposit_in_structure(), and supplies the ID of a deposit object, the function returns a set of structure objects that spatially overlap with the specified deposit object.

Next, the source's local ontology, ONT(CCDB) is described. Here, an ontology ONT(S) consists of a set of *concepts* and inter-concept *relationships*,[4] possibly augmented with additional inference rules and constraints.[5] The ontological grounding ONTG(S) links the object model OM(S) to the source ontology ONT(S). The source ontology serves a number of different purposes.

Creating a Terminological Frame of Reference

For defining the terminology of a specific scientific information source, the source declares its own controlled vocabulary through ONT(S). More precisely, ONT(S) comprises the terms (i.e., concepts) of this vocabulary and the *relationships* among them. The concepts and relationships are often represented as nodes and edges of a directed graph, respectively. Two examples of inter-concept relations are has(co) and has(pm), which are different kinds of part-whole relationships.[6] In Figure 12.4, items ONT1 and ONT2 show fragments of such a concept graph. Once a concept graph is created for a source, one may use it to define additional constraints on object classes and associations.

Semantics of Relationships

The edges in the concept graph of the source ontology represent inter-concept relationships. Often these relationships have their own semantics, which must be specified within ONT(S). Item ONT4 declares two new relationships, tc_has(co) and tc_has(pm). After registration, the mediator interprets this declaration and creates the new (possibly materialized) transitive relations on top of the base

---

4. Most formal approaches (e.g., those based on description logic) consider binary relationships only.

5. For example, ONT4, ONT5 in Figure 12.4 define virtual relations such as *transitive closure* over the base relations.

6. By standards of meronyms, there are different kinds of the has relation, including component-object has(co), portion-mass has(pm), member-collection has(mc), stuff-object has(so), and place-area has(pa) [15].

relations `has(co)` and `has(pm)` provided by the source S. Similarly, the item ONT5 is interpreted by the mediator using a higher-order rule for chaining binary relations:

```
chain(R1,R2)(X,Y)   if   R1(X,Z), R2(Z,Y)
```

With this, ONT5 creates a new relationship `has_co_pm(X,Y)` provided that there is a Z such that `tc_has(co)(X,Z)`, *and* `tc_has(pm)(Z,Y)`.

Ontological Grounding of OM(S)

A local domain constraint specifies additional properties of the given extensional database and thereby establishes an *ontological grounding* ONTG(S) between the local ontology ONT(S) and the object model OM(S) (see Figure 12.3). Items OG1–OG2 in Figure 12.4 refine the domains of the attributes EXPERIMENT.cell_type and STRUCTURE.name from the original type declaration (STRING). The refinement constrains them to take values from those nodes of the concept graph that are *descendants* of the concept cerebellum through the `has(co)` relationship.

This constraint illustrates an important role of the local ontology in a *conceptually lifted* source. By constraining the domain of an attribute to be concept name, C, the corresponding object instance o is *semantically about* C. In addition, this also implies that o is *about* any ancestor concept, C', of C where ancestor is defined via `has(co)` edges only. Similarly, if a specific instance, STRUCTURE.name, has the value `spine process`, it is also about dendrite (ONT2 in Figure 12.4).

In addition to linking attributes to concept names, a constraint may also involve inter-concept relationships. Assume `projects_to(cell, brain_region)` is a relationship in the source ontology ONT(CCDB). A constraint may assert that for all instances *e* of class EXPERIMENT, `projects_to(e.cell_type, 'globus_pallidus')` holds (OG3). The constraint thus *refines* the original relationship `projects_to` to suit the specific semantics of OM(CCDB). Such constraint-defined correspondences between OM(S) and ONT(S) are used in the contextualization process [13].

Intensional Definitions

In the CM wrapper of a source, *S*, one can define virtual classes and associations that can be exported to the mediator as first-class, queriable items by means of an intensional database IDB(S). For example, one can create a new virtual class called DENATURED_PROTEIN in IDB(CCDB) via the rule:

```
DENATURED_PROTEIN(ProtName) if DEPOSIT(ID, ProtName,
   protein, dark, _, _),deposit_in_structure(ID) ≠ Ø
```

Thus, an instance of a DENATURED_PROTEIN is created when a *dark* protein deposit is recorded in an instance of DEPOSIT and there is some structure in which this deposit is found. As a general principle of creating a CM wrapper, such a definition will be supplemented by additional constraints to connect it to the local ontology. For example, assume that ONT(CCDB) already contains a concept called process. Item ONT3 defines denaturation as a specialization of process. The constraint OG4 completes the semantic specification about the new DENATURED_PROTEIN object.

Contextual References

It is a common practice for scientific data sources to tag object instances with attributes from a public standard and to use controlled vocabularies for the values of some of these attributes. For example, the source can specify that the domain of the DEPOSIT.id field can be accessed through an internal method, which, given a protein name, gets its id from a specific database. For example, one can use get_expasy_protein_id to retrieve this information from the Swiss-Prot database on the Web. How the source enforces this integrity constraint is internal to the source and not part of its conceptual export schema.

12.3.3 Interplay Between Mediator and Sources

To address the source registration issue, which components of an existing *n*-source federation that can be seen, or *accessed*, by the new, $n+1^{st}$ source need to be specified. A federation at the mediator consists of: (1) currently *registered conceptual models* CM(S) of each participating source S, (2) one or more *global ontologies* ONT(M) residing at the mediator that have been used in the federation, and (3) *integrated views* IVD(M) defined in a global-as-view (GAV) fashion.

Typical mediator ontologies ONT(M) are *public*, meaning they serve as domain-specific expert knowledge and thus can be used to *glue* conceptual models from multiple sources. Examples of such ontologies are the Unified Medical Language System (UMLS) from the National Library of Medicine[7] and the Biological Process Ontology from the Gene Ontology Consortium.[8] In the presence of multiple ontologies, *articulations*, (mappings between different source ontologies [16])

---

7. The Unified Medical Language System (UMLS) available at *http://www.nlm.nih.gov/research/umls/* is, strictly speaking, a metathesaurus, or a semi-formal ontology with a limited set of pre-defined relationships such as broader-term/narrower-term.

8. See *http://www.geneontology.org/process.ontology* for information about the Biological Process from the Gene Ontology Consortium.

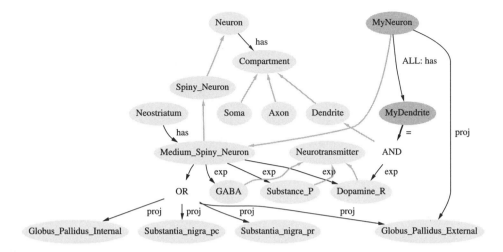

A domain map (DM) after situating new concepts `MyNeuron` and `MyDendrite` (dark).

can be used to register with the mediator information about inter-source relationships. Note that a source, S, usually cannot see all of the previously discussed components (1–3) when defining its conceptual model: Although S sees the mediator's ontologies, `ONT(M)`, and thus can define its own conceptual model, `CM(S)`, relative to the mediator's ontology in a *local-as-view* (LAV) fashion, it cannot directly employ *another* source's conceptual model, `CM(S')`, nor can it query the mediator's integrated view, `IVD(M)`, which is defined *global-as-view* (GAV) on top of the sources. The former is no restriction because S' can register `CM(S')`, in particular `ONT(S')`, with the mediator, at which point S can indirectly refer to registered concepts of S' via `ONT(M)`. The latter guarantees that query processing in this setting does not involve recursion through the Web (i.e., between a source S and the mediator M). The dependency graph in Figure 12.3 is acyclic.[9]

Example 12.3.2   (Contextualization: Local-as-View). Consider the domain map in Figure 12.5. Lighter-colored nodes correspond to concepts that the mediator understands and a source can see. Now assume a source, S, wants to register information about specific neurons and their dendrites, but the mediator ontology (domain map) does not have dedicated names for those specific kinds of neurons and dendrites. In MBM this problem is solved by contextualizing

---

9. At the cost of loss of efficiency, the restriction *no recursion through the Web* could be lifted.

the new local source concepts as views on the mediator's global concepts: In Figure 12.5, the darker-colored source concepts are *hooked* to the mediator's domain map, thereby defining their meaning relative to the mediator's concepts. This is achieved by sending the following first-order axioms (here in description logic syntax) to the mediator:

$$\mathtt{MyDendrite} \equiv \mathtt{Dendrite} \sqcap \exists \mathtt{exp.Dopamine\_R}$$
$$\mathtt{MyNeuron} \sqsubseteq \mathtt{Medium\_Spiny\_Neuron}$$
$$\sqcap \exists \mathtt{proj.Globus\_pallidus\_external}$$
$$\sqcap \forall \mathtt{has.MyDendrite}$$

Thus instances of `MyDendrite` are exactly those dendrites that express Dopamine R(eceptor), and `MyNeuron` objects are medium spiny neurons projecting to Globus Pallidus External and *only* have `MyDendrites`. Assuming properties are *inherited* along the transitive closure of `isa`, it follows that `MyNeuron`, like any `Medium_Spiny_Neuron` projects to certain structures (OR in Figure 12.5). With the newly registered knowledge, it follows that `MyNeuron` definitely projects to *Globus_Pallidus_External*. To specify that it *only* projects to the latter, a *nonmonotonic inheritance* (e.g., using F-logic with well-founded semantics) can be employed.

Note that the intuitive graphical contextualization depicted in Figure 12.5 is not unique; logically equivalent domain maps may have different graphical representations.[10] For domain maps that can be completely axiomatized using a description logic, a reasoning system such as Fast Classification of Terminologies (FaCT) [17] can be employed to compute the deductive closure and, in particular, to derive a unique concept hierarchy and check consistency of a domain map.

12.4 KNOWLEDGE REPRESENTATION FOR MODEL-BASED MEDIATION

This section takes a closer look at the principal mechanisms for specifying glue knowledge: ontologies in the form of domain maps (DMs) and process maps (PMs).

---

10. This is similar to the fact that the same query can have many different syntactic representations. In general, equivalence of first-order (or SQL) queries is not decidable.

12.4.1   Domain Maps

As is standard for ontologies, DMs name and specify relevant concepts by describing the characteristic relationships among them [18]. In this way, DMs provide the basic domain semantics needed to glue data across different sources in multiple-world scenarios. DMs can be depicted more intuitively in the form of labeled, directed graphs. In contrast to many other graph-based notations, however, DMs have a solid formal semantics via a translation to logic rules. The graph form of DMs is defined as follows.

Definition 12.1 Domain Maps

Let C be a set of symbols called *concepts* and $\mathcal{R}$ a set of *roles*. A DM is a directed, labeled graph with nodes C. A concept $C \in \mathcal{C}$ can be understood as denoting a class of objects sharing a set of common properties. To understand how a concept C is defined relative to other concepts, one needs to inspect its outgoing edges. $c \in C$ denotes that c is an *instance* of concept C.[11] Edges are distinguished in DMs as follows:

1. $C \xrightarrow{isa} D$ (short: $C \rightarrow D$) defines that *every C isa D*, that is, $c \in C$ implies $c \in D$.

   The subconcept/subclass relation is very common in DMs, thus the *isa* label is usually omitted and the shorthand notation $C \rightarrow D$ is used instead.

2. $C \xrightarrow{ex:r} D$ defines that for every $c \in C$, there *exists some r*-related $d \in D$.

   Here, $r \in \mathcal{R}$ is a *role*, or, a *binary relation* $r(c, d)$ between instances of C and D.

3. $C \xrightarrow{all:r} D$ defines that for every $c \in C$ and *all* x that are r-related to c (i.e., for which $r(c, x)$ holds), $x \in D$ holds.

4. $C \xrightarrow{r} D$ defines that if $c \in C$ and $d \in D$, then they are r-related, that is, $r(c, d)$ holds.

5. $\text{AND} \rightarrow_i \{D_1, \ldots, D_n\}$ indicates that an AND-node with n outgoing edges to $D_1, \ldots, D_n$, respectively, defines an anonymous concept, the *intersection* of concepts $D_1, \ldots, D_n$.

6. $\text{OR} \rightarrow_i \{D_1, \ldots, D_n\}$, indicates that an OR-node with n outgoing edges to $D_1, \ldots, D_n$, respectively, defines an anonymous concept, the *union* of concepts $D_1, \ldots, D_n$.

---

11. Thus, C and D can be viewed as unary predicates.

7. $C \overset{=}{\to} D$ defines that C is *equivalent* to D, meaning every C *isa* D and vice versa. It could have been denoted also as $C \leftrightarrow D$. However, the directed edge keeps the distinction between C (the definiendum) and its definition D (definiens).

Note that D can be an atomic or a defined concept. When unique, AND nodes are omitted and outgoing arcs directly attached to the concept being defined. In Figure 12.5, unlabeled, grey edges and edges labeled *proj* (*projects-to*) correspond to *isa* edges and *ex*:proj edges, respectively.

Reified Roles as Concepts

In DMs, as in description logics, the concepts are being defined, whereas the roles are only a means to that end. To capture the semantics of roles, or define their properties in terms of each other, they need to be defined in terms of concepts themselves. In logic, this "quoting mechanism" is known as *reification*.

Example 12.4.1 (Roles as Concepts). Consider a DM involving the roles *regulates*, *activates*, and *inhibits*, and assume that in the given domain, *activates* (C, D) and *inhibits* (C, D) are special cases of *regulates* (C, D). Instead of introducing a special notation for *sub-roles*[12] and then defining the mechanics of how roles can be related to one another, roles are turned into first-class citizens by making them concepts using an operator, *make-concept* (mc). The modeling capabilities of DMs can be applied to roles and, for example, simply state that $mc(activates) \overset{isa}{\to} mc(regulates)$.

By modeling roles as concepts, more domain semantics can be formalized, leading to better *knowledge engineering*. In particular, during *query processing*, such formalized knowledge can be automatically employed by the system: Given a DM (formalized as logic rules), an MBM query or view definition involving *activates* and *regulates* knows that the former is a subconcept of the latter. If during query processing a goal `regulates('cAMP',Protein)` is evaluated, the logic rules corresponding to the DM knowledge allow the system to deduce that any result for `activates('cAMP',Protein)` is also an answer for `regulates('cAMP',Protein)`. This is correct because a *substitutability principle* holds, which allows the system to replace a concept, D, with any of its subconcepts, C, that is, for which $C \overset{isa}{\to} D$ holds.

---

12. RDF(S) has such a notion called *subproperty*; see *http://www.w3.org/TR/rdf-schema/*.

Generating the Role Hierarchy

When making a role into a concept, the *isa* hierarchy[13] on concepts induces an *isa* hierarchy on roles.

Domain Maps as Logic Rules

Domain maps borrow from description logics [19] the notions of *concept* and *roles*. Indeed, while some of the previously mentioned constructs of DMs have equivalent formalizations in description logic [20], the fact that additional mechanisms are needed such as *roles as concepts* and *recursive* and *parameterized* roles and concepts, and the fact that executable DMs are wanted during query processing, require a translation into a more general logic framework.

In the following, DMs are formalized in a minimal subset of F-logic [21]. The semantics of DMs could be formalized in other languages, in particular in other deductive database languages. The use of F-logic is convenient because a small subset of it already matches nicely the minimal requirements established for a MBM system [20]. Moreover, implementations of F-logic are readily available [22, 23] and have been used by the authors in different mediator prototypes before [24, 25, 26].

In F-logic, $c : C$ and $C :: D$ denote class membership ($c \in C$) and subclassing ($C \subseteq D$), respectively. Thus, there are logic rules of the form *head* if *body* that express the F-logic semantics of ":" and "::". Say that "::" is a reflexive, transitive, and antisymmetric[14] relation.

Definition 12.2 Compilation of Domain Maps

The mapping $\Psi : \mathrm{DM} \to \mathrm{FL}$ of domain maps to F-logic is defined as follows:

1. $\Psi(C) := \{C : concept\}$, for all atomic concepts $C \in \mathcal{C}$

2. $\Psi(r) := \{r : role\}$, for all roles $r \in \mathcal{R}$

3. $\Psi(C \overset{isa}{\to} D) := \{C :: \Phi D\} \cup \Psi(D)$

4. $\Psi(C \overset{ex:r}{\to} D) :=$

    (a) $\{r(c, \_d), \_d : \Phi D$ if $c : C, \_d = skol_D(c)\} \cup \Psi(D)$

    (b) $\{False$ if $c : C, \neg(r(c, \_d), \_d : \Phi D))\} \cup \Psi(D)$

5. $\Psi(C \overset{all:r}{\to} D) :=$

    (a) $\{d : \Phi D$ if $c : C, r(c, d)\} \cup \Psi(D)$

    (b) $\{False$ if $c : C, r(c, d), \neg d : \Phi(D)\} \cup \Psi(D)$

---

13. Strictly speaking, the *isa* does not have to be a hierarchy but can be any directed acyclic graph.

14. Because concepts are implemented as F-logic classes, this avoids terminological cycles.

6. $\Psi(C \xrightarrow{r} D) := \{r(c, d) \text{ if } c : C, d : \Phi(D)\} \cup \Psi(D)$

7. $\Psi(\text{AND}\rightarrow_i\{D_1, \ldots, D_n\}) :=$
   $\{d : skol_{AND} \text{ if } d : \Phi(D_1), \ldots, d : \Phi(D_n)\} \cup \Psi(D_1) \cup \cdots \cup \Psi(D_n)$

8. $\Psi(\text{OR}\rightarrow_i\{D_1, \ldots, D_n\}) :=$
   $\{d : skol_{OR} \text{ if } d : \Phi(D_1) \vee \ldots \vee d : \Phi(D_n)\} \cup \Psi(D_1) \cup \cdots \cup \Psi(D_n)$

9. $\Psi(C \xrightarrow{=} D) := \{C :: \Phi(D), \Phi(D) :: C \text{ if } \Phi(D)\} \cup \Psi(D)$

Remarks

Here, $\Phi(D)$ is defined similar to $\Psi(D)$, but it returns for a compound concept description D, a new auxiliary symbol $\Phi(D)$ representing the compound. For atomic D, $\Phi(D) = \Psi(D)$ holds simply. The symbols $skol_X$ produce new Skolem function symbols every time they are used in the translation Ψ: For example, in 4(a), a symbolic representation is invented for the existentially quantified variable _d. Note that c, d, _d are logic *variables*, while C, D, D_i, and False are *constants*.[15] The different variants (a) and (b) in the translations of DMs correspond to different intended uses: in 4(a), an anonymous object is *created* for the ∃-quantified variable, in 5(a), all $C.r$ objects are *type coerced* into instances of D. In contrast, the (b) translations only *check* whether the constraints induced by the DM edges are indeed satisfied and signal an inconsistency (False) if they are not.

Example 12.4.2 Roles as Concepts Continued. Consider a DM stating that NProt $\xrightarrow{isa}$ Protein, NProt regulates some Gene, and cfos $\xrightarrow{isa}$ Gene.[16] The role regulates is conceptualized by asserting *mc*(regulates). When making its hidden arguments visible, mc(regulates (C,D)) really denotes a *family* of regulates concepts. The *isa* hierarchy on regulates concepts is derived from the *isa* hierarchy of its arguments. For example:

$$mc(\text{regulates(NProt,cfos)}) \xrightarrow{isa} mc(\text{regulates(NProt,Gene)})$$
$$\xrightarrow{isa} mc(\text{regulates(Protein,Gene)})$$

Deriving the Role Hierarchy

Previously the unary operator, mc, which turns role literals into concepts was introduced. It is implemented in FL as a subclass of the (meta-)class concept by asserting mc::concept and adding further rules for deriving the role hierarchy

---

15. This is reversed from the usual convention used in the rest of the chapter to match this DM notation.

16. Here, NProt stands for *nuclear protein*.

from the concept hierarchy, which are given as set of *mc*-declarations such as
r(C, D) : mc by the user:

```
r(C,D):mc,  r(C',D'):mc,  r(C,D)::r(C',D')
  if (r(C,D):mc ∨ r(C',D'):mc),C::C',  D::D'  (up/down)
r(C,D):mc,  r(C',D'):mc,  r(C,D)::r(C',D')
  if (r(C,D'):mc ∨ r(C',D):mc),  C::C',  D::D'  (mixed)
```

Observe that with these rules, the desired result is obtained in Example 12.4.2.

Recursive Concepts

Consider the *part of* relationship has_a and its interaction with isa. For example,
MyNeuron isa Medium_Spiny_Neuron, which in turn has_a Neostriatum
therefore MyNeuron has_a Neostriatum holds (see Figure 12.5). In the general
case, this gives rise to a recursive rule *if* $C \overset{isa}{\to} D$ *and* $D \overset{has\_a}{\to} E$ *then* $C \overset{has\_a}{\to} E$.
Similarly, one can define that isa and has_a are independently transitive and that
isa is anti-symmetric. For such recursive definitions, an intuitive graph notation
can be devised (e.g., using a dashed edge for the concept being defined to its
recursive definition, see Ludäscher et al. [27]). In a declarative, rule-based query
language like F-logic, an executable specification is:

```
has_a(X,Z)  if  X::Y, has_a(Y,Z).
```

Note that X, Y, Z are concept variables. Such F-logic rules can also be used at the
mediator to handle inductive definitions, such as ONT4 in Figure 12.4, in particular,
when the source does not have the capability to evaluate recursive definitions.

Parameterized Roles and Concepts

Part of relationships such as has_a come in different flavors, F (e.g., $F \in$
{ *member/collection*, *portion/mass*, *phase/activity*, ...}) and transitivity does not
necessarily carry over across flavors [15].[17] This is most naturally modeled by a *parameterized role*, has_a(F), which is transitive *within* each flavor, F, but which
may interact in other ways *across* flavors. Definition 12.2 shows how domain maps
can be formalized as logic rules via a mapping Ψ. This mapping can be extended
for parameterized roles and concepts: For example, assume the parameterized
role has_a(F) should hold between concepts C and D only for some flavors, F,

---

17. For example, *orchestra has_a musician* and *musician has_a arm*, but not *orchestra has_a arm*.

satisfying a condition $\varphi(F)$. One can extend Ψ and compile such a parameterized DM edge into F-logic as follows:

$$\Psi(C \xrightarrow{\text{has\_a(F)}\|\varphi(F)} D) = \{\text{has\_a(F)} c, d \text{ if } c: C, d: \Phi(D), \varphi(F)\} \cup \Psi(D)$$

Note that a parameterized role such as $\text{has\_a}(F)$ has a first-order semantics in F-logic despite its higher-order syntax [28].

12.4.2   Process Maps

PMs provide abstractions of *process knowledge*, that is, temporal and/or causal relationships between events that can be used for situating and linking data across different sources. Like DMs, PMs are directed, labeled graphs, albeit with a very different semantics: Nodes are used to model *states* and edges correspond to *state transitions*, which are labeled with a process name describing the transition. In this way, data providers (e.g., bench scientists) can not only hook their raw data to the (given or refined) DMs but also to processes witnessed in their experimental studies databases (see Figure 12.2 and Figure 12.8).

Initial Process Semantics PM_0

Intuitively, an edge of the form $e_\pi = s \xrightarrow{\{\varphi\}\pi\{\psi\}} s'$ of a PM means that the process π leads from state s to s'; φ is a necessary *precondition* that must hold in s for π to happen, and ψ is a *postcondition*, which holds in s' as a result of π. PM_0 denotes the set of all initial process semantics.

The edge e_π of a PM is called a *process occurrence* of π in PM. Thus, a process occurrence specifies where in a PM a process occurs, and which pre- and postconditions, φ and ψ, this occurrence satisfies. In addition to the semantics implied by the occurrence of e_π in PM, a process π can have an *initial semantics* associated with the process name, π.

To allow for *parameterization* of processes, edge labels where process names are *first-order atoms* (of the form $\pi = \pi(T_1, \ldots, T_n)$ where each term T_i is a logic variable or constant) are considered. For example, consider $\pi = \text{opens(Channel)}$ as describing the opening process of an ion channel. Its initial semantics are defined by the expression:

$$\{\neg \text{open(Channel)}\} \text{ opens(Channel) } \{\text{open(Channel)}\}$$

meaning that *any* transition along a process occurrence of $\pi = \text{opens(Channel)}$ in a PM must be from a state where open(Channel) was *false*. In the successor state, however, (after π has happened), open(Channel) is *true*.

From Process Maps to Domain Maps

The first-order predicates occurring in φ and ψ are called *open(Channel)*, *fluents*, because their truth is state dependent. It is required that the set of *fluent predicate symbols*, $\mathcal{F}$, is disjointed from the set, $\mathcal{P}$, of *process names* and the sets of concept and role names $\mathcal{C}$ and $\mathcal{R}$, respectively. In contrast, the constant parameters used in process occurrences, such as Channel *are* allowed to be concepts from $\mathcal{C}$.

For example, a DM may have that NMDA_receptor $\overset{\text{isa}}{\to}$ Calcium_channel $\overset{\text{isa}}{\to}$ channel in which case the process knowledge about the opening of channels and the static knowledge from a DM are directly linked through the common concept Channel.

Similarly, just as roles are first-class citizens by reifying them into concepts, the same can be done for processes, by specifying additional semantics of processes using domain maps.

Example 12.4.3　(Processes as Concepts). Consider the binds_to(X,Y) process with the initial semantics.

$$\{\neg\texttt{bound(X,Y)}\}\ \texttt{binds\_to(X,Y)}\ \{\texttt{bound(X,Y)}\}$$

Now consider a DM in which processes were reified as concepts as follows:

$$\texttt{dimerizes(X)} \overset{\text{isa}\|\text{X=Y}}{\longrightarrow} \texttt{binds\_to(X,Y)}$$

It is easy to see that this (parameterized) DM edge, when translated into F-logic, allows the system to conclude in the combined knowledge base $(\text{DM} \cup \text{PM}_0)$ that

$$\{\neg\texttt{bound(X,X)}\}\texttt{dimerizes(X)}\ \{\texttt{bound(X,X)}\}.$$

Process Elaboration and Abstraction

The edge, e_π, of a process occurrence can be seen as an *abstraction* of a real process. In addition to its initial semantics, PM_0, and the semantics induced by its concrete occurrence in a specific PM, this abstraction can be *elaborated* by replacing the e_π with a (sub-)process map elab(e_π), whose initial and final states are s and s'. The newly created nodes and edges of the elaboration, elab(e_π), are annotated with the same unique *elaboration identifier eID*. The *eID* includes at least a reference to e_π, indicating the edge being elaborated, and the *author* (data provider) of the elaboration.

The converse of elaboration, *abstraction*, takes a connected subgraph, $\Pi(S, s_0, s_f, E)$, with nodes S, edges E, and distinguished nodes $s_0, s_f \in S$ (initial and final state), and abstracts Π into a single edge $e_\pi = \texttt{abstract}(\Pi(\texttt{S},\texttt{s}_0,\texttt{s}_\texttt{f},\texttt{E}))$. The abstracted edges E of Π are marked with a unique *abstraction identifier aID*, which includes a reference to the new abstraction edge, e_π, and the author of the abstraction.

Definition 12.3 Process Maps

A PM $\Pi(S, s_0, s_f, E)$ is a connected, directed graph with nodes, S, labeled edges, E, and initial and final states $s_0, s_f \in S$. The edges e_π of E are of the form

$$\texttt{s} \xrightarrow{\{\varphi\}\pi\{\psi\}} \texttt{s'} \ (\texttt{e}_\pi)$$

where the *process name* π is a first-order atom and φ and ψ are first-order formulas, called the *precondition* and *postcondition* of e_π, respectively.

Given an edge $e = s_a \dashrightarrow s_b$ of a process map $\Pi(S, s_0, s_f, E)$, the *elaboration*, elab(e), of e is a process map $\Pi'(S', s_a, s_b, E')$ such that (1) the initial and final states are s_a, s_b, (2) $S' \cap S = \{s_a, s_b\}$, and (3) all $e' \in E'$ are linked to e via a common, unique identifier $eid(e', e)$.

A connected subgraph of a PM with distinguished initial and final state is called a *subprocess map* (sub-PM). Given a PM $\Pi(S, s_0, s_f, E)$, the *abstraction* of a sub-PM $\Pi'(S', s_a, s_b, E')$ of Π, denoted abstract (Π'), is a new edge $e_{\pi'} = s_a \dashrightarrow s_b$, where (i) $e_{\pi'} \notin E$, and (ii) all $e' \in E'$ are linked to $e_{\pi'}$ via a common, unique identifier $aid(e', e_{\pi'})$.

Marking edges with elaboration and abstraction identifiers guarantees one-to-one mappings between an edge and its elaboration and similarly, between a sub-PM and its abstraction. In this way, data providers can "double-click" on an edge, e_π, and elaborate the processes into a PM, Π, to provide more precise links to their data. Conversely, they may *collapse* a sub-PM, Π, into a single edge, e_π, if the data does not provide information at the detailed level of Π and hence is more adequately hooked to the overall process, e_π.

Process Maps as Logic Rules

Similarly to DMs, one can translate PMs into a logic representation $\Psi(\text{PM})$. The difference is that for DMs, the formalization in description logic or F-logic yields a first-order logic semantics, whose unique minimal model, $\mathcal{M}(\text{DM})$, interprets concepts and roles as unary and binary predicates over a set of individuals. The model, $\mathcal{M}$, implies that data objects, which are linked as concept instances to a DM, have the properties defined by the domain map (e.g., the neurons in the images linked to MyNeuron in Example 12.3.2 project to Globus_Pallidus_External). In contrast, the logic representation of a PM specifies only some process properties via pre- and postconditions in the PM and the PM's graph structure. The details of the semantics are omitted due to lack of space. The basic idea is that the graph structure of PMs (with its embedded hierarchy of elaborations and abstractions) is formalized via a nested Kripke structure in which the nodes of PM (states) have associated first-order models and in which labeled process edges specify a *temporal accessibility relation* between states.[18] In particular, a process elaboration of an edge, e_π, adds to the initial semantics, PM_0, and the semantics of the pre- and postconditions of the concrete occurrence of e_π in PM, an *elaboration semantics* (i.e., a sequence of intermediate states with first-order constraints along the paths of the elaboration).

12.5   MODEL-BASED MEDIATOR SYSTEM AND TOOLS

At the core of the MBM framework is the KIND mediator system. Other important components are the Spatial Markup and Rendering Tool (SMART) Atlas for annotating, displaying, and relating data with brain atlases, the CCDB, defined in Example 12.3.1 as the primary source of experimental data, and the Knowledge Map Explorer (Know-ME) tool for concept-based navigation of source and mediated views. For a description of Know-ME, see Qian et al. [29]; the other components are described in the following text.

12.5.1   The KIND Mediator Prototype

The architecture of the KIND mediator system is depicted on top in Figure 12.6. At the bottom, a snapshot of the prototype execution is shown: After the user issues a query against the integrated view, the system situates the results on a domain map, in this case ANATOM (simple ontology of brain anatomy). By clicking on the orange diamonds, the user can retrieve the actual result objects, grouped by concept (foreground).

In the first prototype [9, 30] the F-logic implementation FLORA [31] was used as the only query processing and deduction engine. As part of a large, collaborative project [4] the prototype is being re-implemented as a modular, distributed mediator system that includes several additional components, including the following:

✦ *Logic plan generator*: Given a user query, Q, and an integrated view definition IVD, $Q \circ$ IVD is translated into a plan generator program $PG(Q \circ$ IVD$)$ that, when executed, produces an initial logic query plan for $Q \circ$ IVD. Here "$\circ$" denotes query composition.

---

18. See Lausen et al., Section 6 [27] for a formalization of hierarchical processes using nested Kripke structures.

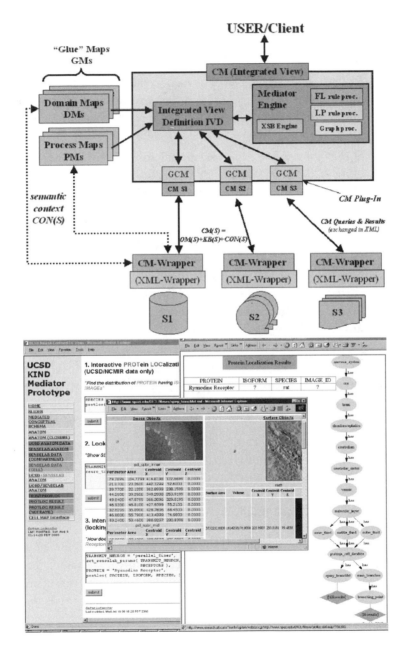

12.6

FIGURE

Top: Architecture of the KIND model-based mediator. **Bottom:** Snapshot of the prototype. Background left shows a mediator shell for issuing *ad hoc* queries against CM(*M*); background right shows a generated subgraph having the requested result data shown in their anatomical context. Clicking on (diamond) result node retrieves the actual result data (see *foreground center*).

✦ *Query rewriter*: This module takes a logic query plan and rewrites it into an executable, distributed plan based on the capabilities of a source (e.g., conjunctive queries with binding patterns or complete SQL).

✦ *Execution plan compiler*: For final execution, the rewritten plan is compiled into a logic program whose run-time execution sends the corresponding subqueries to wrapped sources, retrieves results, and post-processes them (e.g., joins, group-bys, and unions across sources) before sending them to the user.

✦ *SQL plan generator*: For relational sources (those having SQL query capabilities), this wrapper module translates a logic query plan into an equivalent SQL statement, similar to Draxler's tool [32].

A preliminary version of this new system has been recently demonstrated [13] and includes all of the modules previously listed. Plan generation and rewriting is implemented using logic programming technology [33]. The SQL plan generator has been implemented in Java. It is planned that the final system will include specialized inference engines such as FLORA and XSB [34] for handling deductive and object-oriented database capabilities, and FaCT [17] for reasoning tasks over domain maps that are formalized in description logics.

12.5.2   The Cell-Centered Database and SMART Atlas: Retrieval and Navigation Through Multi-Scale Data

The CCDB mentioned earlier, in Example 12.3.1, houses different types of high-resolution, 3D light and electron microscopic reconstructions of cells and subcellular structures produced at the National Center for Microscopy and Imaging Research[19] [14]. It contains structural and protein distribution information derived from confocal, multiphoton, and electron microscopy, including correlated microscopy. Many of the data sets are derived from electron tomography, a powerful technique for deriving 3D information from electron microscopic specimens. Electron tomography is similar in concept to medical imaging techniques like computerized axial tomography (CAT) scans and magnetic resonance imaging (MRI) in that it derives a 3D volume from a series of 2D projections through a structure. In this case, the structures are contained in sections prepared for electron microscopy, which are tilted through a limited angular range. Examples of datasets in the CCDB are shown on the left of Figure 12.7.

---

19. The National Center for Microscopy and Imaging is a research facility specialized in the development of technologies for improving the understanding of biological structure and function relationships spanning the dimensional range from $5nm^3$ to $50\mu m^3$ *(http://www.ncmir.ucsd.edu)*.

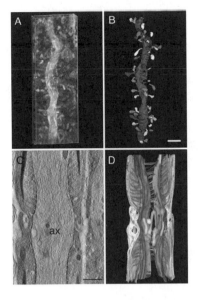

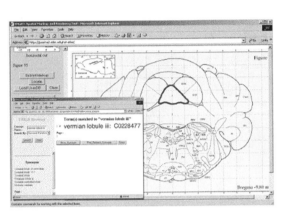

12.7

FIGURE

Left: Examples of tomographic data sets in the CCDB. A and B show a selectively stained spiny dendrite from a Purkinje cell. A is a projection of the volume reconstruction (dendrite appears as white against dark background). B is the segmented dendrite. C and D show a tomographic reconstruction of the node of Ranvier. C is a single computed slice through the volume. D is a surface reconstruction of the various components comprising the node. Scale bar in B = $1\mu m$; in C = $0.5\mu m$. **Right:** Registration of a data set with the Smart Atlas. The user draws a polygon representing the location of a data set, in this case a filled Purkinje neuron. The user specifies the database containing this data, then enters an annotation and selects a concept from the UMLS or some other ontology. The concept ID is stored in the database.

A screenshot of the Smart Atlas tool is shown on the right of Figure 12.7. It is based on a geographic mapping tool [35] and allows users to define polygons on a series of 2D vector images and annotate them with names, relationships, and concept IDs from an ontology such as UMLS. This tool provides another kind of *glue map* (in addition to domain and process maps). First, a brain atlas such as that by Paxinos and Watson [36] is translated into a spatial format, such as Scalable Vector Graphics (SVG). The user then marks up the atlas using the Smart Atlas tool (e.g., with concept names from UMLS). Once the atlas has been (partially) marked up, it can be queried from the same browser: Clicking on any point in the atlas will return the stereotaxic coordinates; clicking on a brain region will return the name of that region, along with any synonyms, and highlight all planes containing that structure. The Smart Atlas can now be used to register a researcher's data to a specific spatial location. This also links the registered data

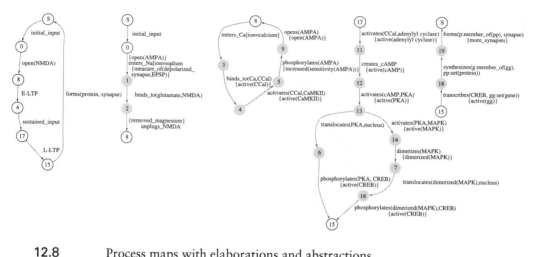

Process maps with elaborations and abstractions.

automatically to the UMLS ontology by virtue of the earlier semantic markup of spatial objects. To register source data, the user draws an arbitrary polygon representing the approximate data location on one of the atlas planes (Figure 12.7, right). The user is then presented with a form that can be used to add annotations or provide additional links to concepts of an ontology. Although the UMLS is used in the examples shown here, the user will eventually be able to use multiple ontologies, including those of their own creation, for semantically indexing data. Tools are also being developed to define new terms and relationships in existing ontologies. Another component of the system has been demonstrated and shows how spatial and conceptual information can be used together in a mediator system [37]; see also Martone et al.'s chapter in *Neuroscience Databases* [38] for further details on the use of the Smart Atlas.

12.6   RELATED WORK AND CONCLUSION

12.6.1   Related Work

Significant progress has been made in the general area of data mediation in recent years, and several prototype mediator architectures have been designed by projects like TSIMMIS [39], SIMS [40], Information Manifold [41], Garlic [42], and MIX [43]. While these approaches focus mostly on structural and schema aspects, the problem of *semantic mediation* has also been addressed: In the DIKE system [44], the focus is on automatic extraction of mappings between semantically analogous elements from different schemas. A global schema is defined in terms of a conceptual model (SDR network), in which the nodes represent concepts and the (directed) edge labels represent their semantic distances, and a score called *semantic relevance* measures the number of instances of the target node that are also instances of the source node. The correspondence between objects is defined in terms of *synonymies*, *homonymies*, and *sub-source similarities*, defined by finding maximal matching between the two graphs.

ODB-Tools [45] is a system developed on top of the MOMIS [46] system for modeling and reasoning about the common knowledge between two to-be-integrated schemas. They present the object-oriented language, ODL_{I^3}, derived from a description logic (OCDL). The language allows a user to create complex objects with finite nesting of values, union and intersection types, integrity constraints, and quantified paths. These constructs are used to define a class in one schema as a *generalization*, *aggregation*, or *equivalent* with respect to another; *subsumption* of a class by another can be inferred. An integrated schema is obtained by clustering schema elements that are close to one another in terms of an affinity metric.

Calvanese et al. [47] perform semantic information integration using a local-as-view approach by expressing the conceptual schema by a description logic language called $\mathcal{DLR}$ and subsequently defining non-recursive Datalog views to express source data elements in terms of the conceptual model. The language $\mathcal{DLR}$ represents concepts, C, relations, R, and a set of assertions of the form $C_1 \subset C_2$ or $R_1 \subset R_2$, where R_1, R_2 are $\mathcal{DLR}$ relations with the same arity. Mediation is accomplished by defining *reconciliation correspondences*, or specifications that a query rewriter uses to match a conceptual-level term to data in different sources.

Recently Peim et al. [48] have proposed an extension to the well-known TAMBIS system [49]. Their approach is similar to ours [18, 50] in that a logic-based ontology (in their case the $\mathcal{ALCQI}$ description logic) interfaces with an *object-wrapped* source. While F-logic [28] is used here as the internal knowledge representation and query language, their work focuses on how a query on the ontology is transformed to monoid comprehensions for semantic query optimization.

12.6.2   Summary: Model-Based Mediation and Reason-Able Meta-Data

MBM was presented as a methodology that supports information integration of scientific data across complex, multiple-world scenarios as found in the neuroscience domain. In this framework, object-oriented models and conceptual models (CM), domain maps (DM), and process maps (PM) all provide means to capture more domain semantics and thus can act as *glue knowledge sources* to link hard-to-correlate sources. Mechanisms to contextualize source data formally were presented. The graph structures thus constructed have been shown to be useful for navigating across related concepts and querying local data during navigation [29].

Logic formalizations of DMs and PMs can be seen as "reason-able" or "executable" "meta-data" (see a paper by Horrocks [51]): Unlike conventional, descriptive meta-data, which are primarily used for data discovery, formal ontologies, such as DMs and PMs, can support much more versatile computational tasks in a mediator system, as illustrated in this chapter. For example, different and apparently unrelated data objects can be associated and retrieved together or even fused by the mediator's integrated view definition (IVD), because IVDs can be defined as deductive rules over DMs and PMs (Figure 12.3). In this way, in model-based mediation (MBM), logic rules play the role of *executable* or *computational* meta-data for scientific data integration. The latter is a challenging application and benchmark for combined database and knowledge representation techniques.

ACKNOWLEDGMENTS

This work has been supported by NIH/NCRR 3 P41 RR08605-08S1 (Biomedical Informatics Research Network [BIRN]) and NSF-NPACI Neurosciences Thrust ACI 9619020. The authors thank their colleagues and students involved in the BIRN project for their contributions, in particular, Xufei Qian, Edward Ross, Joshua Tran, and Ilya Zaslavsky.

REFERENCES

[1]   Y. Papakonstantinou, A. Gupta, and L. M. Haas. "Capabilities-Based Query Rewriting in Mediator Systems." *Distributed and Parallel Databases* 6, no. 1 (1998): 73–110.

[2]   C. Li, R. Yerneni, V. Vassalos, et al. "Capability Based Mediation in TSIMMIS." In *Proceedings of the ACM International Conference on Management of Data (SIGMOD)*, 564–566. 1998.

[3]   National Partnership for Computational Infrastructure (NPACI): Neuroscience Thrust Area, 2001. *http://www.npaci.edu/Thrusts/Neuro/*.

[4]   Biomedical Informatics Research Network Coordinating Center (BIRN-CC). University of California, San Diego. *http://nbirn.net/*, 2001.

[5]   V. Kashyap and A. Sheth. "Semantic and Schematic Similarities Between Database Objects: A Context-Based Approach." *VLDB Journal* 5, no. 4 (1996): 276–304.

[6] D. Calvanese, G. D. Giacomo, M. Lenzerini, et al. "Description Logic Framework for Information Integration." In *Proceedings of the Sixth International Conference on Principles of Knowledge Representation and Reasoning (KR'98)*, 2–13. Morgan Kaufmann, 1998.

[7] O. Bozdagi, W. Shan, H. Tanaka, et al. "Increasing Numbers of Synaptic Puncta During Late-Phase LTP: N-Cadherin is Synthesized, Recruited to Synaptic Sites, and Required for Potentiation." *Neuron* 28, no. 1 (2000): 245–259.

[8] J. Kasahara, K. Fukunaga, and E. Miyamoto. "Activation of Calcium/ Calmodulin-Dependent Protein Kinase IV in Long Term Potentiation in the Rat Hippocampal CA1 Region." *Journal of Biological Chemistry* 276, no. 26 (2001): 24044–24050.

[9] A. Gupta, B. Ludäscher, and M. E. Martone. "An Extensible Model-Based Mediator System with Domain Maps." In Demonstration Session of the 21st International Conference on Data Engineering (ICDE), Heidelberg, Germany, 2001.

[10] S. Chakravarthy, J. Grant, and J. Minker. "Logic-Based Approach to Semantic Query Optimization." *ACM Transactions on Database Systems (TODS)* 15, no. 2 (1990): 162–207.

[11] B. Ludäscher, Y. Papakonstantinou, and P. Velikhov. "Navigation-Driven Evaluation of Virtual Mediated Views." In *Proceedings of the International Conference on Extending Database Technology (EDBT)*, Lecture Notes in Computer Science 1777, 150–165. Springer, 2000.

[12] Y. Papakonstantinou and V. Vassalos. "The Enosys Markets Data Integration Platform: Lessons from the Trenches." In International Conference on Information and Knowledge Management (CIKM), 2001.

[13] A. Gupta, B. Ludäscher, and M. E. Martone. "Registering Scientific Information Sources for Semantic Mediation." In *21st International Conference on Conceptual Modeling (ER)*. Lecture Notes in Computer Science 2503. Springer, 2002.

[14] M. E. Martone, A. Gupta, M. Wong, et al. "A Cell-Centered Database for Electron Tomographic Data." *Journal of Structural Biology* 138 (2002): 145–155. http://ncmir.ucsd.edu/CCDB/.

[15] A. Artale, E. Franconi, N. Guarino, et al. "Part-Whole Relations in Object-Centered Systems: An Overview." *Data and Knowledge Engineering* 20 (1996): 347–383.

[16] P. Mitra, G. Wiederhold, and M. L. Kersten. "A Graph-Oriented Model for Articulation of Ontology Interdependencies." In *Proceedings of the International Conference on Extending Database Technology (EDBT)*, 86–100. 2000.

[17] I. R. Horrocks. "Using an Expressive Description Logic: FaCT or Fiction?" In *KR'98: Principles of Knowledge Representation and Reasoning*, edited by A. G. Cohn, L. Schubert, and S. C. Shapiro, 636–645. San Francisco: Morgan Kaufmann, 1998.

[18] B. Ludäscher, A. Gupta, and M. E. Martone. "Model-Based Mediation with Domain Maps." In *Proceedings of the 17th International Conference on Data Engineering (ICDE)*. New York: IEEE Computer Society, 2001.

[19] D. Calvanese, G. D. Giacomo, M. Lenzerini, et al. "Description Logic Framework for Information Integration." In *Principles of Knowledge Representation and Reasoning*, 2–13. 1998.

[20] B. Ludäscher, A. Gupta, and M. E. Martone. "Model-Based Mediation with Domain Maps." In *Proceedings of the 17th International Conference on Data Engineering (ICDE)*, 81–90. New York: IEEE Computer Society, 2001.

[21] M. Kifer, G. Lausen, and J. Wu. "Logical Foundations of Object-Oriented and Frame-Based Languages." *Journal of the ACM* 42, no. 4 (July 1995): 741–843.

[22] FLORA homepage. http://www.cs.sunysb.edu/~sbprolog/flora/.

[23] FLORID homepage. http://www.informatik.uni-freiberg.de/~dbis/florid/.

[24] B. Ludäscher, A. Gupta, and M. E. Martone. "Model-Based Information Integration in a Neuroscience Mediator System." In *Proceedings of the 26th International Conference on Very Large Data Bases (VLDB)*, 639–642. San Francisco: Morgan Kaufmann, 2000.

[25] R. Himmeröder, G. Lausen, B. Ludäscher, et al. "FLORID: A DOOD-System for Querying the Web." In Demonstration Session at EDBT. Valencia, Spain, 1998.

[26] B. Ludäscher, R. Himmeröder, G. Lausen, et al. "Managing Semistructured Data with FLORID: A Deductive Object-Oriented Perspective." *Information Systems* 23, no. 8 (1998): 589–613.

[27] G. Lausen, B. Ludäscher, and W. May. "On Active Deductive Databases: The Statelog Approach." In *Transactions and Change in Logic Databases*, Lecture Notes in Computer Science 1472, 69–106, edited by B. Freitag, H. Decker, M, Kifer, et al. Springer, 1998.

[28] M. Kifer, G. Lausen, and J. Wu. "Logical Foundations of Object-Oriented and Frame-Based Languages." *Journal of the ACM* 42, no. 4 (July 1995): 741–843.

[29] X. Qian, B. Ludäscher, M. E. Martone, et al. "Navigating Virtual Information Sources with Know-ME." In *EDBT*, Lecture Notes in Computer Science 2287, 739–741, 2002.

[30] B. Ludäscher, A. Gupta, and M. E. Martone. "Model-Based Information Integration in a Neuroscience Mediator System." In *Proceedings of the 26th International Conference on Very Large Data Bases (VLDB)*, 639–642. San Francisco: Morgan Kaufmann, 2000.

[31] G. Yang and M. Kifer. "FLORA: Implementing an Efficient DOOD System Using a Tabling Logic Engine." In Sixth International Conference on Rules and Objects in Databases (DOOD), 2002.

[32] C. Draxler. *A Powerful Prolog to SQL Compiler*, technical report. München, Germany: Centre for Information and Language Processing, Ludwigs-Maximillians-Universität München, 1992.

[33] B. Ludäscher. "Mediator Query Processing with Prolog Technology," technical note, BIRN-DI-TN-2002-01. Biomedical Informatics Research Network, 2002.

[34] K. F. Sagonas, T. Swift, and D. S. Warren. "XSB as an Efficient Deductive Database Engine." In *Proceedings of the ACM International Conference on Management of Data (SIGMOD)*, 442–453. 1994.

[35] I. Zaslavsky. "A New Technology for Interactive Online Mapping." *Cartographic Perspectives* 37 (2000): 65–77.

[36] G. Paxinos and C. Watson. *The Rat Brain in Stereotaxic Coordinates*. San Diego: Academic Press, 1998.

[37] A. Gupta, B. Ludäscher, M. E. Martone, et al. "A System for Managing Alternate Models in Model-Based Mediation." In *Advances in Databases, 19th British National Conference on Databases (BNCOD 19)*, Lecture notes in Computer Science 2405. Springer, 2002.

[38] M. E. Martone, A. Gupta, B. Ludäscher, et al. "Federation of Brain Data through Knowledge-Guided Mediation." In *Neuroscience Databases: A Practical Guide*, edited by R. Kötter, 275–292. Boston: Kluwer Academic Publishers, 2002.

[39] H. Garcia-Molina, Y. Papakonstantinou, D. Quass, et al. "The TSIMMIS Approach to Mediation: Data Models and Languages" (Extended Abstract). In *Next Generation Information Technologies and Systems*. 1995.

[40] C. A. Knoblock, S. Minton, J. L. Ambite, et al. "Modeling Web Sources for Information Integration." In *Proceedings of the Fifteenth National Conference on Artificial Intelligence*. 1998.

[41] A. Y. Levy, A. Rajaraman, and J. J. Ordille. "Querying Heterogeneous Information Sources Using Source Descriptions." In *Proceedings of the International Conference on Very Large Data Bases (VLDB)*, 251–262. 1996.

[42] L. M. Haas, D. Kossmann, E. L. Wimmers, et al. "Optimizing Queries Across Diverse Data Sources." In *Proceedings of the International Conference on Very Large Data Bases (VLDB)*, 276–285. 1997.

[43] C. Baru, A. Gupta, B. Ludäscher, et al. "XML-Based Information Mediation with MIX." In *Proceedings of the ACM SIGMOD International Conference on Management of Data (SIGMOD 1999)*, 597–599. Philadelphia: ACM Press, 1999.

[44] L. Palopoli, G. Terracina, and D. Ursino. "The System DIKE: Towards the Semi-Automatic Synthesis of Cooperative Information Systems and Data Warehouses." In *Proceedings of the ADBIS-DASFAA Symposium*, 108–117. 2000.

[45] D. Beneventano and S. Bergamaschi. "Extensional Knowledge for Semantic Query Optimization in a Mediator-Based System." In *International Workshop on Foundations of Models for Information Integration (FMII-2001)*, 2001.

[46] S. Bergamaschi, S. Castano, and M. Vincini. "Semantic Integration of Semistructured and Structured Data Sources." *SIGMOD Record* 28, no. 1 (1999): 54–59.

[47] D. Calvanese, S. Castano, F. Guerra, et al. "Towards a Comprehensive Methodological Framework for Semantic Integration of Heterogeneous Data Sources." In *Eighth International Workshop on Knowledge Representation Meets Databases (KRDB)*, 2001.

[48] M. Peim, E. Franconi, N. Paton, et al. "Query Processing with Description Logic Ontologies Over Object-Wrapped Databases." In *Proceedings of the International Conference on Scientific and Statistical Database Management (SSDBM)*. 2002.

[49] C. Goble, R. Stevens, G. Ng, et al. "Transparent Access to Multiple Bioinformatics Information Sources." *IBM Systems Journal* 40, no. 2 (2001): 534–551.

[50] A. Gupta, B. Ludäscher, and M. E. Martone. "Knowledge-Based Integration of Neuroscience Data Sources." In *Proceedings of the Twelfth International Conference on Scientific and Statistical Database Management (SSDBM)*, 39–52. IEEE Computer Society, 2000.

[51] I. Horrocks. "DAML+OIL: A Reason-Able Web Ontology Language." In *Proceedings of the International Conference on Extending Database Technology (EDBT)*, 2–13. 2002.

Compared Evaluation of Scientific Data Management Systems

Zoé Lacroix and Terence Critchlow

The variety of biological information systems currently available raises the inevitable question: Which system best meets my needs? To decide which system to choose, or if a custom system is required, a detailed analysis of user needs should be performed. Carefully performing this analysis will identify the best options and clarify the buy-or-build decision.

This chapter outlines several techniques and metrics that can be considered when performing an evaluation. Section 13.1 begins this discussion by defining evaluation techniques. Section 13.2 presents a set of evaluation criteria in detail and describes how they may be applied. Finally, Section 13.3 discusses some of the explicit tradeoffs that can be made and the effects they have on the overall systems. Unfortunately, neither evaluation of the systems described in previous chapters nor comparisons between them are provided because these activities can only be performed within the context of specific user requirements.

13.1 PERFORMANCE MODEL

Whether users are selecting a new system, evaluating an existing system, or determining what requirements a system must meet, they need a performance model. The performance model is used to evaluate the system's ability to meet user requirements and to provide the basis for comparing systems beyond this starting point. The model is composed of a set of specifications and associated metrics that can be used to evaluate a system. Ideally, it identifies the target environment in which the system will actually be deployed and reflects the relative importance of all the system features within that environment. Because of this tight coupling between a model and its environment, the model cannot be directly applied to other environments.

The first and most important step in defining a model is to establish the minimal set of specifications required for a system to be considered of interest. This could be as simple as using previously defined use cases or system requirements, discussed in Section 1.4.1, as the system specification, or it could be as complex as performing a new, detailed evaluation that augments these initial requirements with priorities and a ranking of all desired functionality. This is the most important step because it immediately removes from consideration those systems that do not meet all of the requirements and provides the basis on which all of the systems will be compared.

The list of specifications should be as complete and detailed as possible as two significantly different systems may agree on a small set of specifications while differing on other characteristics. The more complete the specifications, the fewer the number of systems that can meet them and the more likely the solution will meet the users' expectations. For example, in the context of the design of a vehicle if the specification is to transport a person within a town, two possible designs are a car and a bicycle. However, should the specification also include that the vehicle be able to carry heavy objects, only the car satisfies the requirements.

Once the specifications have been identified, they can be translated into a collection of characteristics or metrics that define the areas where the system is going to be evaluated. These criteria can be represented as an evaluation matrix, as outlined in Sections 13.1.1 and 13.2.3. While explicit metrics are useful, evaluating a system must include defining the relative value of appropriate resources or capabilities. Section 13.1.2 outlines several cost models that can be used to compare systems. Finally, Sections 13.1.3 and 13.1.4 present techniques to collect the measurements.

13.1.1   Evaluation Matrix

To measure the performance of a system, two feature sets need to be determined: (1) the perspectives from which the evaluation will occur and (2) the parameters to be measured. Typically, the parameters are derived from the system specifications previously collected. Two obvious perspectives are the developers' perspective and the users' perspective. Each perspective encompasses most, if not all, the evaluation parameters. A matrix representation formed with the perspectives along the x axis and the performance parameters forming the y axis can be used to summarize the evaluation, as illustrated in Section 13.2.3.

13.1.2   Cost Model

There are two common cost measurements used to evaluate systems—time and space. Time can be considered in various ways and at various granularities.

The user response time is the total time needed to answer a query. This time can be split between the actual transaction time (time needed to process the query) and the transmission time (time needed to display the result to the user). For integrated systems, transaction time may be split further: Additional transmissions to several systems may be included, as well as the processing time of each of these systems. While usually less important, pre-processing time should also be considered to ensure the viability of the system. For example, it is important that pre-processing delays do not prevent data from being integrated into the system in a timely manner. The cost function considered for DiscoveryLink is based on time as presented in Section 11.2.3.

Unlike standard models of computation in which uniform access to all data is assumed,[1] models for database management systems must take into consideration the notion of *space*, or the actual location of the data. In particular, as most of the data do not fit into main memory, they are typically kept in secondary storage (e.g., on a magnetic disk). For applications that require the storage of very large data sets, as is often the case for scientific applications, tertiary storage devices such as tape drives are also used. The cost of moving between storage levels is high because data access and writing times take significantly longer. Ultimately, however, the appropriate data need to be in main memory to perform any manipulation, such as returning the results of a query. In database management systems, a *buffer manager* partitions available main memory into buffers, which are regions into which disk blocks can be transferred. All system components that need access to the data interact with the buffer manager.

Space management is tightly coupled to time management. Measuring the space required for the system tasks is a way to evaluate the performance of the system. Efficient space management may significantly decrease the amount of time required to perform required tasks. Space management can be improved by algorithms that minimize data exchanges between tertiary and secondary storage, limit the number of disk accesses, and better exploit data cached in the buffer. In distributed systems, increasing the local storage available for use may decrease the response time because it provides improved caching or eliminates the need for network communication or complex calculations that consume more time than accessing local storage. This complex relationship between time and space highlights the importance of understanding the value of different resources and the tradeoffs made by each system being evaluated.

Of course, there are other ways to measure the cost of a system. One obvious example is the monetary cost to purchase, license, or use it. A less obvious example is the correctness of results provided by the system, as compared to some

---

1. These models are often called random-access models (RAM).

ground truth. These costs can be compared against time, space, and each other to define other tradeoffs. For example, correctness and time can be traded off against each other in a system in which a quick response returns a subset (or superset) of the actual results, and a complete answer is only returned if the query is allowed to run without restriction. Ultimately, the best cost metrics for a given situation depend on which resources are most valuable in that environment.

13.1.3   Benchmarks

A *software benchmark* is a collection of programs used to generate measurements for evaluation of some capability, usually efficiency. Benchmarks guide users in the selection of a system with respect to specific performance considerations. They also offer a good quality assurance test for software developers.

The development of benchmarks is driven either by an application domain (*domain-specific benchmark*) or by the objective to evaluate a general type of system (*generic benchmark*). For example, the SEQUOIA 2000 storage benchmark [1] is a domain-specific benchmark designed to evaluate earth scientific applications, whereas OO7 [2] and XOO7 [3] are generic benchmarks developed to evaluate object-oriented database systems and extensible markup language (XML)-management systems, respectively. While generic benchmarks provide a gauge of performance across a wide range of application domains, the domain-specific benchmarks are able to capture peculiarities of the domain that may not be common in more general applications. Thus, if an appropriate domain-specific benchmark exists, it usually will be preferred.

The selection of a benchmark first requires the analysis of the characteristics the benchmark is expected to evaluate. For example, OO7 aims to capture the ability of an object-oriented database system to perform pointer traversals, updates, and query processing. Once these characteristics are specified, a data set (schema and instance) and queries are designed. The design of the data set needs to be complex enough to reflect the characteristics of the real data. The structure (schema), including data types, and the size of the dataset are the two features that play a role. The data itself is often randomly generated. Typically, queries are grouped in collections that capture the different characteristics. For example, the queries of XOO7 are grouped in three collections: data-driven queries, document-driven queries, and navigational queries.

Benchmarking data management systems involves the evaluation of various tasks such as query execution and transactions (updates). In such contexts, a benchmark consists of: a schema, a set of data corresponding to that schema, and a set of queries. The evaluation of a data management system thus follows these steps: (1) define the data structure (schema) in the system, (2) load the data, and (3) run each of the queries, recording measurements corresponding to the cost model as defined in Section 13.1.2 (typically the time and the space needed to execute the query). The comparison of two systems, therefore, consists in comparing the measurements collected when running the benchmark on each system. The analysis of the collected measurements often requires knowledge of internal implementation decisions, which are typically not available. For more information on benchmarking, refer to Gray's *The Benchmark Handbook* [4].

The design of a benchmark for biological information system is a difficult task. For a user to be satisfied, the benchmark must capture the tasks expected to be performed by the system. Because mediation systems integrate a variety of resources, including Web resources, it is challenging to define a single benchmark that can accurately evaluate an entire system. An alternative is to use a collection of generic benchmarks and combine the results. This has the advantage of being able to leverage existing work, such as relational database and XML benchmarks, where appropriate.

13.1.4   User Survey

Human computer interaction (HCI) is the discipline concerned with the design, evaluation, and implementation of interactive computing systems for human use and the major phenomena surrounding them [5]. Thus, most HCI research focuses on the design and development of user interfaces. A survey is a common approach used by the HCI community for evaluating interfaces. These surveys obtain feedback from a large number of users on a variety of characteristics and allow researchers to identify the strengths and weaknesses of an interface. Their use of surveys to collect requirements and evaluate satisfaction can be exploited to design and evaluate an entire system. Unfortunately, despite previous efforts, there is no commonly accepted characterization of the various tasks that life scientists perform with data management systems. As there is no standard functionality common to bioinformatics applications, each system must be evaluated by a survey designed around the needs of a specific set of users. The evaluation of such a system is further complicated because the expectations of life scientists for computer support evolve significantly over time.

In addition to evaluating individual systems, user surveys can be used to identify and influence trends across an entire scientific domain. For example, consider a survey performed among 31 biologists at Arizona State University, in which 26 agreed that luck was involved in biological discovery, and all agreed on the importance of creativity [6]. Despite being a fairly small survey, it was able to highlight the need for systems that encourage, not discourage, creative exploration of the data. Another classic survey is the set of *unanswerable* queries published in the Department of Energy (DOE) report on Genome Informatics [7] and listed in Figure 8.1. These queries, the result of input from a large number of geneticists, have driven significant improvement in biological data management by motivating the shift from human processing of data to machine processing and manipulation of data. A more recent survey to identify and classify tasks in bioinformatics was conducted interviewing 30 active biologists from academia and industry [8]. Out of the 315 identified tasks, 54% were similarity search, multiple pattern and functional motif search, and sequence retrieval. This emphasizes the need for biological interfaces to combine traditional query languages, such as structured query language (SQL), with searching and analysis tools. Finally, surveys can provide insight into user satisfaction with a system. An example of this type of survey is the Baylor College of Medicine (BCM) Search Launcher User Survey sponsored by the DOE.[2]

13.2   EVALUATION CRITERIA

A consistent set of metrics is needed to compare systems effectively. These metrics should distinguish between the approaches and identify their relative strengths and weaknesses. There are a wide variety of metrics that can be used, each with their advantages and disadvantages. The following six characteristics can be applied using both the computer science (implementation) and the life science (user) perspectives to evaluate genomics data management systems. The metrics are efficiency, extensibility, functionality, scalability, understandability, and usability. These metrics cover a wide range of issues of practical concern. Unfortunately their definitions are vague, and they can be applied with various degrees of vigor. While this makes them difficult to apply consistently, it also provides the flexibility required to differentiate between alternative approaches and a variety of environments. Each user team may refine and specify these criteria with appropriate benchmarks to measure the characteristics of interest with respect to a customized cost model or user survey, as introduced in Section 13.1. Typically, cost models will be used to evaluate implementation performance and user surveys will be used to measure users' satisfaction.

The *implementation perspective* captures the characteristics of the system from the technical point of view. Much of this perspective is driven ultimately by the user requirements. However, it reflects only one of many possible implementations satisfying these requirements. While both views are helpful in understanding a system, and there is significant overlap between them, the true success of a system is determined by whether or not its users are satisfied. Thus, we believe the user perspective is ultimately the more important.

---

2. The BCM Search Launcher User Survey is available at *http://searchlauncher.bcm.tmc.edu/user_survey/user_survey.html*.

13.2.1   The Implementation Perspective

The implementation perspective aims to fulfill the user's expectations, subject to organizational constraints, by optimizing the six metrics subject to the goals and constraints they define.

Efficiency

Implementation efficiency is a combination of query efficiency, data storage size, communication overhead, and data integration overhead. Query efficiency reflects the ability of the system to respond to user queries and reflects factors such as the correlation between the data format and the expected queries. Data storage size can be affected by choices such as using flat files or a relational database and replicating data locally or accessing them remotely. Communication overhead is characterized by data transfer requirements as well as the complexity and frequency of commands executed remotely. The data integration overhead is defined by the complexity of the transformations being performed between the data sources and the user interface.

Each of these four characteristics can be divided between the efficiency during a pre-processing step and the efficiency in response to a query. Often there is a tradeoff between characteristics when deciding at which time to perform a task. For example, the transformation of a data element may be performed as a pre-processing step in a data warehouse, where it is converted prior to being loaded into the warehouse, or at run time in a federated database, where it is converted on the fly in response to a query. This decision may have a dramatic impact on the system's required storage. Although the overall efficiency of the system is important, many systems will seek to reduce the query response time because pre-processing time can be amortized over all of the queries.

Efficiency is a criterion that clearly distinguishes the systems presented in this book. Not surprisingly, mediation systems developed in industry such as discoveryHub, the commercial version of Kleisli (see Chapter 6), or DiscoveryLink (see Chapter 11) usually perform better than the academic ones. This can be explained by the industry's need to provide robust and efficient systems. Simply stated, optimizations are typically designed and developed for industrial systems, and they appear to be less of a priority for academic systems. In general, efficient query processing is not a requirement for the presented academic systems. Some architectural choices lead to systems that are more efficient in certain ways. Simpler data integration platforms, such as link-driven federations, offer efficient query processing because they do not perform data conversion or complex data manipulation. In addition, partially materialized approaches with indices, or completely materialized approaches in data warehouses, are often more efficient than non-materialized ones—as explained in Section 13.3.1. Thus, link-driven federations such as Entrez or SRS (see Chapter 5) often rely on large, pre-computed indices to optimize query processing: SRS queries usually take less than one minute to complete.

Kleisli automatically optimizes queries, and its space management is generally more economical than flat files. However, it has translation time overheads when mapping data between integrated sources and the system. While such translation times may be reduced, they are inevitable in a mediation architecture. DiscoveryLink performs two optimization steps: query rewriting followed by cost-based optimization. Query rewriting transforms the user's query into a semantically equivalent query (i.e., a query that will return the same output) for which more efficient execution plans are possible. Such techniques were described in Chapter 4. Cost-based optimization exploits a broad range of alternative execution strategies, taking input from the wrappers and assessing the cost of functions, scans, and general sub-queries performed at the integrated source [9, 10]. Such methods were introduced in Section 4.4.

Extensibility

Extensibility refers to the ease with which the functionality of the system can be increased. New requirements can result in the need for a new query (in the simplest view) or new tools and types of data. Examples of these types of extensions include the addition of a new similarity search query, the inclusion of expression array data in a system that previously contained only sequence data, and the integration of a new clustering tool. The evaluation of extensibility is complex because most systems are extensible only in certain ways. For example, a given system may simplify adding new tools that use the data already in the system, but integrating new types of data to that same system may be very challenging. While all of these extensions can, in theory, be made to any system, the actual implementation effort varies greatly depending on the system design. This characteristic attempts to quantify the effort required for the system being evaluated.

Many systems presented in this book rely on a mediation architecture that provides a virtual view to users while keeping the data in each integrated source. These approaches typically use wrappers to access and retrieve data from the integrated sources. Extending the system to new applications usually means developing new wrappers and registering them with the system. Systems that exploit meta-information about integrated sources require additional information. Thus, typically these systems are a little less extensible but also more efficient. A system with a semantically consistent view, such as TAMBIS, appears to be less extensible than others because its semantic model must be extended to include all new concepts and relationships in the appropriate way so that queries may be asked against the new data. Within TAMBIS, this implies annotating the sources and services model (see Section 7.4.1). Extensibility may also be more affected by materialized approaches. Partial materialization through indices may be costly to extend, and totally materialized approaches may rely on a management system that is difficult to extend.

Functionality

Functionality reflects the system's ability to perform a wide variety of analysis over the data contained within it. This includes both the types of queries that the system supports and their complexity. For example, a simple keyword search reflects a fairly low level of functionality, a system that supports keyword searches using wildcards across a subset of the attributes would rate better, and a system that augments keyword searches with sequence homology comparisons and data clustering would have much greater functionality.

Functionality of presented systems often depends on their usage. Industrial systems are widely used; therefore, more functionality is provided or made available. Alternatively, academic systems are designed for specific usage in a limited context and thus often provide fewer capabilities.

Scalability

Three basic components comprise scalability: the amount of data the system can handle, the number of users (queries) the system can simultaneously support, and the number of data sources that can be integrated. The amount of data a system can handle is not only limited by available disk space (a limit that is becoming less significant as disks become both larger and cheaper) but also by its ability to effectively manipulate that data. For example, a federated system that dynamically retrieves data from external sources may be designed to respond to queries without needing a disk-based representation. However, if a query returned a large amount of data, it may overload the infrastructure, either because of network bandwidth problems or an inability to hold the results in memory. Similarly, systems need to be designed for multiple users if they are to function beyond a single scientist's desktop. A system may be designed to work for a single user, making the assumption that there would only be one query executing at a time to simplify resource allocation. While this assumption holds, there would not be a problem, but if the system is deployed in a multi-user environment, this will quickly become a limiting factor. Even if a system is designed for concurrent use, there are often real limits on the number of users it can support. These restrictions become important as the system extends its user base from an individual to a lab and eventually to the larger community. Finally, there are several factors affecting the number of data sources that can be integrated realistically into a system, including scarcity of local resources, communication bandwidth, query response times, and the incremental effort required to integrate and maintain each additional data source. These factors are conflicting, and finding an acceptable balance requires understanding your scientists' needs.

The relative importance of each of these three components varies with the underlying approach the system uses and the environment in which the system is expected to be used. For example, values appropriate for a single-user, desktop system in a small company would be very different from those for a large company supporting a community resource. Serious consideration to the number of sources that can be integrated must be given when evaluating systems because scientists constantly want to access more data.

Again, not surprisingly, systems designed and developed in industrial context appear to scale better than academic ones. The main reason is that scalability is often not part of academic systems requirements, whereas it is typically mandatory in industrial context. Kleisli regularly handles hundreds of megabytes and more than 60 types of integrated data sources. However, each invocation of Kleisli only runs one top-level query at a time through the system, generating multiple concurrent sub-queries. In contrast, TAMBIS only provides access to five data sources and offers little explicit support for simultaneous queries and users.

Understandability

Understandability expresses the clarity of the system design. If the system is well designed (e.g., using strong software engineering practices and object-oriented or component-based techniques), as new developers join the project they are able to understand easily the implementation details and quickly begin contributing to the project. If the system is poorly designed and overly complex, people not involved in the original design of the system require additional time before they are able to make significant contributions to the project. In addition, their contributions are likely to further complicate the architecture because identifying the most effective way to implement a specific feature is difficult without a solid understanding of the overall architecture. Unfortunately, without intimate access to implementation details, this characteristic is very hard to determine.

All presented systems offer a clear and understandable design: link-based federation architecture (see Chapter 5), mediation architecture combined with wrappers (see Chapters 6, 7, 8, 11, and 12), or warehousing (see Chapter 10). These designs are described in detail in their respective sections and enable new developers to understand their overall architecture quickly.

Usability

Implementation usability reflects the ability of the user to modify the system behavior and the exposure of underlying system capabilities through some type of application or user interface. Of particular interest is the ability to adapt system capabilities based on specific user needs. For example, while a system may be built on a relational database, it may not provide users full SQL query access, choosing instead to limit the allowable queries by publishing only a simple keyword search interface. There is more to usability, however, than simply expanded capability. While providing a wide variety of complex queries is generally valuable, as we discuss in the next section, it is important to ensure that they are presented in a useful way. It is easy to overwhelm users by providing too many options—forcing people to become experts in your tool before they are able to accomplish anything.

Usability is often not a specification for systems developed in academic contexts. TAMBIS (see Chapter 7) does not provide an API, and its limited extensibility may further restrict this ability. KIND provides an API only for certain features. In contrast, Kleisli provides access to its high-level query languages (CPL and sSQL), an API to SMLNJ function calls, and the Pizzkell suite of JDBC-like interfaces (CPL2Perl and CPL2Java) [11]. SRS requires programming skills in the Icarus language to modify the system. It is likely that the next generation of systems, including future versions of existing ones, will focus on including new interfaces to facilitate their maintenance. For example, a graphical administration tool for SRS that would obviate the use of Icarus is currently under development.

Summary

The implementation perspective can be expressed faithfully with the six metrics. However, it is difficult to optimize all of them concurrently. Indeed, they are far from being independent, with some metrics being positively correlated and others being negatively correlated. For example, the use of a query language such as SQL may improve both extensibility and efficiency. Extensibility could be improved because the system would be able to express a wider variety of queries, and the efficiency could be improved by its ability to perform query optimization. Similarly, a poorly designed system with minimal understandability will be less extensible. On the other hand, a system that is highly functional, with many analysis tools available, is likely to be inefficient by combining these various functionalities. Every system must find the appropriate set of tradeoffs to balance the needs and expectations of its target users. One of the goals of this book is to aid potential developers and users in identifying their needs and choosing the best approach and system for them.

13.2.2   The User Perspective

The user perspective defines the requirements of the system to be developed or chosen. The metrics presented here can be used to identify the users' expectations and needs and evaluate systems accordingly.

Efficiency

From the user's point of view, efficiency is evaluated as the ability of the system to perform a single task in a satisfactory timeframe and its overall ability to support its user base in the succession of their complex tasks. This metric has two components. The first is similar to the implementation perspective but at a slightly higher level: How quickly does the system respond to queries? In effect, this view summarizes implementation efficiency from a purely practical perspective. If the system provides a reasonable response time, the efficiency is deemed acceptable. The faster the response time, the better the ranking. The second is a consolidation of the remaining categories: How effectively can users ask the system the questions they need answers to, get the answers, and continue using those answers in their analysis? Given that the second component is reflected in the remaining characteristics, only the first definition will be considered.

Extensibility

Extensibility expresses the users' ability to ask new questions and customize the system to meet their specific needs. While plug-and-play search and analysis tools are a long way away, a well-designed system often allows the user to extend it in limited ways without becoming intimately familiar with the underlying system. These extensions may be simple variations on previous queries, such as changing the attributes being searched during a keyword search, they may be slightly more complicated variations on a query, such as changing the sequence homology algorithm used to perform a search, or they may be a completely new type of query. Extensibility is not limited to queries; it also reflects the ability of the user to introduce new data, and new data types, into the system. For example, it reflects the user's ability to include data from a new data source. As with implementation extensibility, many systems theoretically can be extended by their users. This characteristic reflects how much effort a user must expend to add new types of queries or data and how much programming skill is required to perform this extension. In a truly extensible system, only minimal programming expertise would be required.

The presented approaches assume that users are not actively involved in system extensions. However, they typically provide users as much flexibility as possible in customizing their queries. For example, when performing a similarity search, the various program parameters and the choice of application version are made available. User extensibility typically is linked to user functionality, and all presented systems perform well for both criteria.

Functionality

User functionality reflects not only the different types of queries that can be asked, but also how those queries can be changed and combined to form new queries. For example, a BLAST search that accepts a protein sequence is useful, but being able to BLAST both protein and nucleotide sequences is better. And the ability to take the results of a BLAST search and use them as input into other BLAST searches, or send them to an analysis program, is even better still. When querying a system, scientists have in mind a specific question they are trying to answer, which may or may not be part of a much larger question. The more of the question that can be answered within a single environment, the more useful that environment is. To completely answer a question, a system must have both the correct types of data and the right capabilities. The more questions a system can answer, the greater its functionality.

All presented systems are generic in that they are not designed to answer a single query or even a small set of queries in a particular context. Instead, they provide query languages that allow the formulation of a variety of queries. The advantages of generic systems are presented in Chapter 4. In addition, some systems such as DiscoveryLink and Kleisli provide access to multiple sources and analytical applications. This variety of integrated resources significantly increases users' functionality.

Scalability

User scalability has some components shared with system scalability: Can the system handle the number of users that it has, and does it provide access to enough data and tools to make it worth using? However, it also has a unique component: the ability of the system to handle large numbers of input objects. This is becoming an increasingly important issue as genomics moves from small-scale science, in which a researcher may focus on a single gene, to large-scale analysis of entire genomes. If a system is only capable of processing a single input at a time, it will not be useful to users who need to analyze hundreds or thousands of these objects. For example, Web-based homology searches often take only a single sequence at a time as input. As a result, these interfaces quickly become of limited use if a scientist has hundreds of clones to analyze. Unless scalability is part of the system requirements, academic systems typically do not scale well.

Understandability

Although a system cannot be considered usable without also being understandable, we separate the concepts for evaluation purposes. For this discussion, a system is understandable if its users not only understand the queries being asked, but also the results being returned. This means that the semantics of both the interface and the data should be clear and well documented so that when a question is asked, what that question was is known exactly. This sounds obvious, but it is often overlooked. For even simple queries such as keyword searches, the semantics may not be clear. Questions such as "Is the text of the entire system searched for that keyword or are only certain objects or attributes searched?" must be answered. Failure to understand the semantics of a query may lead to asking the wrong question or misinterpreting the results. Furthermore, it may appear that the semantics of a system is always well defined. Unfortunately, this is not always the case. Many systems do not precisely define the data they contain and instead rely on the user's domain expertise to guide them. This situation is even worse in integrated systems in which returned data have been obtained from a variety of data sources. In these systems, there may be many subtle semantic inconsistencies. Even in systems that claim to provide consistent views of the data, the precise semantics associated with some aspects of the data may be elusive. These types of implicit semantics dramatically reduce the understandability of the results.

Most presented systems were designed to be understandable to their scientific users. Some systems put a creative emphasis on this criteria and provide original solutions. TAMBIS focused on a transparent access to data through an ontology reflecting the scientists' view of the data. KIND returns outputs in the context of a *domain map*—a graphical representation of ontological knowledge of the scientific domain. In contrast, other systems, such as Kleisli and DiscoveryLink, rely on the data organization provided by the integrated sources, assuming that this organization is appropriately known and understood by users.

Usability

Usability is probably the most important feature of a system and one of the most difficult to obtain. While focusing on facets of the other characteristics, it is remarkably easy to develop a system that, despite providing all of the required efficiency, extensibility, functionality, and scalability, is unusable by its target audience. This occurs because programmers often design a system for themselves, forgetting that their users are not programmers and have no desire to become programmers. An ideal system provides an intuitive query interface, directly supporting only the queries that need to be executed and returning the results in the most useful format.

In reality, the number of scientists and types of queries a system needs to support makes this impossible. One approach to increasing the usability of a system is to provide multiple interfaces targeting different user groups. For example, a graphical interface can help novice users comfortably interact with a system, but it may be too slow for experts. To increase its usability, a system could either add shortcut keys or have a separate command line interface for experts. While requiring additional development effort, this would allow users familiar with the system to perform their queries quickly, while not forcing novices to learn the more advanced interface immediately.

13.3 TRADEOFFS

This section explores some of the tradeoffs to be considered when evaluating systems and some of the unique characteristics of biological data management systems that complicate their design and evaluation. As with the evaluation metrics, there is no clearly best approach, but rather the user requirements and system constraints need to be included in the evaluation. The purpose of the following sections is simply to call out certain characteristics and encourage the readers to consider them. This is meant to be an illustrative, not exhaustive, list. There are many tradeoffs and considerations that are not discussed but that may be important for evaluating a system within a specific environment.

13.3.1 Materialized vs. Non-Materialized

Materialized approaches usually are faster than non-materialized ones for query execution. This makes intuitive sense because the data is stored in a single location and in a format supportive of the queries. To confirm intuition, tests were run in 1995 with several implementations of the query: "*Retrieve the HUGO names, accession numbers, and amino acid sequences of all known human genes mapped to chromosome c*" [12]. These tests were performed using the Genomic Unified Schema (GUS) warehouse as the materialized source and the K2/Kleisli[3] system as the non-materialized source. The query requires integrating data from the Genome DataBase (GDB), the Genome Sequence DataBase (GSDB), and GenBank. Measures showed that for all implementations, the warehouse is significantly faster. In certain cases, queries executed by K2 as part of this evaluation failed to complete due to network timeouts. The expression of the query (using semi-joins

---

3. The version of K2 used for this comparison is much earlier than the version presented in Chapter 8.

rather than nested loop iterations) also affected the performance of the execution of the query.

In addition to the communication overhead, the middleware between the user interface and the remote data may introduce computational overhead. Recently, tests have been performed at IBM to determine whether or not a middleware approach such as DiscoveryLink (presented in Chapter 11) affects the access costs when interacting with a single database. They conducted two series of tests in which DiscoveryLink was compared to a production database at Aventis [13]. The results show that, in the tested context, for a single user, the middleware did not affect the performance. None of the tested queries involved the manipulation of large amounts of data; however, they presented many sub-queries and unions. In some cases, accessing the database through a middleware and a wrapper was even faster than the direct access to the database system. The load test shows that both configurations scale well, and the response times for both approaches are comparable to the single-user case.

There are a variety of factors to be considered beyond the execution cost. Materialized databases are generally more secure because queries can be performed entirely behind a firewall. Non-materialized approaches have the advantage that they always return the most up-to-date information available from the sources, which can be important in a highly dynamic environment. They also require significantly less disk space and can be easier to maintain (particularly if the system does not resolve semantics conflicts).

13.3.2   Data Distribution and Heterogeneity

Many systems presented in the previous chapters are mediation systems. Mediation systems integrate fully autonomous, distributed, heterogeneous data sources such as various database systems (relation, object-relational, object, XML, etc.) and flat files. In general, the performance characteristics of distributed database systems are not well understood [14]. There are not enough distributed database applications to provide a framework for evaluation and comparison. In addition, the performance models of distributed database systems are not sufficiently developed, and it is not clear that the existing benchmarks to test the performance of transaction processing applications in pure database contexts can be used to measure the performance of distributed transaction management. Furthermore, because the resources are not always databases, the mediation approach is more complex than the multi-database and other distributed database architectures typically studied in computer science.

For many bioinformatics systems, issues related to data distribution and heterogeneity are considerable and significantly affect the performance. As a result, they typically integrate only the minimal number of sources required to perform a given task, even when additional information could be useful. The complexity of this domain and the lack of objective information favor domain-specific evaluation approaches over generic ones for this characteristic.

13.3.3 Semi-Structured Data vs. Fully Structured Data

Previous chapters have pointed out that scientific data are usually complex, and their structures can be fluid. For these reasons, a system relying on a semi-structured framework rather than a fully structured approach, such as a relational database, seems more adequate. Although there are systems that utilize meta-level capabilities within relational databases to develop and maintain flexibility, they are usually not scalable enough to meet the demands of modern genomics. The success of XML as a self-describing data representation language for electronic information interchange makes it a good candidate for biological data representation. The design of a generic benchmark for evaluating XML management systems is a non-trivial task in general, and it becomes much more challenging when combined with data management and performance issues inherent to genomics.

Some attempts have been made to design an XML generic benchmark. Three XML generic benchmarks limited to locally stored data and in a single machine or single user environment have been designed: XOO7 [3], XMach-1 [15], and XMark [16]. XOO7 attempts to harness the similarities in data models of XML and object-oriented approaches. The XMach-1 benchmark [15] is a multi-user benchmark designed for business-to-business applications, which assumes the data size is rather small (1 to 14 KB). XMark [16] is a newer benchmark for XML data stores. It consists of an application scenario that models an Internet auction site and 20 XQuery queries designed to cover the essentials of XML query processing.

XOO7 appears to be the most comprehensive benchmark. Both XMark and XMach-1 focus on a data-centric usage of XML. All three benchmarks provide queries to test relational model characteristics such as selection, projection, and reduction. Properties such as transaction processing, view manipulation, aggregation, and update, are not yet tested by any of the benchmarks. XMach-1 covers delete and insert operations, although the semantics of such operations are not yet clearly defined for the XML query model. Additional information about XML benchmarks can be found in Bressan et al.'s *XML Management System Benchmarks* [17].

Native XML systems have been compared to XML-enabled systems (relational systems that provide an XML interface that allows users to view and query their data in XML) with three collections of queries: data-driven, document-driven, and navigational queries [18]. Tests confirm that XML-enabled management systems perform better than XML native systems for data-driven queries. However, XML native systems outperform XML-enabled ones on document-driven and navigational queries. This is not unexpected because enabled systems are tuned to optimize the execution of relational queries. However, they do not efficiently represent nested or linked data. Thus, navigational queries within enabled systems are rather slow; whereas native systems are able to exploit the concise representation of data in XML. Finally, document queries may use the implicit order of elements within the XML file. This ordering is not typically represented in relational databases, therefore defining an appropriate representation is a tedious task and negatively affects performance.

The type of system that is most appropriate depends heavily on the types of queries expected, the data being integrated, and the tools with which the system must interact. Scientific queries exploit all characteristics of XML queries: data, navigation, and document. An XML biological information system will need to perform well in all these contexts. An XML biological benchmark will be needed to evaluate XML biological information systems.

13.3.4   Text Retrieval

For many tasks, scientists access their data through a document-based interface. Indeed, a large amount of the data consists of textual annotations. Life scientists extensively use search engines to access data and navigation to explore the data. Unlike database approaches, structured models cannot be used to represent a document or many queries over document sets (e.g., given a document, find other documents that are similar to it). The evaluation of a textual retrieval engine typically relies on the notion of *relevance* of a document. A document is relevant if it satisfies the query. The notion of relevance is subjective because retrieval engines typically provide users with a limited query language consisting of Boolean expression of keywords or phrases (strings of characters). In such context, the query often does not express the user's intent, and thus, the notion of relevance is used to capture the level of satisfaction of the user rather than the validation of the query. Relevance is considered to have two components: *recall* and *precision*. *Recall* is the ratio of the number of relevant documents retrieved by the engine to the total number of relevant documents in the entire data set. A recall equal to one means all relevant documents were retrieved, whereas a recall of zero means no relevant document was retrieved. A recall of one does not guarantee the satisfaction of the user; indeed, the engine may have retrieved numerous non-relevant documents (noise).

Precision is the ratio of the number of relevant documents retrieved by the engine to the total number of retrieved documents and thus reflects the noise in the response. A precision equal to one means all retrieved documents are relevant, whereas a precision of zero means no retrieved document is relevant. Ideally, a document would have both a precision and a recall of one, returning exactly the set of documents desired. Unfortunately, state-of-the-art text query engines are far from that ideal. Currently, recall and precision are inversely related in most systems, and a balance is sought to obtain the best overall performance while not being overly restrictive.

13.3.5 Integrating Applications

System requirements usually include the ability to use sophisticated applications to access and analyze scientific data. The more applications that are available, the better functionality the system has. However, integrating applications such as BLAST may significantly affect the system performance in unanticipated and unpredictable ways. For example, a call to blastp against a moderate size data set will return a result within seconds, whereas a call to tblastn against a large data set may require hours. The evaluation of the performance of the overall integration approach must include information about the stand-alone performance of the integrated resources. This information, including the context in which optimal performance can be obtained, is often poorly documented. This is partially because many of the useful analysis tools are developed in academic contexts where little effort is made to characterize and advertise their performance. Readers who are involved in tool development are invited to better characterize the performance of these tools for systems to better integrate them.

13.4 SUMMARY

Each of the systems described in this book was designed to address specific user needs, and these requirements led to vastly different approaches. These systems represent the wide spectrum of tradeoffs that may be made. Ideally, a table or other mechanism would summarize their characteristics with respect to the variety of parameters presented in Section 13.2 and would allow readers to identify the system or approach that best meets their needs quickly. Unfortunately, such a comparison is not possible without significantly more insight into, on one hand, the users' requirements and, on the other hand, the systems implementation and feedback on user satisfaction. In particular, it would require familiarity with the environment in which the system was to be used, the users who would be working with it, the value of various resources in the environment, and how the system would be expected to interact with other tools. Although it would be possible to invent example users and evaluate some of the systems with respect to them, this would involve vast simplifications and would be a disservice to systems targeting different users. It is safe to say, however, that given the tradeoffs that must be made when developing a system, there is no approach that is obviously better than all the others. Instead, each user group could analyze carefully the specific requirements corresponding to their needs and use the approach presented in this chapter to select the approach and system that best meets them. When the requirements are identified, contacting the systems' designers and asking them how their approach performs in such context will allow each user team to compile their own comparison matrix and select an appropriate approach and system. When performing this evaluation, it is important to consider all of the users' requirements because focusing on only a few could lead to the selection of a less desirable approach. For that purpose, the contact information for each of the presented systems is provided in the System Information section.

While evaluating systems using the metrics proposed in Section 13.2 is somewhat subjective, when applied consistently, they form a reasonable basis for identifying the strengths and weaknesses of disparate systems. In addition, their flexible definitions allow them to be refined as needed to obtain the proper level of detail with respect to a particular evaluation's requirements. For example, if efficiency is an important consideration, a more detailed evaluation could be performed, resulting in specific information about query efficiency, data storage size, communication overhead, and data integration overhead. Similarly, metrics can be combined if only a high-level overview of a system is desired. Finally, readers should feel free to introduce new metrics to capture other properties of systems if you determine them to be important to an evaluation. There is nothing sacred about this evaluation matrix that can be refined and extended to meet one's needs.

REFERENCES

[1]   M. Stonebraker, J. Frew, K. Gardels, et al. "The Sequoia 2000 Benchmark." In Proceedings of the *ACM SIGMOD International Conference on Management Data*, 2–11. ACM Press, 1993.

[2]   M. Carey, D. DeWitt, and J. Naughton. "The OO7 Benchmark." In Proceedings of the *ACM SIGMOD International Conference on Management Data*, 12–21. ACM Press, 1993.

[3]   S. Bressan, G. Dobbie, Z. Lacroix, et al. "XOO7: Applying OO7 Benchmark to XML Query Processing Tool." In Proceedings of the *ACM International Conference on Information and Knowledge Management (CIKM)*, 167–174. 2001.

[4]   J. Gray, ed. *The Benchmark Handbook: For Database and Transaction Processing Systems*, second ed. San Francisco: Morgan Kaufmann.

[5]   T. T. Hewett, R. Baecker, S. Card, et al. "Human-Computer Interaction." In *Curricula For Human-Computer Interaction*. New York: ACM Press, *http://sigchi.org/cdg/cdg2.html*.

[6]   A. E. Lawson: "What Do Biologists Think about the Nature of Biology?" *Science and Education* (2003), not yet published.

[7]   R. J. Robbins, ed: *Report of the Invitational DOE Workshop on Genome Informatics*, April 1993, 26–27. Baltimore, MD: 1993.

[8]   R. Stevens, C. Goble, P. Baker, et al. "A Classification of Tasks in Bioinformatics." *Bioinformatics* 17, no. 2 (2001): 180–188.

[9]   L. M. Haas, D. Kossmann, E. L. Wimmers, et al. "Optimizing Queries Across Diverse Data Sources." In Proceedings of the *23rd International Conference on Very Large Databases (VLDB)*, 276–328. Morgan Kaufmann, 1997.

[10] M. T. Roth, F. Ozcan, and L. M. Haas. "Cost Models DO Matter: Providing Cost Information for Diverse Data Sources in a Federated System." In Proceedings of the *23rd International Conference on Very Large Databases (VLDB)*, 599–610. Morgan Kaufmann, 1999.

[11] L. Wong. "Kleisli: Its Exchange Format, Supporting Tools, and an Application in Protein Interaction Extraction." In *IEEE International Symposium on Bioinformatics and Biomedical Engineering (BIBE)*, 21–28. IEEE Computer Society, 2000.

[12] S. B. Davidson, V. Tannen, J. Crabtree, et al. "K2/Kleisli and GUS: Experiments in Integrated Access to Genomic Data Sources." IBM Systems Journal 40, no. 2 (2001): 512–531.

[13] L. Haas, P. Scharz, P. Kodali, et al. "DiscoveryLink: A System for Integrated Access to Life Sciences Data Sources." *IBM Systems Journal* 40, no. 2 (2001): 489–511.

[14] T. Oszu and P. Valduriez. *Principles of Distributed Database Systems*, 2nd ed. Upper Saddle River, NJ: Prentice Hall, 1999.

[15] T. Böhme and E. Rahm. "Multi-user Evaluation of XML Data Management systems with Xmach-1. In *Efficiency and Effectiveness of XML Tools and Techniques and Data Integration over the Web, VLDB 2002 Workshop EEXTT and CAiSE 2002 Workshop DIWeb*, edited by S. Bressan, A. B. Chaudhri, M. L. Lee, et al., 148–158. Springer, 2003.

[16] A. R. Schmidt, F. Waas, M. L. Kerste, et al. "Assessing XML Data Management with Xmark." In *Efficiency and Effectiveness of XML Tools and Techniques and Data Integration over the Web, VLDB 2002 Workshop EEXTT and CAiSE 2002 Workshop DIWeb*, edited by S. Bressan, A. B. Chaudhri, M. L. Lee, et al., 144–145. Springer, 2003.

[17] S. Bressan, M. L. Lee, Y. G. Li, et al. "XML Management System Benchmarks." In *XML Data Management: Native XML and XML-Enabled Database Systems*, edited by A. Chaudri, A. Rashid, and R. Zicari. Addison Wesley, 2003.

[18] U. Nambiar, Z. Lacroix, S. Bressan, et al. "Current Approaches to XML Management." *IEEE Internet Computing Journal* 6, no. 4 (July–August 2002): 43–51.

Concluding Remarks

As the first book focusing on management systems for biological data, this material is a detailed introduction to the variety of problems and issues facing data integration and the presentation of numerous systems. The major issue these systems are trying to address is the large number of distributed, semantically disparate data sources that need to be combined into a useful and usable system for geneticists and biologists to perform their research. This issue is complicated by the variety of data formats, inconsistent semantics, and custom interfaces supported by these sources—as well as the highly dynamic nature of these characteristics and the data themselves. Ideally, a data integration system would provide consistent access to all of the data and tools needed by scientists. However, no single system meets this ideal for all users. This final section provides a brief summary and a peek into the future of bioinformatics.

SUMMARY

The introductory chapters establish a terminology shared by computer scientists and life scientists. They focus on the different steps in the design of a system and highlight the differences between the problems faced by those in bioinformatics and other facets of these respective disciplines. Upon first glance, these differences may seem insignificant, but understanding them is the first step in understanding the realities of the environment in which bioinformatics solutions must work. The desire to simplify this environment is common in people starting out in this domain, but overcoming it is critical to successfully addressing the problems being faced. Many of the challenges in bioinformatics are derived from the inherent complexity of the domain, and failure to embrace this results in approaches that, while acceptable in theory, are not workable in the real, complex world in which bioinformatics solutions must be applied.

Once a common background has been established, the following chapters present several bioinformatics systems that are currently in use. The wide variety of systems described in this book provides significant insight into the complexity of performing data integration in the rapidly changing domain of genomics. The fact that these systems are still evolving indicates that none of these approaches has yet led to an ideal solution for all applications. This is a testament to the difficulty of creating a bioinformatics solution that addresses the needs of all users. Most of these systems evolved independently, and many began as attempts at addressing specific challenges facing scientists in a particular organization. The challenges focused on by a given solution are generally the most important problems facing the associated organization or its customers. While each system presented here has met its original goals, as the scope of its usage evolved, it has encountered new challenges.

As discussed in Chapter 13, evaluating a system requires detailed knowledge of the environment in which the system will be deployed. Part of the reason no single approach is clearly better than another is that the bioinformatics community places conflicting goals on systems. As a simple example, notice that although providing a semantically consistent view of the data greatly improves the usability of the system, it also places practical limits on the number of data sources to which the system can provide access. This is because each data source provides its own unique semantics for the data it contains, and an expert is required to perform the mapping from these semantics to the global ones. However, the more sources to which a system provides access, the more valuable it is in general. As scalability and semantic consistency are mutually exclusive goals, a system can excel in only one of them, providing at best marginal performance in the other. Whether such a system is better than another depends on the users' values. This example illustrates only one of the tradeoffs bioinformatics systems strive to meet. Because of these conflicting constraints, it is currently impossible for a single system to provide *the* bioinformatics solution that meets every scientist's needs. Although this is a discouraging realization, it is not a situation unique to bioinformatics. Indeed, it appears to be a characteristic of any rapidly evolving scientific domain, and as such, the techniques used by bioinformaticians are more generally applicable than typically thought.

LOOKING TOWARD THE FUTURE

As one becomes familiar with the problems facing bioinformatics and the approaches being pursued to address them, it is easy to become disenchanted. The problems are daunting, and there is no clear path that will lead to a unifying solution. Some issues, such as query optimization and data caching, are just now being investigated seriously in this context. Other issues appear as the result of applying existing technology in new ways and the development of new technology. Indeed, sometimes it feels as if we are moving in the wrong direction: As it becomes increasingly easy to distribute data via the Web, the number and heterogeneity of data sources containing information relevant to scientists keeps increasing. Unfortunately, a lack of community standards results in each source publishing its own distinct semantics and interfaces. The number of tools available to researchers, and their complexity, continues to increase without significant progress at making them interoperable. Multimedia data is becoming more common as genomics research continues to move onto computers and out of the wet-lab, which causes problems for data integration systems that are expecting textual data. Large-scale data are also becoming more common as access to powerful computers and related infrastructures increases. This changes the value of bandwidth and requires rethinking many assumptions about the underlying data. Grid technology is emerging and will likely soon allow data and computation to be spread transparently among a large number of machines. How this technology will be used is not entirely clear, but it will likely have a significant impact on computational biology.

While each of these issues raises significant data integration and access challenges, they also provide new opportunities to solve existing bioinformatics problems and, in turn, to advance the state of genomics research. For example, grid technology may be able to minimize the impact of large data sets by moving the computation to the place where the data resides. Thus, there is still hope that we will achieve the goal of providing scientists with intuitive access to all the relevant data they need.

One of the more promising emerging trends is an effort to define data semantics precisely through ontologies. A possible, although not necessarily probable, result of this effort is a single unifying ontology that is able to identify accurately the information contained in all data sources. Having this global ontology would allow mappings between related concepts to be easily identified, and thus would greatly reduce the burden placed on integration systems. Unfortunately, this vision may take decades to be realized, if it happens at all. The major reason for this is that life science is an inherently complex domain, and there is a lot of information that is not yet understood. Thus, the ability to correctly define the semantics between these complex concepts is severely limited by this lack of comprehension. Because of this difficulty, the ontologies currently being developed are generally small and define semantic concepts only for a specific sub-community of life science. The creation and adoption of these smaller ontologies are likely to occur over the next few years. Although a less than ideal solution, these ontologies could be extremely useful to bioinformatics by reducing the number of semantic definitions that need to be integrated.

Integrating data from multiple resources also raises challenging issues related to data provenance, data ownership, data quality, privacy, and security, which will need to be addressed in the short future. Indeed, integrated data is often composed of several data items, each coming from a different resource. Tracking data provenance is critical to scientific applications as it enables users to know where each data item comes from. This knowledge is relevant to data ownership and quality. For example, when exploiting data, it is important to give credit to the researcher who has generated or annotated the data. In addition, data provenance may affect the expected quality of the data (e.g., when they are not curated or validated) and thus the way it should be exploited. But if scientific integration systems evolve to track down data provenance, they might also enable to reconstitute the original datasets, which raises privacy and security issues as scientific discovery will need to integrate more and more clinical data. Biological integration systems may have to comply with regulations such as the privacy provisions and the standards for the security of electronic health information of the U.S. federal law, the Health Insurance Portability and Accountability Act of 1996 (HIPAA).

Which trends will continue and impact the bioinformatics community as a whole has yet to be seen. The only thing certain is that bioinformatics will continue to be an exciting and evolving discipline for years to come. As comprehensive as we have tried to be, this book provides only an introduction to the fascinating world of bioinformatics data integration. Furthermore, while the challenges outlined herein are daunting, addressing them is only the first step the evolving, multidisciplinary field of bioinformatics must take. Once these challenges have been overcome, there is still a huge amount of work to be done to use that information effectively to understand the mechanics of life. Despite the tremendous amount of work still to do, the path is fascinating and the rewards for successfully unraveling the mysteries of the genome are unparalleled. We hope that this book has provided not just insight into the challenges currently being addressed in bioinformatics, but also inspiration to help overcome them.

Appendix: Biological Resources

Useful biological resources, databases, organizations, and applications are listed in three tables. The tables include resources cited in the book. The acronyms commonly used to refer to the resources are spelled out, and current URLs are provided. Additional resources are the Public Catalog of Databases available at INFOBIO-GEN (*http://www.infobiogen.fr/services/dbcat*) and the Biocatalog available at the European Bioinformatics Institute (*http://www.ebi.ac.uk/biocat*).

Category	Databases and URLs
Comprehensive Data Center: Broad content including sequence, structure, function, etc.	EBI (European Bioinformatics Institute): *http://www.ebi.ac.uk/*
	EMBL (European Molecular Biology Laboratory): *http://www.embl-heidelberg.de/*
	ExPaSy (Expert Protein Analysis System—Swiss Institute of Bioinformatics): *http://us.expasy.org*
	The INFOBIOGEN Deambulum: *http://www.infobiogen.fr/services/deambulum/english/menu.html*
	Institut Pasteur: *http://bioweb.pasteur.fr/docs/gendocdb/banques.html*
	NCBI (National Center for Biotechnology and Information): *http://www.ncbi.nlm.nih.gov/*
	TIGR (The Institute of Genome Research): *http://www.tigr.org/*
	Whitehead/MIT (Massachusetts Institute of Technology) Genome Center: *http://www-genome.wi.mit.edu/*

A.1

TABLE

Biological databases.

Category	Databases and URLs
DNA or Protein Sequence	DDBJ (DNA Data Bank of Japan): *http://www.ddbj.nig.ac.jp/*
	dbEST (Expressed Sequence Tags Database): *http://www.ncbi.nih.gov/dbEST*
	EMBL (Nucleotide Sequence Database): *http://www.ebi.ac.uk/embl/index.html*
	GenBank and the NCBI Nucleotide Database: *http://www.ncbi.nlm.nih.gov/Genbank*
	GenPept (protein database translated from the last release of GenBank): *ftp://www.infobiogen.fr/pub/db/genpept/* or *ftp://ftp.ncbi.nih.gov/genbank/*
	GSDB (Genome Sequence DataBase): *http://wehih.wehi.edu.au/gsdb/gsdb.html*
	PIR (Protein Information Resource): *http://pir.georgetown.edu/*
	RefSeq (comprehensive integrated non-redundant set of sequences): *http://www.ncbi.nlm.nih.gov/LocusLink/refseq.html*
	Swiss-Prot (protein knowledgebase): *http://us.expasy.org/sprot/*
Genomes: Complete genome sequences and related information for specific organisms	EBI complete genomes: *http://www.ebi.ac.uk/genomes/*
	EcoCyc (Genome of *Escherichia coli*): *http://biocyc.org/ecocyc*
	FlyBase (Database of *Drosophila* Genome): *http://flybase.bio.indiana.edu/*
	GDB (Genome database): *http://www.gdb.org*
	Institut Pasteur complete genomes: *http://www.pasteur.fr/actu/presse/com/dossiers/GBgenomics/GBintro.html*
	MGD (Mouse Genome Database): *http://www.informatics.jax.org/*
	NCBI complete genomes: *http://www.ncbi.nlm.nih.gov/Genomes/index.html*
	RatMap (Rat Genome Database): *http://ratmap.gen.gu.se*
	SGD (Saccharomyces Genome Database): *http://genome-www.stanford.edu/Saccharomyces/*
	UCSC Genome Bioinformatics: *http://genome.ucsc.edu/*
	WormBase (Genome and Biology of *C. Elegans*): *http://www.wormbase.org/*

A.1 Continued.

TABLE

Category	Databases and URLs
Genetics: Gene mapping, mutations, and diseases	AllGenes (predicted human and mouse genes): *http://www.allgenes.org*
	GeneCards (human genes): *http://bioinfo.weizmann.ac.il/cards*
	GeneLynx (human genes): *http://www.genelynx.org*
	Genew (database of approved HUGO symbols): *http://www.gene.ucl.ac.uk/nomenclature/*
	GDB (Genome Database): *http://gdbwww.gdb.org/gdb/*
	HGMD (Human Gene Mutation Database): *http://archive.uwcm.ac.uk/uwcm/mg/hgmd0.html*
	OMIM (Online Mendelian Inheritance in Man): *http://www.ncbi.nlm.nih.gov/entrez/query.fcgi?db=OMIM*
Gene Expression: Microarray and cDNA gene expression	ArrayExpress (microarray data): *http://www.ebi.ac.uk/arrayexpress*
	BodyMap (expression information about human and mouse genes): *http://bodymap.ims.u-tokyo.ac.jp/*
	dbEST (Expressed Sequence Tag Database): *http://www.ncbi.nlm.nih.gov/dbEST/index.html*
	GeneX (gene expression database): *http://www.ncgr.org/genex*
	GEO (Gene Expression Omnibus): *http://www.ncbi.nlm.nih.gov/geo/*
	MGED (Microarray Gene Expression Database): *http://www.mged.org*
	UniGene (partition of GenBank into clusters that contain the sequences that represent a unique gene): *http://www.ncbi.nlm.nih.gov/UniGene/*
Structure: Three-dimension structures of small molecules, proteins, DNA	CSD (Cambridge Structural Database): *http://www.ccdc.cam.ac.uk/prods/csd/csd.html*
	HSSP (database of Homology-derived Secondary Structure of Proteins): *http://www.hgmp.mrc.ac.uk/Bioinformatics/Databases/hssp-help.html*
	NDB (Nucleic Acid Database): *http://ndbserver.rutgers.edu/NDB/ndb.html*
	PDB (Protein Data Bank): *http://www.rcsb.org/pdb/index.html*

TABLE A.1 Continued.

Category	Databases and URLs
Classification of Protein Family and Protein Domains	Blocks database (protein blocks): *http://www.blocks.fhcrc.org/*
	CATH (Protein Structure Classification Database): *http://www.biochem.ucl.ac.uk/bsm/cath_new/index.html*
	InterPro (resource for whole genome analysis): *http://www.ebi.ac.uk/interpro/index.html*
	Pfam (database of protein families): *http://pfam.wustl.edu/*
	PRINTS (Protein Fingerprint Database): *http://www.bioinf.man.ac.uk/dbbrowser/PRINTS/*
	ProDom (protein domain families): *http://prodes.toulouse.inra.fr/prodom/2002.1/html/home.php*
	PROSITE (database of protein families and domains): *http://www.expasy.ch/prosite/*
	SCOP (Structure Classification of Proteins): *http://scop.mrc-lmb.cam.ac.uk/scop/*
Protein Pathway Protein–Protein Interactions and Metabolic Pathway	BIND (Biomolecular Interaction Network Database): *http://www.binddb.org/*
	DIP (Database of Interacting Proteins): *http://dip.doe-mbi.ucla.edu/*
	EcoCyc (Encyclopedia of *E. coli* Genes and Metabolism): *http://biocyc.org/ecocyc*
	KEGG (Kyoto Encyclopedia of Genes and Genomes): *http://www.genome.ad.jp/kegg/kegg2.html#pathway*
	WIT (Metabolic Pathway): *http://wit.mcs.anl.gov/WIT2/*
Proteomics: Proteins, Protein family	AfCS (Alliance for Cellular Signaling): *http://cellularsignaling.org/*
	JCSG (Joint Center for Structural Genomics): *http://www.jcsg.org/scripts/prod/home.html*
	PKR (Protein Kinase Resource): *http://pkr.sdsc.edu/html/index.shtml*
Pharmacogenomics, Pharmacogenetics, Single Nucleotide Polymorphism (SNP), Genotyping	ALFRED (Allele Frequency Database): *http://alfred.med.yale.edu/alfred/index.asp*
	CEPH (Centre d'Etude du Polymorphisme Humain genotype database): *http://www.cephb.fr/cephdb/*

TABLE A.1   Continued.

Category	Databases and URLs
	dbSNP (Single Nucleotide Polymorphism Database): *http://www.ncbi.nlm.nih.gov/SNP/*
	LocusLink: *http://www.ncbi.nlm.nih.gov/LocusLink*
	PharmGKB (Pharmacogenetics Knowledge Base): *http://pharmgkb.org*
	SNP consortium: *http://snp.cshl.org*
Tissues, Organs, and Organisms	BRAID (Brain Image Database): *http://braid.rad.jhu.edu/interface.html*
	NeuroDB (Neuroscience Federated Database): *http://www.npaci.edu/DICE/Neuro/*
	Visible Human Project: *http://www.nlm.nih.gov/research/visible/visible_human.html*
	Whole Brain Atlas: *http://www.med.harvard.edu/AANLIB/home.html*
Literature Reference	PubMed (MEDLINE bibliographic database): *http://www.ncbi.nlm.nih.gov/entrez/*
	USPTO (US Patent and Trademark Office): *http://www.uspto.gov/*

Organization	Descriptions
Human Genome Organization (HUGO) Gene Nomenclature Committee (HGNC) *http://www.gene.ucl.ac.uk/nomenclature/*	HGNC is responsible for the approval of a unique symbol for each gene and designates descriptions of genes. Aliases for genes are also listed in the database.
Gene Ontology Consortium (GO) *http://www.geneontology.org*	GO is to develop ontologies describing the molecular function, biological process, and cellular component of genes and gene products for eukaryotes. Members include genome databases of fly, yeast, mouse, worm, and *Arabidopsis*.
Plant Ontology Consortium *http://plantontology.org*	Produce structured, controlled vocabularies applied to plant-based database information.
Microarray Gene Expression Data Society (MGED) *http://www.mged.org/*	The MGED group facilitates the adoption of standards for DNA-microarray experiment annotation and data representation, as well as the introduction of standard experimental controls and data normalization methods.
NBII (National Biological Information Infrastructure) *http://www.nbii.gov/disciplines/systematics.html*	NBII provides links to taxonomy sites for all biological disciplines.
ITIS (Integrated Taxonomic Information System) *http://www.itis.usda.gov/*	ITIS provides taxonomic information on plants, animals, and microbes of North America and the world.
MeSH (Medical Subject Headings) *http://www.nlm.nih.gov/mesh/meshhome.html*	National Library of Medicine (NLM) controlled vocabulary used for indexing articles, cataloging books and other holdings, and searching MeSH-indexed databases, including MEDLINE.
SNOMED (Systematized Nomenclature of Medicine) *http://www.snomed.org/*	SNOMED is recognized globally as a comprehensive, precise controlled terminology created for the indexing of the entire medical record.
International Classification of Diseases, Ninth Revision, Clinical Modification (ICD-9-CM) *http://www.cdc.gov/nchs/about/otheract/icd9/abticd9.htm*	ICD-9-CM is the official system of assigning codes to diagnoses and procedures associated with hospital utilization in the United States. It is published by the U.S. National Center for Health Statistics.

A.2 Biological ontology resources.

TABLE

Organization	Descriptions
International Union of Pure and AppliedChemistry (IUPAC) International Union of Biochemistry and Molecular Biology (IUBMB) Nomenclature Committee *http://www.chem.qmul.ac.uk/iubmb/*	IUPAC and IUBMB make recommendations on organic, biochemical, and molecular biology nomenclature, symbols, and terminology.
PharmGKB (Pharmacogenetics Knowledge Base) *http://pharmgkb.org/*	PharmGKB develops an ontology for pharmacogenetics and pharmacogenomics.
mmCIF (The macromolecular Crystallographic Information File): *http://pdb.rutgers.edu/mmcif/* or *http://www.iucr.ac.uk/iucr-top/cif/index.html*	mmCIF is sponsored by IUCr (International Union of Crystallography) to provide a dictionary for data items relevant to macromolecular crystallographic experiments.
LocusLink *http://www.ncbi.nlm.nih.gov/LocusLink/.*	LocusLink contains gene-centered resources including nomenclature and aliases for genes.
RiboWeb *http://riboweb.stanford.edu/riboweb/login-frozen.html*	The RiboWeb provides access to a knowledge base containing a standardized representation of ribosomal structural information.
ENZYME *http://us.expasy.org/enzyme/*	ENZYME is a repository of information relative to the nomenclature of enzymes.
IMGT (ImMunoGeneTics information system) *http://imgt.cines.fr/*	IMGT is a high-quality integrated information system specializing in immunoglobulins (IG), T-cell receptors (TR), major histocompatibility complex (MHC), and related proteins of the immune system of human and other vertebrate species.

Category	Application Names and URLs
Microarray analysis	MAS (Affymetrix MicroArray Suite): *http://www.affymetrix.com/products/software/specific/mas.affx*
	ImaGene (BioDiscovery): *http://www.biodiscovery.com/*
Sequence similarity search	BLAST (Basic Local Alignment Search Tool): *http://www.ncbi.nlm.nih.gov/BLAST/blast_overview.html*
	FASTA (Sequence similarity and homology search): *http://www.ebi.ac.uk/fasta33/index.html* and *http://www.ebi.ac.uk/fasta33/genomes.html*
	SMART (Simple Modular Architecture Research Tool): *http://smart.embl-heidelberg.de/*
	WU-BLAST (Washington University BLAST) *http://blast.wustl.edu/blast/README.html*
Multiple sequence alignment	CAP (Contig Assembly Program): *http://fenice.tigem.it/bioprg/interfaces/cap3.html* ClustalW: *http://www-igbmc.u-strasbg.fr/BioInfo/ClustalW* ClustalX: *http://www-igbmc.u-strasbg.fr/BioInfo/ClustalX*
	LASSAP (LArge Scale Sequence compArison Package), also known as BioFacet: *http://www.gene-it.com/index.html*
	MEGA: *http://www.megasoftware.net/*
	MultAlin: *http://prodes.toulouse.inra.fr/multalin/multalin.html*
	PAUP (Phylogenetic Analysis Using Parsimony): *http://paup.csit.fsu.edu/paupfaq/faq.html*
	Phylip: *http://evolution.genetics.washington.edu/phylip.html*
	TMAP: *http://www.mbb.ki.se/tmap/*
Analysis	EMBOSS (European Molecular Biology Open Software Suite): *http://www.hgmp.mrc.ac.uk/Software/EMBOSS/*
	HMMER (Profile hidden Markov models for biological sequence analysis): *http://hmmer.wustl.edu/*
	GeneSpring (Silicon Genetics): *http://www.silicongenetics.com/cgi/SiG.cgi/Products/GeneSpring/index.smf*
	PSORT: *http://psort.nibb.ac.jp/*
	Spotfire: *http://spotfire.com*
	StackPACK: *http://www.sanbi.ac.za/Dbases.html*
	Wise (Genewise): *http://www.ebi.ac.uk/Wise2/index.html*
Sequence folding	Mfold: *http://www.bioinfo.rpi.edu/applications/mfold/*
Structure prediction	NNPREDICT (Protein Secondary Structure Prediction): *http://www.cmpharm.ucsf.edu/~nomi/nnpredict.html*

A.3 Biological tools and systems.

TABLE

Category	Application Names and URLs
Pattern recognition	Partek: *http://www.partek.com*
Retrieval systems	BioRS (Biomax): *http://www.biomax.de/index.html*
	DBGET/LinkDB: *http:www-genome.ad.jp/dbget*
Database systems	AceDB: *http://www.acedb.org/*
	GUS (Genomics Unified Schema platform): *http://www.gusdb.org*
	GIMS (Genome Information Management System): *http://www.cs.man.uk/img/gims/*
	Informax: *http://www.informaxinc.com*
	MySQL (open source DBMS): *http://www.mysql.com/*
	SeqStore: *http://www.accelrys.com/dstudio/ds_seqstore/*
	Tripos: *http://www.tripos.com/*

Glossary

AADM The Affymetrix Analysis Data Model. The relational database schema the Affymetrix *LIMS* and MicroDB systems use to store GeneChip expression results.

aggregation A computation whose result value depends on a stream of input values, such as an average, sum, or standard deviation.

API (application programming interface) This is composed of any set of routines generally available for use by programmers to provide portable code. The programmer only has to worry about the call and its parameters and not the details of implementation, which may vary from system to system.

ASN1 Abstract Syntax Notation One. This is an *ISO* standard for open systems interconnection.

automatic summary table A special table created to cache the results of a specific query against other tables. Subsequently, when another query is submitted, the query processor may be able to deduce that the new query can be rewritten as a query against the cached result. Using the pre-computed result in this way can have a large performance benefit. The downside is that the automatic summary table must be maintained as the underlying tables are updated.

autonomy of databases Degree of control of the database into an integration architecture that includes what transactions are permissible, how it executes transactions, and so on. In the context of integration of distributed resources, examples of integration affecting the autonomy of each resource are tight integration, semi-autonomous integration, and total isolation. They respectively characterize a lack of control, moderate control, and total control. Autonomy is the second characterization of integration with *distribution*.

bag A data type that represents a homogeneous collection of objects such that the order of appearance of these objects in the collection is unimportant, but the number of occurrences is important. Unlike a set, a value may occur multiple times in a bag.

bindjoin A federated join algorithm in which the federated server ships values of the join column(s) from one of the tables to the remote data source that stores the other table. The remote source searches its table for rows with matching values, and returns these to the federated server.

BLAST The Basic Local Alignment Search Tool. Used to compare a gene or protein sequence against other sequences.

blastn An implementation of BLAST used for nucleotide–nucleotide comparisons.

blastp An implementation of BLAST used for protein–protein comparisons.

BLOB Binary Large Object. A data type for representing a long string of binary data (e.g., an image or a video) whose internal structure is unknown to the database management system. Due to their potential for great size, database systems typically manage BLOB data with special techniques to eliminate unnecessary copying and allow random access to sub-pieces. Unlike a *CLOB*, a BLOB is not associated with a particular character set or encoding.

Boolean circuit A family of Boolean circuits is an infinite collection of acyclic Boolean circuits made up of AND, OR, and NOT gates.

box plots An excellent tool for conveying location and variation information in data sets, particularly for detecting and illustrating location and variation changes between different groups of data.

browsing The act of accessing information available on the World Wide Web. This is typically an interactive process, with a person examining Web pages and following links.

bulk data type Refers to data types that are collections of objects. Examples of bulk data types are sets, bags, lists, and arrays.

CDATA Textual portion of an XML document that is ignored by the parser.

cDNA Complementary DNA. DNA copies of the mRNA expressed in a specified tissue.

CDS Coding sequences.

CGI Common Gateway Interface.

CLI Call-Level Interface. A general-purpose interface to IBM DB2 that conforms to *ODBC* 2.0 level 2 and ODBC 3.0 level 1, but can be used without an ODBC driver. It also supports some ODBC 3.0 level 2 functions, as well as some DB2-specific functions.

CLOB Character Large Object. A data type for representing a long string of character data (e.g., a text document or genomic sequence) whose internal structure is unknown to the database management system. Due to their potential for great size, database systems typically manage CLOB data with special techniques to eliminate unnecessary copying and allow random access to sub-pieces. A CLOB is associated with a specific character set, and character codes will be translated appropriately when it is retrieved.

CNS tissue Central nervous system tissue.

co-clustered fragment Gene fragments derived from the same UniGene cluster or consensus sequence cluster.

CPL (Collection Programming Language) A high-level query language based on the comprehension syntax and supported by Kleisli.[1]

comparative genomics The study of human genetics by comparisons with model organisms such as mice, fruit flies, and the bacterium *E. coli.*

complex value data Data whose type system includes not only simple types such as strings, and numbers, but also arbitrarily nested sets, lists, bags, records, and variants.

---

1. L. Wong. "Kleisli: A Functional Query System." *Journal of Functional Programming* 10, no. 1 (2000): 19–56.

Conceptual Model (CM) An abstraction of the objects represented in an application, as well as their properties and their relationships, that provides a conceptual representation of the application. CMs typically capture the class and object structure as well as domain-specific relationships of the modeled world. CMs can be expressed in a variety of ways such as through entity-relationship diagrams (*ER*), class diagrams in the Unified Modeling Language (UML), or by using formal approaches based on first-order predicate logic.

CORBA Common Object Request Broker Architecture. An *OMG* standard for an architecture and infrastructure that allows computer applications to work together over networks.

CPU Central Processing Unit.

database (DB) A collection of information organized in such a way that a computer program can quickly select desired pieces of data (see *database management system*).

database management system A collection of programs that enables storing, accessing, modifying, and extracting information in a database.

data cleansing (Also called data scrubbing) This is the process of amending or removing data in a database that is incorrect, incomplete, improperly formatted, or duplicated. See also *data curation*.

data curation The process of storing and checking the accuracy of data so they remain accessible indefinitely. When applied in the context of multiple data sources, this also implies the reconciliation of semantic conflicts that may arise from conflicting information.

data fusion The process of deriving insight from information acquired from multiple sources (sensor, databases, information gathered by human, etc.) of which data integration is a key step. The term was first used by the military to correlate and analyze information in time and space, to identify and track individual objects (equipment and units), to assess the situation, to determine threats, and to detect patterns in activity.

data integration A process that combines data from multiple, possibly heterogeneous and inconsistent, data sources into a single, consistent source.

data mining Analyzing, exploring, or clustering a data set with statistical techniques.

data model Provides the means for specifying particular data structures, for constraining the data associated with these structures, and for manipulating the data within a database system. To handle data outside the database system, this traditional definition is extended to include a data exchange format, which is a means for bringing data outside the database system into it and also for moving data inside the database system to the outside.

data-shipping Within the client/server context, data-shipping consists of transferring the data from the client to the server and performing the execution of the query at the server. (See *query-shipping* for an alternate approach.)

data source Any data repository (e.g., database, flat files).

data type Classifies a particular type of information. Examples of data types are: integer, floating point, number, character, and string. (See *bulk data type*.)

data warehouse A collection of data integrated from multiple sources and contained within a unique system, usually a database. Data needs to be translated to a common format, cleansed, and reconciled before being integrated into the data warehouse. It constitutes a subject-oriented, integrated, time variant, and nonvolatile data repository.

Datalog A query language that allows users to access and manipulate data contained in predicates through if-then-else rules.

DBMS (*See database management system.*)

description logics Knowledge representation languages tailored for expressing knowledge about concepts and concept hierarchies.

distributed database systems. A collection of logically interrelated databases, distributed at multiple sites and connected by a computer network such that each database has autonomous processing capability and participates in the execution of queries that are split across multiple sites. Distribution of databases characterizes the fact that the data are split over several databases. Distribution is the second characterization of databases with autonomy.

DNA Deoxyribonucleic acid. A linear nucleic acid polymer composed of four kinds of nucleotides: Adenine, Thymine, Guanine, Cytosine. In native form inside the nucleus, it is a double-helix of two anti-parallel strands held together by hydrogen bonds. DNA is the carrier of genetic information for many species.

DNA microarray A mechanism for massively parallel gene expression and gene discovery studies in which probes (or oligonucleotide sequences) with known identity are placed on glass or nylon substrates and used to determine complementary binding through hybridization. A synonym for this is probe array.

DNA sequencing The experimental process of determining the nucleotide sequence of a region of DNA. This is done by labelling each nucleotide (A, C, G, or T) with either a radioactive or fluorescent marker that identifies it. There are several methods of applying this technology, each with its advantages and disadvantages. For more information, refer to a current textbook. High throughput laboratories frequently use automated sequencers, which are capable of rapidly reading large numbers of templates. Sometimes the sequences may be generated more quickly than they can be characterized.

domain map (DM) A kind of ontology to denote semantic networks of terms and their relationships. A precise meaning can be associated to DMs via a logic formalization. DMs are used to express terminological knowledge.

EDB Extensional database.

ER model Entity-relationship model. A data model consisting of entity classes and relationships traditionally used to describe relational database schema.

enzyme A biological macromolecule, usually a protein, that acts as a catalyst. Enzyme Nomenclature Committee classifies these molecular activities by assigning a unique Enzyme Catalogue (EC) number.

EST sequence Expressed Sequence Tags. Short sequence fragments (<200 base pairs) that are known to express collectively in a given tissue or a pool of tissue. Clusters of these sub-fragments assembled into consensus sequences act as identifiers of genes or transcripts expressed in that tissue.

extensional database (EDB) Refers to (i) the set of tuples (i.e., "facts") stored in a database and/or (ii) the relational schema of the tuples/facts which are stored directly in a database.

FDM Functional Data Model.

federation A collection of semi-autonomous, distributed databases in which each database has significant autonomy while still providing the capability to access integrated resources in a unified manner.

First Order logic (FO) is the logic that can only quantify over sets of values. Second-order logic can quantify over functions, and higher-order logic can quantify over any type of entity.

foreign key This is a field in a relational table that matches the *primary key* column of another table. The foreign key can be used to cross-reference tables.

FTP File Transfer Protocol.

functional genomics The study of genes, their resulting proteins, and the role played by the proteins in the body's biochemical processes.

functional programming languages Programming languages that emphasize a particular paradigm of programming technique known as "functional programming." In this paradigm, all programs are expressed as mathematical functions and are generally free from side effects. Examples of functional programming languages are LISP, Haskell, and SML.

gene An abstract entity that is the fundamental physical and functional unit of heredity. A gene is an ordered sequence of nucleotides located in a particular position on a particular chromosome that encodes a specific functional product (i.e., a protein or *RNA* molecule).

GeneChip Affymetrix whole genome arrays or dynamic custom arrays.

gene expression The process by which a gene's coded information is converted into the structures present and operating in the cell. Expressed genes include those that are transcribed into *mRNA* and then translated into protein and those that are transcribed into RNA but not translated into protein (e.g., transfer and ribosomal RNAs).

gene fragment An abstract sub-sequence fragment of a representative target transcript (*mRNA*) from which the individual probes or oligo sequences are derived.

Synonyms include composite target sequence, probe set, sequence fragment, and target sequence.

gene product The biochemical material, either RNA or protein, resulting from expression of a gene. The amount of gene product is used to measure how active a gene is; abnormal amounts can be correlated with disease-causing alleles.

genome All the genetic material in the chromosomes of a particular organism; its size is generally given as its total number of base pairs.

genome project See *Human Genome Initiative.*

genomics The study of genes and their function.

gene chip microarray technology Development of *cDNA* microarrays from a large number of genes. Used to monitor and measure changes in gene expression for each gene represented on the chip.

global schema A single, unifying, semantically consistent view of data contained in multiple, distributed, heterogeneous data sources.

global-as-view (GAV) An integration approach is global-as-view (GAV) when the *global schema* is expressed with respect to the source schemas (the schemas of the integrated sources). Queries asked against the global schema are easily translated into source queries by replacing the meaning of each relation and attribute of the global schema with its definition in terms of the source schemas. GAV is an alternative to LAV.

grid Grid computing is a form of distributed computing that involves coordinating and sharing computing, application, data, storage, or network resources across dynamic and geographically dispersed organizations.

GUI Graphical User Interface.

heterogeneous databases Used in the context of distributed databases when systems differ in some way, such as data representation, query language, or semantics.

host variable The *SQL* representation for an application program variable. A host variable can be the container for data inserted into or retrieved from the database, or it can represent a query parameter whose value will be supplied by the application just prior to execution of the query.

HTTP Hypertext Transfer Protocol.

Human Genome Initiative Collective name for several projects begun in 1986 by the U.S. Department of Energy to create an ordered set of DNA segments from known chromosomal locations, to develop new computational methods for analyzing genetic map and DNA sequence data, and to develop new techniques and instruments for detecting and analyzing DNA. This DOE initiative is now known as the Human Genome Program. The joint national effort, led by DOE and National Institutes of Health, is now known as the Human Genome Project.

Human Genome Project (HGP) Formerly titled *Human Genome Initiative.*

hybridization The biochemical process by which two complementary, single-stranded nucleic acid chains form a stable, double-stranded helix chain. *DNA microarrays* use hybridization reactions to assay target transcripts extracted from the samples.

IDB Intensional database.

intensional database (IDB) Refers to (i) the set of tuples (i.e., virtual relations) in a database, which are defined by means of logic rules (e.g., Datalog formulas or SQL "create view" statements) and/or (ii) the relational schema of the virtual relations defined by those rules.

ISA relationship When between two entities, captures the notion of generalization The opposite or inverse of generalization is called specialization. If "A" ISA "B", then "B" is the more generic concept and A is the specific concept. The most significant property of an ISA relationship is that of inheritance. All that is specified to be true about the generic concept is also true for the specific concept. That means that all attributes, their values, and constraint (rules) are inherited from the more generic level concept down to the more specific level concept as are all relationships in which the more generic level concept participates.

ISDK InSilico Discovery Kit describes experimental steps carried out in computers the same way an experimental protocol describes the steps carried out in a wet laboratory.

ISO International Organization for Standardization. An international standards-making body.

JDBC Java DataBase Connectivity. JDBC technology is an application programming interface (*API*) that provides cross-database connectivity to a wide range of relational database systems from the Java programming language. It also provides access to other tabular data sources, such as spreadsheets or flat files.

K2MDL The K2 mediator definition language, a high-level language that extends *ODMG's ODL* with *OQL* definitions and variants.

KEGG Kyoto Encyclopedia of Genes and Genomes. An effort to computerize current knowledge of molecular and cellular biology in terms of the information pathways that consist of interacting molecules or genes and to provide links from the gene catalogs produced by genome sequencing projects.

known gene Refers to officially approved genes by the model organism nomenclature committee. For example, HUGO is for humans.

Kripke structure Modal logics provide a general framework for reasoning about what is necessarily or possibly true, in particular when dealing with several "possible worlds" that are reachable from one another by a temporal accessibility relation. Kripke structures are families of conventional first-order logic structures (one for each "possible world"), which may be reachable from one another as described by the accesibility relation R.

LAV See *local-as-view*.

LIMS Laboratory Information Management System. Software that helps manage the workflow and data associated with a laboratory.

link-driven federation of databases This type of federation allows a set of data sources to be browsed by a user who asks a single retrieval query and then explores the output by browsing from one source to the other via hyperlinks. An example of a link-driven federation is SRS.

LISP This is a programming language invented by John McCarthy in the late 1950s as a formalism for reasoning about the use of recursion equations as a model for computation.

list A data type that represents a homogeneous collection of objects such that both the order of appearance and the number of occurrences of objects in the collection are important.

local-as-view (LAV) An integration approach is local-as-view when the source schemas are expressed by means of the *global schema*. Queries asked against the global schema are translated into source queries by replacing the sub-query–defining schema components of source schemas. An alternative approach to LAV is *global-as-view*.

LOGSPACE The class of problems solvable in deterministic logarithmic space.

materialized query table See *automatic summary table*.

materialized view The cached result of a query against a database. The query can restructure or be intended to load data into a data warehouse. (See also *automatic summary table*.)

mediator A middleware component of a database integration infrastructure that translates data from fully autonomous distributed heterogeneous data sources to a semantically consistent representation. Mediators do not assume that integrated sources will all be relational databases; instead, they can be various database systems (relational, object-relational, object, XMLetc.), flat files, and so on.

MGED The Microarray Gene Expression Data society. An international organization for facilitating the sharing of microarray data from functional genomics and proteomics experiments.

MIAME Minimum information about a microarray experiment. This is a set of guidelines developed by the MGED Society to outline the minimum information required to unambiguously interpret microarray data and subsequently to allow independent verification of this data at a later stage if required.

microarray See *DNA microarray*.

middleware Connectivity software that consists of a set of enabling services that allow multiple processes running on one or more machines to interact across a network.

ML (meta language) A functional programming language.

model-based mediation (MBM) A wrapper/mediator approach and architecture for information integration in which representations of domain semantics (domain maps, process maps, and semantic integrity constraints) are used to facilitate queries across sources.

mRNA A single-stranded ribonucleic acid molecule derived from the DNA template of a gene when the gene is transcribed during the gene expression process, which takes place in the nucleus of the cell. mRNA specifies the order of the amino acids to be coded in a protein by the translation process which takes place inside the cytoplasm of cell. Its role is to transmit instructions from DNA sequences in the nucleus to the protein-making machinery in the cytoplasm of the cell.

multi-database A system consisting of fully autonomous distributed databases. The integration component is in charge of providing the user with a query language to query integrated resources, executing the query by collecting needed data from each integrated resource, and returning the result to the user.

non-materialized view The result of a query that is not cached and restructure a database. The query that defines the non-materialized view, usually is stored as a functional definition of the data contained within it, and it is this function that is used to recreate the view dynamically on demand.

NP (NPTIME) The set of problems solvable in non-deterministic polynomial time that cannot be solved deterministically in polynomial time.

NP-complete A set of problems in NP such that any NP problem reduces to it.

NRC Nested Relational Calculus.

object-oriented model An approach to programming and data storage in which objects are the primary concepts. In this approach, data and functionality are tightly coupled. Methods are associated with an object and are the only way to manipulate or access the data contained within that object. This approach also makes use of concepts such as object inheritance, which may not be available in other models.

Object Query Language (OQL) A query language that allows users to access and manipulate data contained in object-oriented databases such as those formalized by the *ODMG*.

ODBC Open DataBase Connectivity. A widely accepted application programming interface for database access. It is based on the call-level interface specifications from X/Open and *ISO*/IEC for database *APIs,* and *SQL* is its database access language.

ODL Object Definition Language. A standard for object definition specified by the *ODMG.*

ODMG Object Data Management Group. A standard-making body for object-oriented databases.

OIL (Ontology Inference Layer) A proposal for a Web-based representation and inference layer for ontologies, which combines the widely used modeling primitives from frame-based languages with the formal semantics and reasoning services provided by description logics.

OLAP (on-line analytical processing) OLAP transforms raw data so that it reflects the real dimensionality of the enterprise as understood by the user.

OMG Object Management Group. A standard-setting body focused on developing standards for interoperable enterprise applications.

one-world/multiple-world scenarios Here "world" means a coherent fragment of an application domain (i.e., classes of objects and their relationships, that naturally belong together and form a coherent domain, and where the relationships among the objects and classes is evident). Thus, a one-world mediation scenario can be solved without additional cross-world knowledge, while a multiple-world scenario often requires specialized knowledge to bridge semantic gaps.

ontology A description of concepts and relationships that exist among the concepts for a particular domain of knowledge. In the world of structured information and databases, ontologies in life science provide controlled vocabularies for terminology as well as specifying object classes, relations, and functions.

OQL Object Query Language. A standard for querying object-oriented databases specified by the *ODMG.*

primary key Primary key of a relational table uniquely identifies each record in the table. It can either be a normal attribute that is guaranteed to be unique (such as a Social Security number in a table with no more than one record per person), or it can be generated by the *DBMS* (such as a globally unique identifier, or GUID, in Microsoft SQL Server).

probe The individual 25mer sub-sequences that are tiled on a microarray. These are derived from the gene fragments that collectively detect the target transcript. Synonyms used are oligonucleotide sequence and target sequence.

process map (PM) A kind of ontology used in model-based mediation to describe semantic networks of procedural knowledge (i.e., the processes of a domain and how they influence and depend on each other). (See also *domain maps*.)

proteome Proteins expressed by a cell or organ at a particular time and under specific conditions.

proteomics The study of the full set of proteins encoded by a genome.

pharmacogenomics The study of the interaction of an individual's genetic makeup and response to a drug.

P (PTIME) The class of problems solvable in deterministic polynomial time.

query A program written in a database *query language* for retrieving and transforming information in a database.

querying Accessing and manipulating a data source using a *query language*.

query language A language that enables users to access and manipulate data, usually stored within a database management system. Examples of query languages are the relational algebra, the Structural Query Language (*SQL*), the database logic (*Datalog*), the Object Query Language (*OQL*), and *XQuery*.

query-shipping In the client/server context, query-shipping consists of partially or completely performing a query at the client site and sending only the results to the server. (See *data-shipping* for an alternate approach.)

RDF The Resource Description Framework (RDF) integrates a variety of applications including library catalogs and World Wide Web directories; syndication and aggregation of news, software, and content; and personal collections of music, photos, and events using *XML* as an interchange syntax. The RDF specifications provide a lightweight ontology system to support the exchange of knowledge on the Web.

record A data type that represents an object comprising several data fields. Each data field has a name and a value.

relational algebra A query language that allows users to access and manipulate data contained in relations with algebraic operators: union, intersection, difference, selection, projection, Cartesian product, join, and renaming.

relational model The standard data model used in commercial database management systems. This data model is based on the relational algebra and presents data as a collection of tables. Each table represents a complex data type, and each column represents an attribute. Each row in a table contains an instance of that type.

RNA Ribonucleic acid.

schema The physical data representation in a database system. It characterizes the way the data is organized in the system (e.g., tables, relations, classes, entities, concepts, etc.).

schema integration The process of mapping source schemas to a global, integrated schema. It consists in (1) identifying the components of a database that are related to one another, (2) selecting the best representation for the global schema, and (3) mapping and integrating the components.

searching Accessing a data source through a phrase (string of characters, keyword, DNA sequence, etc.) or a Boolean expression of phrases. The output is the set of strings (documents, sequences, etc.) that are similar to the given phrase.

semantic mediation. See *model-based mediation.*

Semantic Web This is a collaborative effort led by the W3C aiming to represent data on the World Wide Web. It is based on the Resource Description Framework (*RDF*), which integrates a variety of applications using *XML* for syntax and *URIs* for naming.

set A data type that represents a homogeneous collection of objects such that the order of appearance and the number of occurrences of these objects in the collection are unimportant.

SGML Standard Generalized Markup Language.

Skolem functions Or more precisely, Skolem function symbols (after Albert Thoralf Skolem, 1987–1963) are used to create symbolic names when eliminating existential quantifiers in first-order logic statements. For example the formula:

$$\forall P \ person(P) \Rightarrow \exists M \ person(M) \wedge mother(P,M)$$

states that each person P has a mother M. One can obtain a formula that is equivalent with respect to satisfiability by replacing the existential quantifier by a new unary function symbol $f\_m(X)$ denoting the mother of X:

$$\forall P \ person(P) \Rightarrow person(f\_m(P)) \wedge mother(P, f\_m(P))$$

In general, a Skolem function depends on those universal quantifiers in whose scope it occurs (here: $\forall P$).

SML Standard Markup Language. A programming language based on the functional programming paradigm. In this paradigm, all programs are expressed as mathematical functions and are generally free from side effects.

SNP Single nucleotide polymorphism.

SOAP The Simple Object Access Protocol (SOAP) Version 1.2 provides the definition of the *XML*-based information which can be used for exchanging structured and typed information between peers in a decentralized, distributed environment.

sSQL Simplified SQL. An SQL-like query language supported by Kleisli. It extends SQL to the nested relational data model and to multiple, heterogeneous, distributed data sources.

stored procedure A piece of application code, typically including one or more database accesses, that is invoked by the client-side portion of a database application but is executed on the database server. Stored procedures typically are used to reduce client-server communication when multiple accesses to the database are required between interactions with the user.

Structured Query Language (SQL) The standard query language for expressing queries and transformation on relational database management systems. It allows access and manipulation of data with select-from-where statements.

systems biology A new field in biology that is attempting to develop a system-level understanding of biological systems. System-level understanding requires understanding the structures and behaviors of systems as a whole, as well as how to control and design them.

table expression A query language construct that represents data whose value is a table, rather than a scalar value or row. Examples include a reference to a database table, the result of a table function, and a subquery.

target sequence See *gene fragment.*

target transcript. See *mRNA.*

TC° This is the class of those languages recognized by polynomial-size, bouded-depth, unbounded *fan-in* (e.g., maximum number of inputs) Boolean circuits augmented by *threshold gates* (i.e., unbounded fan-in gates that output 1 if and only if more than half of their outputs are non-zero).

transcription The process of synthesizing *mRNA* from a sequence of *DNA* (a gene) template.

transcriptome The full complement of activated genes, *mRNAs*, or transcripts in a particular tissue at a particular time.

translation The process by which the genetic code carried by *mRNA* directs the synthesis of proteins from amino acids.

UML (Unified Modeling Language) The industry-standard language for specifying, visualizing, constructing, and documenting the artifacts of software systems.

UMLS Unified Medical Language System.

URI Uniform Resource Identifiers (also known as *URLs*).

URL Uniform Resource Locators (also known as *URIs*) are short strings that identify resources in the Web: documents, images, downloadable files, services, electronic mailboxes, and other resources. They make resources available under a variety of naming schemes and access methods such as *HTTP, FTP*, and Internet mail addressable in the same simple way.

variant A data type representing an object that is one of several types. Variant types enables to consider data of different types within the same composed type.

view A structured presentation of the data contained within a database. The default view of data is the view of the data as defined by the global schema. However, alternative views (e.g., data summaries) may be presented to provide additional insight into the data.

view integration A virtual integration of multiple data sources.

XA An industry-standard interface for transaction management that is based on the X/Open specification. XA allows multiple compliant data managers to cooperate in a single transaction and ensures that all updates in the transaction are either committed or rolled back as a group, regardless of which data manager made each change.

XML Extensible Markup Language. A simple, very flexible text format derived from *SGML* that is a standard format for structured documents and data on the World Wide Web.

XQuery A standard query language that allows users to access and manipulate data contained in *XML* documents.

W3C The World Wide Web (WWW) Consortium.

warehouse See *data warehouse*.

workflow Workflows are used in business applications to assess, analyze, model, define, and implement the core business processes of an organization (or other business entity). A workflow approach automates the business procedures where documents, information, or tasks are passed between participants according to a defined set of rules to achieve, or contribute to, an overall business goal. In the context of scientific applications, a workflow approach may address overall collaborative issues among scientists, as well as the physical integration of scientific data and tools.

wrapper A wrapper is generally used within a mediator-wrapper architecture for integrating multiple data sources. Each data source typically is accessed through an existing interface program. However, the mediator program is unable to communicate directly with this existing interface program, often because of some input–output format incompatibility. A wrapper is a program that handles this incompatibility so the mediator program can communicate with the interface program.

System Information

Chapter 5	SRS
Name and Version of System	SRS (Sequence Retrieval System), version 7.0. (Additional information is available at *http://www.lionbioscience.com/solutions/products/srs.*)
Status of Development and Maintenance	SRS is a commercial system and is being further developed by LION bioscience with 2 major releases/year and around four maintenance releases/year. SRS is available to academics free of charge.
Contact	srs@lionbioscience.com

Chapter 6	Kleisli
Name and Version of System	discoveryHub (version 5) Available for Solaris, Linux, and Windows platforms.
Status of Development and Maintenance	Kleisli is being developed, maintained, and commercialized by geneticXchange Inc. Academic licenses are available (e.g., Stanford University is an academic customer).
Contact Person	Brian Donnelly GeneticXchange Inc. 713 Santa Cruz Avenue Menlo Park, California 94025-4519, USA Tel: +1 (650) 321-9573 Email: info@geneticxchange.com URL: *http://www.geneticxchange.com* European Operations: Tel: +44 (0)1296 660348 Email: infoeurope@geneticxchange.com Asia-Pacific Operations: Tel: +61 (0)2 6281 7655 Email: infoapac@geneticxchange.com

Chapter 7	TAMBIS
Name and Version of System	TAMBIS 0.96 A demonstrational Java applet and video examples are publicly accessible. (Additional information is available at *http://imgproj.cs.man.ac.uk/tambis/index.html*.)
Status of Development and Maintenance	TAMBIS is a public system developed at the University of Manchester in the UK with the support of the Bioinformatics programme of the British Biotechnology and Biological Sciences Research Council (BBSRC) in partnership with the Engineering and Physical Sciences Research Council (EPSRC) and Zeneca Pharmaceuticals. TAMBIS is no longer maintained. An academic license may be obtained.
Contact	tambis-help@cs.man.ac.uk

Chapter 8	K2
Name and Version of System	K2 0.5 alpha K2 is implemented in pure Java, under JDK 1.2. It is provided as a .jar file, about 850 K, and requires ORO's Perl module for doing regular expression matching and JGL for handling collections. Its OQL/ODL implementation is based on the ODMG 2.0 specification, with some additions, and a few portions that are not yet implemented. (Additional information is available at *http://db.cis.upenn.edu/K2/*.)
Status of Development and Maintenance	K2/KLEISLI was developed at the University of Pennsylvania and is currently maintained by Scott Harker. Academic licenses are available.
Contact Person	Dr. Val Tannen Department of Computer and Information Science University of Pennsylvania 200 South 33rd Street Philadelphia, Pennsylvania 19104-6389, USA Tel: +1 (215) 898-2665 FAX: +1 (215) 898-0587 Email: val@cis.upenn.edu

Chapter 9	P/FDM Mediator
Name and Version of System	P/FDM Mediator (Additional information is available at *http://www.csd.abdn.ac.uk/~gjlk/mediator/.*)
Status of Development and Maintenance	The P/FDM Mediator, described in Chapter 9, is a research prototype developed with support from the British Biotechnology and Biological Sciences Research Council (BBSRC) in partnership with the Engineering and Physical Sciences Research Council (EPSRC). This system is not currently developed or maintained.
Contact Person	Dr. Graham J. L. Kemp Department of Computing Science, Chalmers University of Technology SE-412 96, Göteborg, Sweden Tel: (+46) 31-772-5411 FAX: (+46) 31-165655 Email: kemp@cs.chalmers.se URL: *http://www.cs.chalmers.se/~kemp/*

Chapter 10	GeneExpress
Name and Version of System	GX Software System 1.4.2, Genesis 1.1 (both as of November 2002) (Additional information is available at *http://www.genelogic.com/products.cfm.*)
Status of Development and Maintenance	GX is a commercial system developed at Gene Logic Inc. Continuous maintenance and software upgrades are provided. The next major release: GX 2.0/Genesis 2.0 is planned for summer 2003. Academic licenses are available.
Contact Person	Dr. Victor M. Markowitz Gene Logic, Inc. 2001 Center Street Berkeley, California 94704, USA Tel: +(510) 981-3141 URL: *http://www.genelogic.com/products.cfm*

Chapter 11	DiscoveryLink
Name and Version of System	DiscoveryLink is an IBM services offering based on DB2 UDB V7.2 and higher version numbers. (Additional information is available at *http://www.ibm.com/discoverylink.*)
Status of Development and Maintenance	DB2 UDB is supported via IBM's normal customer support channels. Additionally, customers may contract for services to install, set up, configure, tune, and/or maintain the system, as well as to write new wrappers. These services are optional. DiscoveryLink is available through IBM's scholars program, which offers free licenses for qualifying academic purposes. See *http://www-3.ibm.com/sofware.info/university/* for more information.
Contact	ls@us.ibm.com, or visit the Web site at *http://www.ibm.com/discoverylink*

Chapter 12	KIND
Name and Version of System	KIND Mediator (Knowledge-Based Integration of Neuroscience Data) version 1.01 (Additional information is available at *http://www.nbirn.net.*)
Status of Development and Maintenance	KIND is under development at the University of California, San Diego, for the Biomedical Informatics Research Network (BIRN), an initiative of the National Center for Research Resources (NCRR), a component of the National Institutes of Health (NIH). Earlier prototypes (KIND versions 0.X) have been demonstrated at various conferences, including the Human Brain Project meetings 2000 and 2001. A demonstration is available at *http://www.npaci.edu/DICE/Neuro/* but is no longer actively maintained. Currently the system is completely redesigned and maintained. The system will be public but access is currently limited to the BIRN research group. In the future, a license may be available.

Chapter 12	KIND
Contact People	Dr. Bertram Ludäscher San Diego Supercomputer Center University of California, San Diego 9500 Gilman Drive, MC 0505 La Jolla, California 92093-0505, USA Tel: +1 (858) 822-0864 FAX: +1 (858) 534-5113 Email: ludaesch@sdsc.edu Dr. Amarnath Gupta San Diego Supercomputer Center University of California, San Diego 9500 Gilman Drive La Jolla, California 92093, USA Tel: +1 (858) 822-0994 FAX: +1 (858) 534-5113 Email: gupta@sdsc.edu

Index